The Numerical Method of Lines and Duality Principles Applied to Models in Physics and Engineering

Fabio Silva Botelho
Federal University of Santa Catarina – UFSC
Department of Mathematics, Campus Trindade
Florianopolis – SC – Brazil

CRC Press
Taylor & Francis Group
Boca Raton London New York

CRC Press is an imprint of the
Taylor & Francis Group, an **informa** business

A SCIENCE PUBLISHERS BOOK

First edition published 2024
by CRC Press
2385 NW Executive Center Drive, Suite 320, Boca Raton FL 33431

and by CRC Press
4 Park Square, Milton Park, Abingdon, Oxon, OX14 4RN

CRC Press is an imprint of Taylor & Francis Group, LLC

Library of Congress Cataloging-in-Publication Data (applied for)

ISBN: 978-1-032-19209-3 (hbk)
ISBN: 978-1-032-19210-9 (pbk)
ISBN: 978-1-003-25813-1 (ebk)

DOI: 10.1201/9781003258131

Typeset in Times New Roman
by Radiant Productions

Preface

The first part of this text develops theoretical and numerical results concerning the original conception of the generalized method of lines as well as more recent advances relating to such a method.

We recall that, for the generalized method of lines, the domain of the partial differential equation in question is discretized in lines (or more generally, in curves), and the concerning solution is written on these lines as functions of the boundary conditions and the domain boundary shape.

Beyond some improvements related to its original conception, we develop an approximate proximal approach that has been successful for a large class of models in physics in engineering, including results also suitable for a large class of domain shapes and concerning boundary conditions.

In its first chapter, the text presents applications to a Ginzburg-Landau type equation. The results include an initial description of the numerical procedures and, in a second step, we present the related software either in MATHEMATICA or MATLAB® and examples of the corresponding line expressions for each problem considered.

In the third chapter, we develop numerical results for the time independent, incompressible Navier-Stokes system in fluid mechanics. The results presented include a description of the numerical method in question, the corresponding software and the related line expressions obtained for each case considered. Furthermore, the establishment of an equivalent elliptic system is also presented. Here we highlight the procedure to obtain such a system is rather standard and well known, however, we emphasise the novelty is the identification of the correct concerning boundary conditions that lead to the mentioned equivalence between these two mentioned systems.

In the second text part, we present basic topics on the calculus of variations, duality theory, and constrained optimization, including new results in Lagrange multipliers for non-smooth optimization in a Banach space context. Furthermore, a generalization of the Ekeland variational principle is also developed.

Finally, in the third text part, we develop applications connecting the generalized method of lines and the duality theory. We recall that the dual and primal dual formulations are fundamentally important specially for systems that have no optimal global solution in a classical sense. In such a case, the critical points of the dual formulations correspond to weak cluster points of minimizing sequences for the primal ones, and reflect the average behavior of such sequences close to the corresponding global infima values for the primal models addressed. Such evaluations are important from both theoretical and practical points of view, since they may serve as a reliable tool for projects close to global optimality for a concerning primal model.

At this point we start to describe a summary of each chapter content.

Summary of each book chapter content

Chapter 1—The Generalized Method of Lines Applied to a Ginzburg-Landau Type Equation

This chapter develops an improvement concerning the original format of the generalized method of lines. We recall that for the generalized method of lines, the domain of the partial differential equation in question is discretized in lines (or in curves), and the concerning solution is developed on these lines, as functions of the boundary conditions and the domain boundary shape. In this text, we introduce a slight change in the way we truncate the series solution generated through the application of the Banach fixed point theorem to obtain the relation between two adjacent lines. With such a new improvement, we have got very good results even as a typical parameter is too much small, decreasing substantially the numerical solution error for such a class of small parameters. In the last sections, we present numerical examples and results related to a Ginzburg-Landau type equation.

Chapter 2—An Approximate Proximal Numerical Procedure Concerning the Generalized Method of Lines

This chapter develops an approximate proximal approach for the generalized method of lines. We recall again that for the generalized method of lines, the domain of the partial differential equation in question is discretized in lines (or in curves), and the concerning solution is developed on these lines, as functions of the boundary conditions and the domain boundary shape. Considering such a context, along the text we develop an approximate numerical procedure of proximal nature applicable to a large class of models in physics and engineering. Finally, in the last sections, we present numerical examples and results related to a Ginzburg-Landau type equation.

Chapter 3—Approximate Numerical Procedures for the Navier-Stokes System through the Generalized Method of Lines

This chapter develops applications of the generalized method of lines to numerical solutions of the time-independent, incompressible Navier-Stokes system in fluid mechanics. Once more, we recall that for such a method, the domain of the partial differential equation in question is discretized in lines (or more generally in curves), and the concerning solutions are written on these lines as functions of the boundary conditions and the domain boundary shape. More specifically, in this text we present softwares and results for a concerning approximate proximal approach as well as results based on the original conception of the generalized method of lines.

Chapter 4—An Approximate Numerical Method for Ordinary Differential Equation Systems with Applications to a Flight Mechanics Model

This chapter develops a new numerical procedure suitable for a large class of ordinary differential equation systems found in models in physics and engineering. The main numerical procedure is analogous to those concerning the generalized method of lines, originally published in the referenced books of 2011 and 2014 [22, 12], respectively. Finally, in the last section, we apply the method to a model in flight mechanics.

Chapter 5—Basic Topics on the Calculus of Variations

This chapter develops the main concepts on the basic calculus of the variations theory. Topics addressed include the Gâteaux variation, sufficient conditions for extremals of convex functionals, natural and essential boundary conditions, and the second Gâteaux variation.

The chapter finishes with a formal proof of the Gâteaux variation formula for the scalar case.

Chapter 6—More Topics on the Calculus of Variations

In this chapter, we present some more advanced topics on the calculus of variations, including Fréchet differentiability, the Legendre-Hadamard condition, the Weierstrass condition, and the Weierstrass-Erdmann corner conditions.

Chapter 7—Convex Analysis and Duality Theory

This chapter develops basic and advanced concepts on convex analysis and duality theory. Topics such as convex functions, weak lower semi-continuity, polar functionals, and the Legendre transform are developed in detail.

Among the topics on duality theory, we highlight the min-max theorem and the Ekeland variational principle.

Chapter 8—Constrained Variational Optimization

In this chapter, we introduce the main definitions and results on constrained variational optimization.

Topics addressed include the duality theory for the convex case, the Lagrange multiplier theorem for equality, and equality/inequality restrictions in a Banach spaces context.

Chapter 9—On Lagrange Multiplier Theorems for Non-Smooth Optimization for a Large Class of Variational Models in Banach Spaces

This chapter develops optimality conditions for a large class of non-smooth variational models. The main results are based on standard tools of functional analysis and calculus of variations. Firstly, we address a model with equality constraints and, in a second step, a more general model with equality and inequality constraints, always in a general Banach space context. We highlight, the results in general are well known, however, some novelties are introduced related to the proof procedures, which are in general softer than those concerning the present literature.

Chapter 10—A Convex Dual Formulation for a Large Class of Non-Convex Models in Variational Optimization

This chapter develops a convex dual variational formulation for a large class of models in variational optimization. The results are established through basic tools of functional analysis, convex analysis, and duality theory. The main duality principle is developed as an application to a Ginzburg-Landau type system in superconductivity in the absence of a magnetic field.

Chapter 11—Duality Principles and Numerical Procedures for a Large Class of Non-Convex Models in the Calculus of Variations

This chapter develops duality principles and numerical results for a large class of non-convex variational models. The main results are based on the fundamental tools of convex analysis, the duality theory, and calculus of variations. More specifically, the approach is established for a class of non-convex functionals similar to those found in some models in phase transition. Finally, in the last section we present a concerning numerical example and the respective software.

Chapter 12—Dual Variational Formulations for a Large Class of Non-Convex Models in the Calculus of Variations

This chapter develops dual variational formulations for a large class of models in variational optimization. The results are established through the basic tools of functional analysis, convex analysis, and, the duality theory. The main duality principle is developed as an application to a Ginzburg-Landau type system in superconductivity in the absence of a magnetic field. In the first sections, we develop new general dual convex variational formulations, more specifically, dual formulations with a large region of convexity around the critical points which are suitable for the non-convex optimization for a large class of models in physics and engineering. Finally, in the last section, we present some numerical results concerning the generalized method of lines applied to a Ginzburg-Landau type equation.

Chapter 13—A Note on the Korn's Inequality in a n-Dimensional Context and a Global Existence Result for a Non-Linear Plate Model

In the first part of this chapter, we present a new proof for the Korn inequality in a n-dimensional context. The results are based on standard tools of real and functional analysis. For the final result, the standard Poincaré inequality plays a fundamental role. In the second text part, we develop a global existence result for a non-linear model of plates. We address a rather general type of boundary conditions and the novelty here is the more relaxed restrictions concerning the external load magnitude.

Acknowledgments

I would like to thank my colleagues and Professors from Virginia Tech - USA, where I got my Ph.D. degree in Mathematics in 2009. At Virginia Tech, I had the opportunity of working with many exceptionally qualified people. I am especially grateful to Professor Robert C. Rogers for his excellent work as advisor. I would like to thank the Department of Mathematics for its constant support and this opportunity of studying mathematics at an advanced level. Among many other Professors, I particularly thank Martin Day (Calculus of Variations), William Floyd and Peter Haskell (Elementary Real Analysis), James Thomson (Real Analysis), Peter Linnell (Abstract Algebra), and George Hagedorn (Functional Analysis) for the excellent lecture courses. Finally, special thanks to all my Professors at I.T.A. (Instituto Tecnológico de Aeronáutica, SP-Brasil) my undergraduate and masters school. Specifically about I.T.A., among many others, I would like to express my gratitude to Professors Leo H. Amaral, Tânia Rabelo, and my master thesis advisor Antônio Marmo de Oliveira, also for their valuable work.

Fabio Silva Botelho

December 2022   Florianópolis - SC, Brazil

Contents

SECTION II: CALCULUS OF VARIATIONS, CONVEX ANALYSIS AND RESTRICTED OPTIMIZATION

SECTION III: DUALITY PRINCIPLES AND RELATED NUMERICAL EXAMPLES THROUGH THE GENERALIZED METHOD OF LINES

THE GENERALIZED METHOD OF LINES

I

Chapter 1

The Generalized Method of Lines Applied to a Ginzburg-Landau Type Equation

1.1 Introduction

This chapter develops an improvement concerning the original format of the generalized method of lines. We recall that for the generalized method of lines, the domain of the partial differential equation in question is discretized in lines (or in curves), and the concerning solution is developed on these lines, as functions of the boundary conditions and the domain boundary shape. In this text, we introduce a slight change in the way we truncate the series solution generated through the application of the Banach fixed point theorem to obtain the relation between two adjacent lines. With such a new improvement, we have got very good results even though a typical parameter is very small, decreasing substantially the numerical solution error for such a class of small parameters. In the last sections, we present numerical examples and results related to a Ginzburg-Landau type equation.

We start with a kind of matrix version of the Generalized Method of Lines. Applications are developed for models in physics and engineering.

1.2 On the numerical procedures for Ginzburg-Landau type ODEs

We first apply Newton's method to a general class of ordinary differential equations. The solution here is obtained similarly to the generalized method of lines procedure. See the next sections for details on such a method for PDEs.

For a C^1 class function f and a continuous function g, consider the second order equation,

$$\begin{cases} u'' + f(u) + g = 0, \text{ in } [0,1] \\ \\ u(0) = u_0, \ u(1) = u_f. \end{cases} \tag{1.1}$$

In finite differences we have the approximate equation:

$$u_{n+1} - 2u_n + u_{n-1} + f(u_n)d^2 + g_n d^2 = 0.$$

Assuming such an equation is non-linear and linearizing it about a first solution $\{\tilde{u}\}$, we have (in fact, this is an approximation),

$$u_{n+1} - 2u_n + u_{n-1} + f(\tilde{u}_n)d^2 + f'(\tilde{u}_n)(u_n - \tilde{u}_n)d^2 + g_n d^2 = 0.$$

Thus we may write

$$u_{n+1} - 2u_n + u_{n-1} + A_n u_n d^2 + B_n d^2 = 0,$$

where,

$$A_n = f'(\tilde{u}_n),$$

and

$$B_n = f(\tilde{u}_n) - f'(\tilde{u}_n)\tilde{u}_n + g_n.$$

In particular for $n = 1$ we get

$$u_2 - 2u_1 + u_0 + A_1 u_1 d^2 + B_1 d^2 = 0.$$

Solving such an equation for u_1, we get

$$u_1 = a_1 u_2 + b_1 u_0 + c_1,$$

where

$$a_1 = (2 - A_1 d^2)^{-1}, \ b_1 = a_1, \ c_1 = a_1 B_1.$$

Reasoning inductively, having

$$u_{n-1} = a_{n-1} u_n + b_{n-1} u_0 + c_{n-1},$$

and

$$u_{n+1} - 2u_n + u_{n-1} + A_n u_n d^2 + B_n d^2 = 0,$$

we get

$$u_{n+1} - 2u_n + a_{n-1}u_n + b_{n-1}u_0 + c_{n-1} + A_n u_n d^2 + B_n d^2 = 0,$$

so that

$$u_n = a_n u_{n+1} + b_n u_0 + c_n,$$

where,

$$a_n = (2 - a_{n-1} - A_n d^2)^{-1},$$

$$b_n = a_n b_{n-1},$$

and

$$c_n = a_n (c_{n-1} + B_n d^2),$$

$\forall n \in 1, ..., N-1$.
    We have thus obtained

$$u_n = a_n u_{n+1} + b_n u_0 + c_n \equiv H_n(u_{n+1}), \forall n \in \{1, ..., N-1\},$$

and in particular,

$$u_{N-1} = H_{N-1}(u_f),$$

so that we may calculate,

$$u_{N-2} = H_{N-2}(u_{N-1}),$$

$$u_{N-3} = H_{N-3}(u_{N-2}),$$

and so on, up to finding,

$$u_1 = H_1(u_2).$$

    The next step is to replace $\{\tilde{u}_n\}$ by the $\{u_n\}$ calculated, and repeat the process up to the satisfaction of an appropriate convergence criterion. We present numerical results for the equation,

$$\begin{cases} u'' - \frac{u^3}{\varepsilon} + \frac{u}{\varepsilon} + g = 0, \text{ in } [0,1] \\ u(0) = 0, \ u(1) = 0, \end{cases} \tag{1.2}$$

where, and

$$g(x) = \frac{1}{\varepsilon},$$

    The results are obtained for $\varepsilon = 1.0$, $\varepsilon = 0.1$, $\varepsilon = 0.01$ and $\varepsilon = 0.001$. Please see Figures 1.1, 1.2, 1.3 and 1.4 respectively.

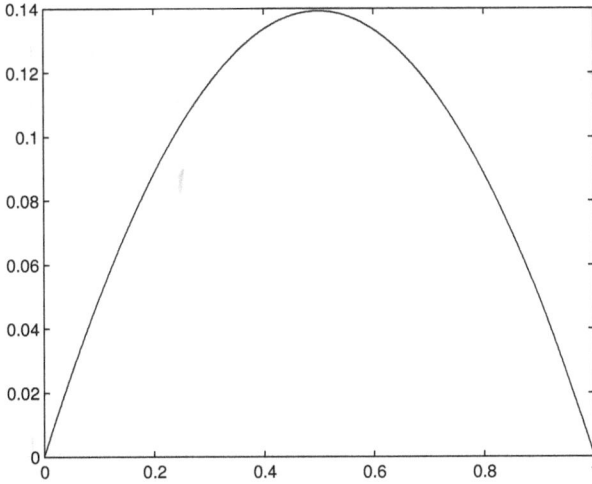

Figure 1.1: The solution $u(x)$ by Newton's method for $\varepsilon = 1$.

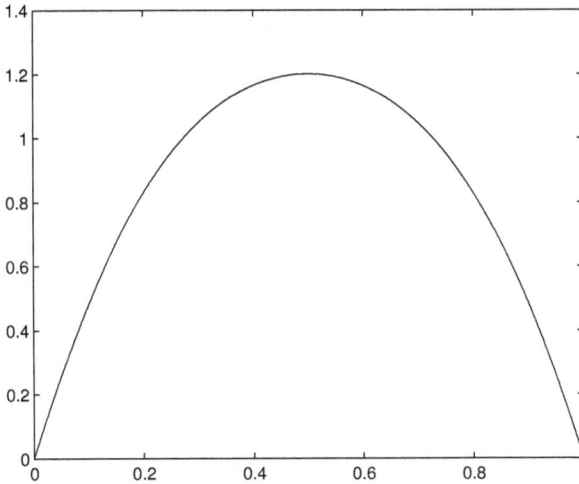

Figure 1.2: The solution $u(x)$ by Newton's method for $\varepsilon = 0.1$.

1.3 Numerical results for related P.D.E.s

1.3.1 A related P.D.E on a special class of domains

We start by describing a similar equation, but now in a two dimensional context. Let $\Omega \subset \mathbb{R}^2$ be an open, bounded, connected set with a regular boundary denoted by $\partial\Omega$. Consider a real Ginzburg-Landau type equation (see [4], [9], [52], [53]

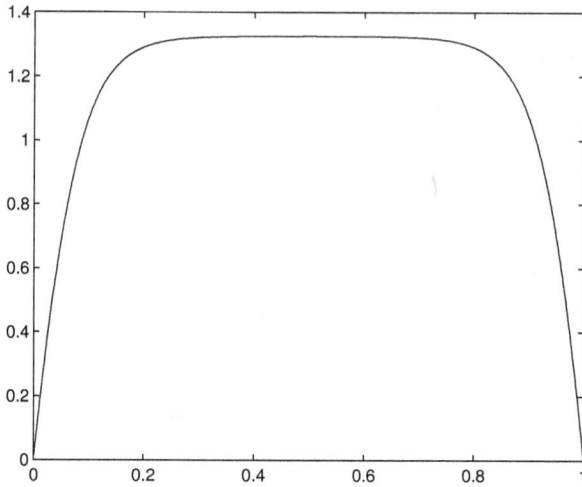

Figure 1.3: The solution $u(x)$ by Newton's method for $\varepsilon = 0.01$.

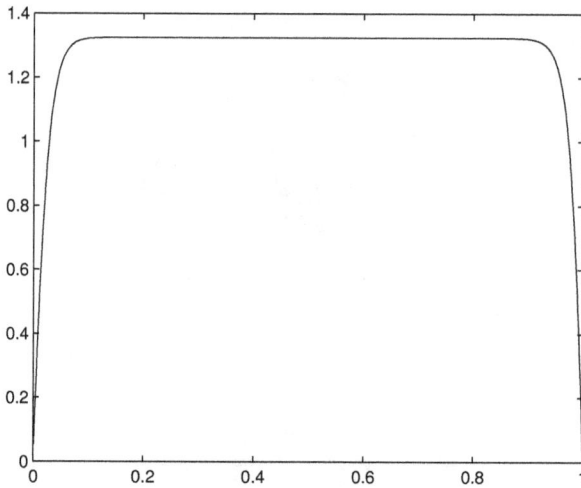

Figure 1.4: The solution $u(x)$ by Newton's method for $\varepsilon = 0.001$.

for details about such an equation), given by

$$
\begin{cases}
\varepsilon \nabla^2 u - \alpha u^3 + \beta u = f, & \text{in } \Omega \\
u = 0, & \text{on } \partial\Omega,
\end{cases}
\tag{1.3}
$$

where α, β, $\varepsilon > 0$, $u \in U = W_0^{1,2}(\Omega)$, and $f \in L^2(\Omega)$. The corresponding primal variational formulation is represented by $J : U \to \mathbb{R}$, where

$$J(u) = \frac{\varepsilon}{2} \int_{\Omega} \nabla u \cdot \nabla u \, dx + \frac{\alpha}{4} \int_{\Omega} u^4 \, dx - \frac{\beta}{2} \int_{\Omega} u^2 \, dx + \int_{\Omega} f u \, dx.$$

1.3.2 About the matrix version of G.M.O.L.

The generalized method of lines was originally developed in [22]. In this work we address its matrix version. Consider the simpler case where $\Omega = [0,1] \times [0,1]$. We discretize the domain in x, that is, in $N + 1$ vertical lines obtaining the following equation in finite differences (see [71] for details about finite differences schemes).

$$\frac{\varepsilon(u_{n+1} - 2u_n + u_{n-1})}{d^2} + \varepsilon M_2 u_n / d_1^2 - \alpha u_n^3 + \beta u_n = f_n, \quad (1.4)$$

$\forall n \in \{1, ..., N-1\}$, where $d = 1/N$ and u_n corresponds to the solution on the line n. The idea is to apply the Newton's method. Thus choosing a initial solution $\{(u_0)_n\}$ we linearize (1.4) about it, obtaining the linear equation:

$$u_{n+1} - 2u_n + u_{n-1} \quad + \quad \tilde{M}_2 u_n - \frac{3\alpha d^2}{\varepsilon}(u_0)_n^2 u_n$$
$$+ \quad \frac{2\alpha}{\varepsilon}(u_0)_n^3 d^2 + \frac{\beta d^2}{\varepsilon} u_n - f_n \frac{d^2}{\varepsilon} = 0, \quad (1.5)$$

where $\tilde{M}_2 = M_2 \frac{d^2}{d_1^2}$ and

$$M_2 = \begin{bmatrix} -2 & 1 & 0 & 0 & \cdots & 0 \\ 1 & -2 & 1 & 0 & \cdots & 0 \\ 0 & 1 & -2 & 1 & \cdots & 0 \\ \vdots & \vdots & \vdots & \vdots & \ddots & \vdots \\ 0 & 0 & \cdots & 1 & -2 & 1 \\ 0 & 0 & \cdots & \cdots & 1 & -2 \end{bmatrix}, \quad (1.6)$$

with N_1 lines corresponding to the discretization in the y axis. Furthermore $d_1 = 1/N_1$.

In particular for $n = 1$ we get

$$u_2 - 2u_1 \quad + \quad \tilde{M}_2 u_1 - \frac{3\alpha d^2}{\varepsilon}(u_0)_1^2 u_1$$
$$+ \quad \frac{2\alpha}{\varepsilon}(u_0)_1^3 d^2 + \frac{\beta d^2}{\varepsilon} u_1 - f_1 \frac{d^2}{\varepsilon} = 0. \quad (1.7)$$

Denoting

$$M_{12}[1] = 2I_d - \tilde{M}_2 + 3\frac{\alpha d^2}{\varepsilon}(u_0)_1^2 I_d - \frac{\beta d^2}{\varepsilon} I_d,$$

where I_d denotes the $(N_1 - 1) \times (N_1 - 1)$ identity matrix,

$$Y_0[1] = \frac{2\alpha d^2}{\varepsilon} (u_0)_1^3 - f_1 \frac{d^2}{\varepsilon},$$

and $M_{50}[1] = M_{12}[1]^{-1}$, we obtain

$$u_1 = M_{50}[1]u_2 + z[1].$$

where

$$z[1] = M_{50}[1] \cdot Y_0[1].$$

Now for $n = 2$ we get

$$u_3 - 2u_2 + u_1 \;\; + \;\; \tilde{M}_2 u_2 - \frac{3\alpha d^2}{\varepsilon} (u_0)_2^2 u_2$$
$$+ \;\; \frac{2\alpha}{\varepsilon} (u_0)_2^3 d^2 + \frac{\beta d^2}{\varepsilon} u_2 - f_2 \frac{d^2}{\varepsilon} = 0, \qquad (1.8)$$

that is,

$$u_3 - 2u_2 + M_{50}[1]u_2 + z[1] \;\; + \;\; \tilde{M}_2 u_2 - \frac{3\alpha d^2}{\varepsilon} (u_0)_2^2 u_2$$
$$+ \;\; \frac{2\alpha}{\varepsilon} (u_0)_2^3 d^2 + \frac{\beta d^2}{\varepsilon} u_2 - f_2 \frac{d^2}{\varepsilon} = 0, \quad (1.9)$$

so that denoting

$$M_{12}[2] = 2I_d - \tilde{M}_2 - M_{50}[1] + 3\frac{\alpha d^2}{\varepsilon} (u_0)_2^2 I_d - \frac{\beta d^2}{\varepsilon} I_d,$$

$$Y_0[2] = \frac{2\alpha d^2}{\varepsilon} (u_0)_2^3 - f_2 \frac{d^2}{\varepsilon},$$

and $M_{50}[2] = M_{12}[2]^{-1}$, we obtain

$$u_2 = M_{50}[2]u_3 + z[2],$$

where

$$z[2] = M_{50}[2] \cdot (Y_0[2] + z[1]).$$

Proceeding in this fashion, for the line n we obtain

$$u_{n+1} - 2u_n \;\; + \;\; M_{50}[n-1]u_n + z[n-1] + \tilde{M}_2 u_n - \frac{3\alpha d^2}{\varepsilon} (u_0)_n^2 u_n$$
$$+ \;\; \frac{2\alpha}{\varepsilon} (u_0)_n^3 d^2 + \frac{\beta d^2}{\varepsilon} u_n - f_n \frac{d^2}{\varepsilon} = 0, \qquad (1.10)$$

so that denoting

$$M_{12}[n] = 2I_d - \tilde{M}_2 - M_{50}[n-1] + 3\frac{\alpha d^2}{\varepsilon}(u_0)_n^2 I_d - \frac{\beta d^2}{\varepsilon}I_d,$$

and also denoting

$$Y_0[n] = \frac{2\alpha d^2}{\varepsilon}(u_0)_n^3 - f_n\frac{d^2}{\varepsilon},$$

and $M_{50}[n] = M_{12}[n]^{-1}$, we obtain

$$u_n = M_{50}[n]u_{n+1} + z[n],$$

where

$$z[n] = M_{50}[n] \cdot (Y_0[n] + z[n-1]).$$

Observe that we have

$$u_N = \theta,$$

where θ denotes the zero matrix $(N_1 - 1) \times 1$, so that we may calculate

$$u_{N-1} = M_{50}[N-1] \cdot u_N + z[N-1],$$

and,

$$u_{N-2} = M_{50}[N-2] \cdot u_{N-1} + z[N-2],$$

and so on, up to obtaining

$$u_1 = M_{50}[1] \cdot u_2 + z[1].$$

The next step is to replace $\{(u_0)_n\}$ by $\{u_n\}$ and thus to repeat the process until convergence is achieved.

This is the Newton's Method, what seems to be relevant is the way we inverted the big matrix $((N_1 - 1) \cdot (N-1)) \times ((N_1 - 1) \cdot (N-1))$, in fact instead of inverting it directly we have inverted $N - 1$ matrices $(N_1 - 1) \times (N_1 - 1)$ through an application of the generalized method of lines.

1.3.3 Numerical results for the concerning partial differential equation

We solve the equation

$$
\begin{cases}
\varepsilon \nabla^2 u - \alpha u^3 + \beta u + 1 = 0, & \text{in } \Omega = [0,1] \times [0,1] \\
u = 0, & \text{on } \partial\Omega,
\end{cases} \tag{1.11}
$$

through the algorithm specified in the last section. We consider $\alpha = \beta = 1$. For $\varepsilon = 1.0$ see Figure 1.5, for $\varepsilon = 0.0001$ see Figure 1.6.

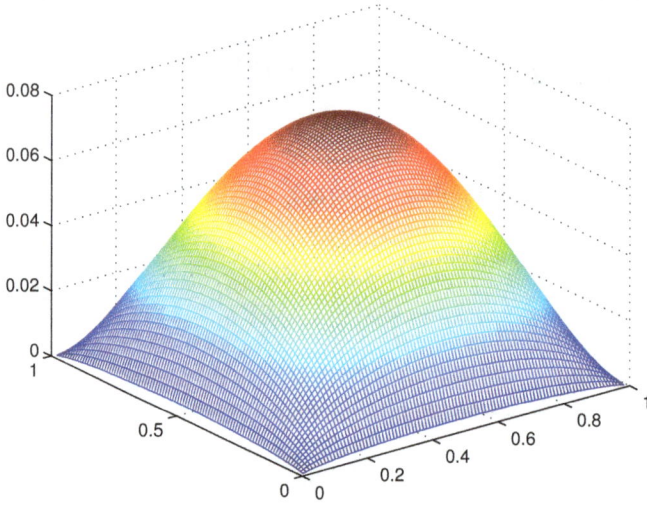

Figure 1.5: The solution $u(x,y)$ for $\varepsilon = 1.0$.

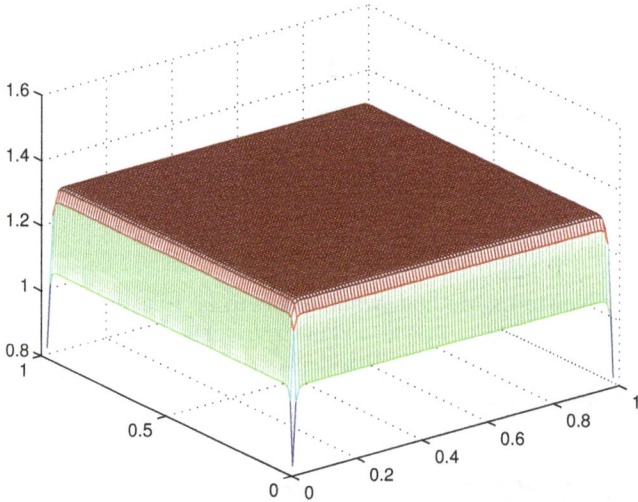

Figure 1.6: The solution $u(x,y)$ for $\varepsilon = 0.0001$.

1.4 A numerical example concerning a Ginzburg-Landau type equation

We start by recalling that the generalized method of lines was originally introduced in the book entitled "Topics on Functional Analysis, Calculus of Variations and Duality" [22], published in 2011.

Indeed, the present results are extensions and applications of previous ones which have been published since 2011, in books and articles such as [22, 17, 12, 13]. About the Sobolev spaces involved, we would mention [1, 2]. Concerning the applications, related models in physics are addressed in [4, 52].

We also emphasize that, in such a method, the domain of the partial differential equation in question is discretized in lines (or more generally, in curves), and the concerning solution is written on these lines as functions of boundary conditions and the domain boundary shape.

In fact, in its previous format, this method consists of an application of a kind of partial finite differences procedure combined with the Banach fixed point theorem to obtain the relation between two adjacent lines (or curves).

In the present article, we propose an improvement concerning the way we truncate the series solution obtained through an application of the Banach fixed point theorem to find the relation between two adjacent lines. The results obtained are very good even as a typical parameter $\varepsilon > 0$ is very small.

In the next lines and sections, we develop in detail such a numerical procedure.

1.4.1 About the concerning improvement for the generalized method of lines

Let $\Omega \subset \mathbb{R}^2$ where

$$\Omega = \{(r,\theta) \in \mathbb{R}^2 \ : \ 1 \le r \le 2, 0 \le \theta \le 2\pi\}.$$

Consider the problem of solving the partial differential equation

$$
\begin{cases}
-\varepsilon \left(\frac{\partial^2 u}{\partial r^2} + \frac{1}{r}\frac{\partial u}{\partial r} + \frac{1}{r^2}\frac{\partial^2 u}{\partial \theta^2} \right) + \alpha u^3 - \beta u = f, & \text{in } \Omega, \\
u = u_0(\theta), & \text{on } \partial\Omega_1, \\
u = u_f(\theta), & \text{on } \partial\Omega_2.
\end{cases}
\tag{1.12}
$$

Here

$$\Omega = \{(r,\theta) \in \mathbb{R}^2 \ : \ 1 \le r \le 2, 0 \le \theta \le 2\pi\},$$

$$\partial\Omega_1 = \{(1,\theta) \in \mathbb{R}^2 \ : \ 0 \le \theta \le 2\pi\},$$

$$\partial\Omega_2 = \{(2,\theta) \in \mathbb{R}^2 \ : \ 0 \le \theta \le 2\pi\},$$

$\varepsilon > 0$, $\alpha > 0, \beta > 0$, and $f \equiv 1$, on Ω.

In a partial finite differences scheme, such a system stands for

$$
-\varepsilon \left(\frac{u_{n+1} - 2u_n + u_{n-1}}{d^2} + \frac{1}{t_n}\frac{u_n - u_{n-1}}{d} + \frac{1}{t_n^2}\frac{\partial^2 u_n}{\partial \theta^2} \right) + \alpha u_n^3 - \beta u_n = f_n,
$$

$\forall n \in \{1, \cdots, N-1\}$, with the boundary conditions

$$u_0 = 0,$$

and

$$u_N = 0.$$

Here N is the number of lines and $d = 1/N$.
In particular, for $n = 1$ we have

$$-\varepsilon \left(\frac{u_2 - 2u_1 + u_0}{d^2} + \frac{1}{t_1} \frac{(u_1 - u_0)}{d} + \frac{1}{t_1^2} \frac{\partial^2 u_1}{\partial \theta^2} \right) + \alpha u_1^3 - \beta u_1 = f_1,$$

so that

$$u_1 = \left(u_2 + u_1 + u_0 + \frac{1}{t_1}(u_1 - u_0)\,d + \frac{1}{t_1^2} \frac{\partial^2 u_1}{\partial \theta^2} d^2 + (-\alpha u_1^3 + \beta u_1 - f_1)\frac{d^2}{\varepsilon} \right) /3.0,$$

We solve this last equation through the Banach fixed point theorem, obtaining u_1 as a function of u_2.
Indeed, we may set

$$u_1^0 = u_2$$

and

$$\begin{aligned} u_1^{k+1} &= \left(u_2 + u_1^k + u_0 + \frac{1}{t_1}(u_1^k - u_0)\,d + \frac{1}{t_1^2} \frac{\partial^2 u_1^k}{\partial \theta^2} d^2 \right. \\ &\quad \left. + (-\alpha(u_1^k)^3 + \beta u_1^k - f_1)\frac{d^2}{\varepsilon} \right) /3.0, \end{aligned} \tag{1.13}$$

$\forall k \in \mathbb{N}$.
Thus, we may obtain

$$u_1 = \lim_{k \to \infty} u_1^k \equiv H_1(u_2, u_0).$$

Similarly, for $n = 2$, we have

$$\begin{aligned} u_2 &= \left(u_3 + u_2 + H_1(u_2, u_0) + \frac{1}{t_1}(u_2 - H_1(u_2, u_0))\,d + \frac{1}{t_1^2} \frac{\partial^2 u_2}{\partial \theta^2} d^2 \right. \\ &\quad \left. + (-\alpha u_2^3 + \beta u_2 - f_2)\frac{d^2}{\varepsilon} \right) /3.0, \end{aligned} \tag{1.14}$$

We solve this last equation through the Banach fixed point theorem, obtaining u_2 as a function of u_3 and u_0.
Indeed, we may set

$$u_2^0 = u_3$$

and

$$u_2^{k+1} = \left(u_3 + u_2^k + H_1(u_2^k, u_0) + \frac{1}{t_2}(u_2^k - H_1(u_2^k, u_0)) \, d + \frac{1}{t_2^2} \frac{\partial^2 u_2^k}{\partial \theta^2} d^2 \right.$$
$$\left. + (-\alpha(u_2^k)^3 + \beta u_2^k - f_2) \frac{d^2}{\varepsilon} \right) / 3.0, \tag{1.15}$$

$\forall k \in \mathbb{N}$.

Thus, we may obtain

$$u_2 = \lim_{k \to \infty} u_2^k \equiv H_2(u_3, u_0).$$

Now reasoning inductively, having

$$u_{n-1} = H_{n-1}(u_n, u_0),$$

we may get

$$u_n = \left(u_{n+1} + u_n + H_{n-1}(u_n, u_0) + \frac{1}{t_n}(u_n - H_{n-1}(u_n, u_0)) \, d + \frac{1}{t_n^2} \frac{\partial^2 u_n}{\partial \theta^2} d^2 \right.$$
$$\left. + (-\alpha u_n^3 + \beta u_n - f_n) \frac{d^2}{\varepsilon} \right) / 3.0, \tag{1.16}$$

We solve this last equation through the Banach fixed point theorem, obtaining u_n as a function of u_{n+1} and u_0.

Indeed, we may set

$$u_n^0 = u_{n+1}$$

and

$$u_n^{k+1} = \left(u_{n+1} + u_n^k + H_{n-1}(u_n^k, u_0) + \frac{1}{t_n}(u_n^k - H_{n-1}(u_n^k, u_0)) \, d + \frac{1}{t_n^2} \frac{\partial^2 u_n^k}{\partial \theta^2} d^2 \right.$$
$$\left. + (-\alpha(u_n^k)^3 + \beta u_n^k - f_n) \frac{d^2}{\varepsilon} \right) / 3.0, \tag{1.17}$$

$\forall k \in \mathbb{N}$.

Thus, we may obtain

$$u_n = \lim_{k \to \infty} u_n^k \equiv H_n(u_{n+1}, u_0).$$

We have obtained $u_n = H_n(u_{n+1}, u_0)$, $\forall n \in \{1, \cdots, N-1\}$.
In particular, $u_N = u_f(\theta)$, so that we may obtain

$$u_{N-1} = H_{N-1}(u_N, u_0) = H_{N-1}(0) \equiv F_{N-1}(u_N, u_0) = F_{N-1}(u_f(\theta), u_0(\theta)).$$

Similarly,

$$u_{N-2} = H_{N-2}(u_{N-1}, u_0) = H_{N-2}(H_{N-1}(u_N, u_0)) = F_{N-2}(u_N, u_0) = F_{N-1}(u_f(\theta), u_0(\theta)),$$

an so on, up to obtaining

$$u_1 = H_1(u_2) \equiv F_1(u_N, u_0) = F_1(u_f(\theta), u_0(\theta)).$$

The problem is then approximately solved.

1.4.2 Software in Mathematica for solving such an equation

We recall that the equation to be solved is a Gingurg-Landau type one, where

$$\begin{cases} -\varepsilon \left(\frac{\partial^2 u}{\partial r^2} + \frac{1}{r}\frac{\partial u}{\partial r} + \frac{1}{r^2}\frac{\partial^2 u}{\partial \theta^2} \right) + \alpha u^3 - \beta u = f, & \text{in } \Omega, \\ u = 0, & \text{on } \partial\Omega_1, \\ u = u_f(\theta), & \text{on } \partial\Omega_2. \end{cases} \tag{1.18}$$

Here

$$\Omega = \{(r, \theta) \in \mathbb{R}^2 \ : \ 1 \le r \le 2, \ 0 \le \theta \le 2\pi\},$$

$$\partial\Omega_1 = \{(1, \theta) \in \mathbb{R}^2 \ : \ 0 \le \theta \le 2\pi\},$$

$$\partial\Omega_2 = \{(2, \theta) \in \mathbb{R}^2 \ : \ 0 \le \theta \le 2\pi\},$$

$\varepsilon > 0$, $\alpha > 0, \beta > 0$, and $f \equiv 1$, on Ω. In a partial finite differences scheme, such a system stands for

$$-\varepsilon \left(\frac{u_{n+1} - 2u_n + u_{n-1}}{d^2} + \frac{1}{t_n}\frac{u_n - u_{n-1}}{d} + \frac{1}{t_n^2}\frac{\partial^2 u_n}{\partial \theta^2} \right) + \alpha u_n^3 - \beta u_n = f_n,$$

$\forall n \in \{1, \cdots, N-1\}$, with the boundary conditions

$$u_0 = 0,$$

and

$$u_N = u_f[x].$$

Here N is the number of lines and $d = 1/N$.

At this point we present the concerning software for an approximate solution. Such a software is for $N = 10$ (10 lines) and $u_0[x] = 0$.

****

1. $m8 = 10$; ($N = 10$ *lines*)

2. $d = 1/m8$;

3. $e_1 = 0.1$; ($\varepsilon = 0.1$)

4. $A = 1.0$;

5. $B = 1.0$;

6. $For[i = 1, i < m8, i++, f[i] = 1.0]$; ($f \equiv 1$, on Ω)

7. $a = 0.0$;

8. $For[i = 1, i < m8, i++,$
   $Clear[b, u]$;
   $t[i] = 1 + i * d$;
   $b[x\_] = u[i+1][x]$;

9. $For[k = 1, k < 30, k++,$ (we have fixed the number of iterations)
   $$z = \left(u[i+1][x] + b[x] + a + \frac{1}{t[i]}(b[x] - a) * d \right.$$
   $$\left. + \frac{1}{t[i]^2} D[b[x], \{x, 2\}] * d^2 + (-A * b[x]^3 + B * u[x] + f[i]) * \frac{d^2}{e_1} \right) / 3.0;$$
   $z =$
   $Series[z, \{u[i+1][x], 0, 3\}, \{u[i+1]'[x], 0, 1\}, \{u[i+1]''[x], 0, 1\},$
   $\{u[i+1]'''[x], 0, 0\}, \{u[i+1]''''[x], 0, 0\}]$;
   $z = Normal[z]$,
   $z = Expand[z]$;
   $b[x\_] = z]$;

10. $a_1[i] = z$;

11. $Clear[b]$;

12. $u[i+1][x\_] = b[x]$;

13. $a = a_1[i]$];

14. $b[x\_] = u_f[x]$;

15. $For[i = 1, i < m8, i++,$
    $A_1 = a_1[m8 - i]$;
    $A_1 = Series[A_1, \{u_f[x], 0, 3\}, \{u'_f[x], 0, 1\}, \{u''_f[x], 0, 1\}, \{u'''_f[x], 0, 0\},$
    $\{u''''_f[x], 0, 0\}]$;

$A_1 = Normal[A_1];$

$A_1 = Expand[A_1];$

$u[m8 - i][x_-] = A_1;$

$b[x_-] = A_1;$

$Print[u[m8/2][x]];$

****

The numerical expressions for the solutions of the concerning N lines are given by

$$
\begin{aligned}
u[1][x] \;=\;& 0.47352 + 0.00691u_f[x] - 0.00459u_f[x]^2 + 0.00265u_f[x]^3 \\
& + 0.00039(u_f'')[x] - 0.00058u_f[x](u_f'')[x] + 0.00050u_f[x]^2(u_f'')[x] \\
& - 0.000181213u_f[x]^3(u_f'')[x] \tag{1.19}
\end{aligned}
$$

$$
\begin{aligned}
u[2][x] \;=\;& 0.76763 + 0.01301u_f[x] - 0.00863u_f[x]^2 + 0.00497u_f[x]^3 \\
& + 0.00068(u_f'')[x] - 0.00103u_f[x](u_f'')[x] + 0.00088u_f[x]^2(u_f'')[x] \\
& - 0.00034u_f[x]^3(u_f'')[x] \tag{1.20}
\end{aligned}
$$

$$
\begin{aligned}
u[3][x] \;=\;& 0.91329 + 0.02034u_f[x] - 0.01342u_f[x]^2 + 0.00768u_f[x]^3 \\
& + 0.00095(u_f'')[x] - 0.00144u_f[x](u_f'')[x] + 0.00122u_f[x]^2(u_f'')[x] \\
& - 0.00051u_f[x]^3(u_f'')[x] \tag{1.21}
\end{aligned}
$$

$$
\begin{aligned}
u[4][x] \;=\;& 0.97125 + 0.03623u_f[x] - 0.02328u_f[x]^2 + 0.01289u_f[x]^3 \\
& + 0.00147331(u_f'')[x] \\
& - 0.00223u_f[x](u_f'')[x] + 0.00182uf[x]^2(u_f'')[x] \\
& - 0.00074u_f[x]^3(u_f'')[x] \tag{1.22}
\end{aligned}
$$

$$
\begin{aligned}
u[5][x] \;=\;& 1.01736 + 0.09242u_f[x] - 0.05110u_f[x]^2 + 0.02387u_f[x]^3 \\
& + 0.00211(u_f'')[x] - 0.00378u_f[x](u_f'')[x] + 0.00292u_f[x]^2(u_f'')[x] \\
& - 0.00132u_f[x]^3(u_f'')[x] \tag{1.23}
\end{aligned}
$$

$$
\begin{aligned}
u[6][x] \;=\;& 1.02549 + 0.21039u_f[x] - 0.09374u_f[x]^2 + 0.03422u_f[x]^3 \\
& + 0.00147(u_f'')[x] - 0.00634u_f[x](u_f'')[x] + 0.00467u_f[x]^2(u_f'')[x] \\
& - 0.00200u_f[x]^3(u_f'')[x] \tag{1.24}
\end{aligned}
$$

$$
\begin{aligned}
u[7][x] \;=\; & 0.93854 + 0.36459 u_f[x] - 0.14232 u_f[x]^2 + 0.04058 u_f[x]^3 \\
& + 0.00259 (u_f'')[x] - 0.00747373 u_f[x](u_f'')[x] + 0.0047969 u_f[x]^2 (u_f'')[x] \\
& - 0.00194 u_f[x]^3 (u_f'')[x]
\end{aligned}
\tag{1.25}
$$

$$
\begin{aligned}
u[8][x] \;=\; & 0.74649 + 0.57201 u_f[x] - 0.17293 u_f[x]^2 + 0.02791 u_f[x]^3 \\
& + 0.00353 (u_f'')[x] - 0.00658 u_f[x](u_f'')[x] + 0.00407 u_f[x]^2 (u_f'')[x] \\
& - 0.00172 u_f[x]^3 (u_f'')[x]
\end{aligned}
\tag{1.26}
$$

$$
\begin{aligned}
u[9][x] \;=\; & 0.43257 + 0.81004 u_f[x] - 0.13080 u_f[x]^2 + 0.00042 u_f[x]^3 \\
& + 0.00294 (u_f'')[x] - 0.00398 u_f[x](u_f'')[x] + 0.00222 u_f[x]^2 (u_f'')[x] \\
& - 0.00066 u_f[x]^3 (u_f'')[x]
\end{aligned}
\tag{1.27}
$$

1.5 The generalized method of lines for a more general domain

This section develops the generalized method of lines for a more general type of domain. In our previous publications [22, 17], we highlighted that the method addressed there may present a relevant error as a parameter $\varepsilon > 0$ is too small, that is, as ε is about 0.01, 0.001, or even smaller.

In the present section, we develop a solution for such a problem through a proximal formulation suitable for a large class of non-linear elliptic PDEs.

At this point, we reintroduce the generalized method of lines, originally presented in F. Botelho [22]. In the present context, we add new theoretical and applied results to the original presentation. Especially the computations are all completely new. Consider first the equation

$$
\varepsilon \nabla^2 u + g(u) + f = 0, \text{ in } \Omega \subset \mathbb{R}^2,
\tag{1.28}
$$

with the boundary conditions

$$
u = 0 \text{ on } \Gamma_0 \text{ and } u = u_f, \text{ on } \Gamma_1.
$$

From now on we assume that u_f, g and f are smooth functions (we mean C^∞ functions), unless otherwise specified. Here Γ_0 denotes the internal boundary of Ω and Γ_1 the external one. Consider the simpler case where

$$
\Gamma_1 = 2\Gamma_0,
$$

and suppose there exists $r(\theta)$, a smooth function such that

$$\Gamma_0 = \{(\theta, r(\theta)) \mid 0 \leq \theta \leq 2\pi\},$$

being $r(0) = r(2\pi)$.

In polar coordinates the above equation may be written as

$$\frac{\partial^2 u}{\partial r^2} + \frac{1}{r}\frac{\partial u}{\partial r} + \frac{1}{r^2}\frac{\partial^2 u}{\partial \theta^2} + g(u) + f = 0, \text{ in } \Omega, \tag{1.29}$$

and

$$u = 0 \text{ on } \Gamma_0 \text{ and } u = u_f, \text{ on } \Gamma_1.$$

Define the variable t by

$$t = \frac{r}{r(\theta)}.$$

Also defining $\bar{u}$ by

$$u(r, \theta) = \bar{u}(t, \theta),$$

dropping the bar in $\bar{u}$, equation (1.28) is equivalent to

$$\frac{\partial^2 u}{\partial t^2} + \frac{1}{t}f_2(\theta)\frac{\partial u}{\partial t}$$
$$+ \frac{1}{t}f_3(\theta)\frac{\partial^2 u}{\partial \theta \partial t} + \frac{f_4(\theta)}{t^2}\frac{\partial^2 u}{\partial \theta^2}$$
$$+ f_5(\theta)(g(u) + f) = 0, \tag{1.30}$$

in Ω. Here $f_2(\theta)$, $f_3(\theta)$, $f_4(\theta)$ and $f_5(\theta)$ are known functions.

More specifically, denoting

$$f_1(\theta) = \frac{-r'(\theta)}{r(\theta)},$$

we have:

$$f_2(\theta) = 1 + \frac{f_1'(\theta)}{1 + f_1(\theta)^2},$$

$$f_3(\theta) = \frac{2f_1(\theta)}{1 + f_1(\theta)^2},$$

and

$$f_4(\theta) = \frac{1}{1 + f_1(\theta)^2}.$$

Observe that $t \in [1,2]$ in Ω. Discretizing in t (N equal pieces which will generate N lines) we obtain the equation

$$\frac{u_{n+1} - 2u_n + u_{n-1}}{d^2} + \frac{(u_n - u_{n-1})}{d} \frac{1}{t_n} f_2(\theta)$$

$$+ \frac{\partial(u_n - u_{n-1})}{\partial\theta} \frac{1}{t_n d} f_3(\theta) + \frac{\partial^2 u_n}{\partial\theta^2} \frac{f_4(\theta)}{t_n^2}$$

$$+ f_5(\theta) \left(g(u_n) \frac{1}{\varepsilon} + f_n \frac{1}{\varepsilon} \right) = 0, \tag{1.31}$$

$\forall n \in \{1, ..., N-1\}$. Here, $u_n(\theta)$ corresponds to the solution on the line n.

At this point we introduce a proximal approach, initially around the point $\{(U_0)_n\}$ through a constant $K > 0$ to be specified.

In partial finite differences, such a system stands for

$$\frac{u_{n+1} - 2u_n + u_{n-1}}{d^2} + \frac{(u_n - u_{n-1})}{d} \frac{1}{t_n} f_2(\theta)$$

$$+ \frac{\partial(u_n - u_{n-1})}{\partial\theta} \frac{1}{t_n d} f_3(\theta) + \frac{\partial^2 u_n}{\partial\theta^2} \frac{f_4(\theta)}{t_n^2}$$

$$+ f_5(\theta) \left(g(u_n) \frac{1}{\varepsilon} + f_n \frac{1}{\varepsilon} \right)$$

$$- \frac{K}{\varepsilon} u_n + \frac{K}{\varepsilon} (U_0)_n = 0, \tag{1.32}$$

$\forall n \in \{1, ..., N-1\}$. Here, $u_n(\theta)$ corresponds to the solution on the line n.

Thus denoting $\varepsilon = e_1$ we may obtain

$$u_{n+1} - 2u_n + u_{n-1} - \frac{K}{e_1} u_n + T_n + \hat{f}_n = 0,$$

where

$$T_n = \frac{(u_n - u_{n-1})}{d} \frac{1}{t_n} f_2(\theta) d^2$$

$$+ \frac{\partial(u_n - u_{n-1})}{\partial\theta} \frac{1}{t_n d} f_3(\theta) d^2 + \frac{\partial^2 u_n}{\partial\theta^2} \frac{f_4(\theta)}{t_n^2} d^2$$

$$+ f_5(\theta)(g(u_n)) \frac{d^2}{e_1} \tag{1.33}$$

and

$$\hat{f}_n = (f_5(\theta) f_n + K(U_0)_n) \frac{d^2}{e_1},$$

$\forall n \in \{1, \cdots, N-1\}$.

In particular, for $n = 1$, we have

$$u_2 - 2u_1 + u_0 - K\, u_1 \frac{d^2}{e_1} + T_1 + \hat{f}_1.$$

Hence

$$u_1 = a_1 u_2 + b_1 u_0 + c_1 T_1 + F_1 + E_{r_1},$$

where

$$a_1 = \left(2 + K\frac{d^2}{e_1}\right)^{-1},$$

$$b_1 = a_1$$

$$c_1 = a_1$$

$$F_1 = a_1\, \hat{f}_1.$$

$$E_{r_1} = 0.$$

With such a result for $n = 2$ we have

$$u_3 - 2u_2 + u_1 - K\, u_2 \frac{d^2}{e_1} + T_2 + \hat{f}_2.$$

Hence

$$u_2 = a_2 u_3 + b_2 u_0 + c_2 T_2 + F_2 + E_{r_2},$$

where

$$a_2 = \left(2 + K\frac{d^2}{e_1} - a_1\right)^{-1},$$

$$b_2 = a_2 b_1$$

$$c_1 = a_2(c_1 + 1)$$

$$F_1 = a_2(\hat{f}_2 + F_1)$$

$$E_{r_2} = a_2(c_1(T_1 - T_2)).$$

Reasoning inductively, having

$$u_{n-1} = a_{n-1}u_n + b_{n-1}u_0 + c_{n-1}T_{n-1} + F_{n-1} + E_{r_{n-1}},$$

for the line n, we have

$$u_{n+1} - 2u_n + u_{n-1} - K\, u_{n-1}\frac{d^2}{e_1} + T_n + \hat{f}_n.$$

Hence

$$u_n = a_n u_{n+1} + b_n u_0 + c_n T_n + F_n + E_{r_n},$$

where

$$a_n = \left(2 + K\frac{d^2}{e_1} - a_{n-1}\right)^{-1},$$

$$b_n = a_n b_{n-1}$$

$$c_n = a_n(c_{n-1} + 1)$$

$$F_n = a_n(\hat{f}_n + F_{n-1})$$

$$E_{r_n} = a_n(E_{r_{n-1}} + c_{n-1}(T_{n-1} - T_n)),$$

$\forall n \in \{1, \cdots, N-1\}$.

In particular we have $u_N = u_f$.

Thus, for $n = N - 1$ we have got

$$u_{N-1} \approx a_{N-1}u_f + b_{N-1}u_0 + c_{N-1}T_{N-1}(u_{N-1}, u_f) + F_{N-1}.$$

This last equation is an ODE in u_{N-1} which may be easily solved with the boundary conditions

$$u_{N-1}(0) = u_{N-1}(2\pi)$$

and

$$u'_{N-1}(0) = u'_{N-1}(2\pi).$$

Having u_{N-1} we may obtain u_{N-2} through $n = N - 2$.

Observe that,

$$u_{N-2} \approx a_{N-2}u_{N-1} + b_{N-2}u_0 + c_{N-2}T_{N-2}(u_{N-2}, u_{N-1}) + F_{N-2}.$$

This last equation is again an ODE in u_{N-2} which may be easily solved with the boundary conditions

$$u_{N-2}(0) = u_{N-2}(2\pi)$$

and

$$u'_{N-2}(0) = u'_{N-2}(2\pi).$$

We may continue inductively with such a reasoning up to obtaining u_1.

The next step is to replace $\{(U_0)_n\}$ by $\{u_n\}$ and repeat the process until an appropriate convergence criterion is satisfied.

The problem is then solved.

At this point we present a software in MATHEMATICA based on this last algorithm. Some changes has been implemented concerning this last previous conception, in particular in order to make it suitable as a software in MATHE-MATICA.

Here the concerning software.

****

1. $m8 = 10$;
   $d = 1.0/m8$;
   $K = 5.0$;
   $e1 = 1.0$;
   $A = 1.0$;
   $B = 1.0$;
   $Clear[t3]$;

2. $For[i = 1, i < m8 + 1, i + +,$
   $uo[i] = 0.0]$;

3. $For[k = 1, k < 20, k + +,$
   $Print[k]$;
   $a[1] = 1/(2.0 + K * d^2/e1)$;
   $b[1] = a[1]$;
   $c[1] = a[1] * (K * uo[1] + 1.0 * t3 * f5[x]) * d^2/e1$;

4. $For[i = 2, i < m8, i + +,$
   $a[i] = 1/(2.0 + K * d^2/e1 - a[i - 1])$;
   $b[i] = a[i] * (b[i - 1] + 1)$;
   $c[i] = a[i] * (c[i - 1] + (K * uo[i] + 1.0 * t3 * f5[x]) * d^2/e1)]$;
   $u[m8] = uf[x]$;

5. $For[i = 1, i < m8, i + +,$
   $Print[i]$;
   $t[m8 - i] = 1 + (m8 - i) * d$;
   $A1 = a[m8 - i] * u[m8 - i + 1] +$
   $b[m8 - i] * (-A * u[m8 - i + 1]^3 + B * u[m8 - i + 1]) * d^2/e1) * f5[x] * t3 +$
   $c[m8 - i]$
   $+ d^2 * b[m8 - i] * (f4[x] * D[u[m8 - i + 1], x, 2]/t[m8 - i]^2) * t3$
   $+ f3[x] * t3 * b[m8 - i]/t[m8 - i] * D[(uo[m8 - i + 1] - uo[m8 - i])/d, x] * d^2$
   $+ f2[x] * t3 * 1/t[m8 - i] * b[m8 - i] * d^2(uo[m8 - i + 1] - uo[m8 - i])/d$;
   $A1 = Expand[A1]$;
   $A1 = Series[A1, \{t3, 0, 2\}, \{uf[x], 0, 3\}, \{uf'[x], 0, 1\},$

$\{uf''[x], 0, 1\}, \{uf'''[x], 0, 0\}, \{uf''''[x], 0, 0\},$
$\{f5[x], 0, 2\}, \{f4[x], 0, 2\}, \{f3[x], 0, 2\},$
$\{f2[x], 0, 2\}, \{f5'[x], 0, 0\}, \{f4'[x], 0, 0\},$
$\{f3'[x], 0, 0\}, \{f2'[x], 0, 0\}, \{f5''[x], 0, 0\},$
$\{f4''[x], 0, 0\}, \{f3''[x], 0, 0\}, \{f2''[x], 0, 0\}];$
$A1 = Normal[A1];$
$u[m8 - i] = Expand[A1]];$

6. $For[i = 1, i < m8 + 1, i++,$
$uo[i] = u[i]];$
$Print[Expand[u[m8/2]]];]$
t3 = 1; For[i = 1, i ¡ m8, i++, Print["u[", i, "]=", u[i]]]

****

Here we present the lines expressions for the lines $n = 2$, $n = 4$, $n = 6$ and $n = 8$ of a total of $N = 10$ lines,

$$
\begin{aligned}
u[2] \ = \ & 0.08f5[x] - 0.00251535f2[x]f5[x] + 0.00732449f5[x]^2 + 0.2uf[x] \\
& +0.0534377f2[x]uf[x] - 0.00202221f2[x]^2uf[x] + 0.0503925f5[x]uf[x] \\
& +0.0035934f2[x]f5[x]uf[x] + 0.00580877f5[x]^2uf[x] \\
& -0.00741623f5[x]^2uf[x]^2 - 0.0276371f5[x]uf[x]^3 \\
& -0.00455457f2[x]f5[x]uf[x]^3 - 0.0103934f5[x]^2uf[x]^3 \\
& +0.0534377f3[x](uf')[x] - 0.00404441f2[x]f3[x](uf')[x] \\
& +0.0035934f3[x]f5[x](uf')[x] - 0.00506561f3[x]f5[x]uf[x]^2(uf')[x] \\
& -0.00202221f3[x]^2(uf'')[x] + 0.0206072f4[x](uf'')[x] \\
& +0.00161163f2[x]f4[x](uf'')[x] + 0.00479841f4[x]f5[x](uf'')[x] \\
& -0.00784991f4[x]f5[x]uf[x]^2(uf'')[x] \quad\quad\quad (1.34)
\end{aligned}
$$

$$
\begin{aligned}
u[4] \;=\;& 0.12f5[x] - 0.0115761f2[x]f5[x] + 0.0101622f5[x]^2 + 0.4uf[x] \\
&+0.0747751f2[x]uf[x] - 0.00755983f2[x]^2uf[x] + 0.0868571f5[x]uf[x] \\
&-0.00156623f2[x]f5[x]uf[x] + 0.00847093f5[x]^2uf[x] \\
&-0.013132f5[x]^2uf[x]^2 - 0.0534765f5[x]uf[x]^3 \\
&-0.00387899f2[x]f5[x]uf[x]^3 - 0.0177016f5[x]^2uf[x]^3 \\
&+0.0747751f3[x](uf')[x] - 0.0151197f2[x]f3[x](uf')[x] \\
&-0.00156623f3[x]f5[x](uf')[x] + 0.00341553f3[x]f5[x]uf[x]^2(uf')[x] \\
&-0.00755983f3[x]^2(uf'')[x] + 0.0323379f4[x](uf'')[x] \\
&+0.000516904f2[x]f4[x](uf'')[x] + 0.0063856f4[x]f5[x](uf'')[x] \\
&-0.0116439f4[x]f5[x]uf[x]^2(uf'')[x]
\end{aligned} \tag{1.35}
$$

$$
\begin{aligned}
u[6] \;=\;& 0.12f5[x] - 0.0178517f2[x]f5[x] + 0.00790756f5[x]^2 + 0.6uf[x] \\
&+0.0704471f2[x]uf[x] - 0.010741f2[x]^2uf[x] + 0.0978209f5[x]uf[x] \\
&-0.00934984f2[x]f5[x]uf[x] + 0.00692381f5[x]^2uf[x] \\
&-0.0131832f5[x]^2uf[x]^2 - 0.0703554f5[x]uf[x]^3 \\
&+0.00171175f2[x]f5[x]uf[x]^3 - 0.0174038f5[x]^2uf[x]^3 \\
&+0.0704471f3[x](uf')[x] - 0.0214821f2[x]f3[x](uf')[x] \\
&-0.00934984f3[x]f5[x](uf')[x] + 0.0199603f3[x]f5[x]uf[x]^2(uf')[x] \\
&-0.010741f3[x]^2(uf'')[x] + 0.0329474f4[x](uf'')[x] \\
&-0.00316522f2[x]f4[x](uf'')[x] + 0.00478074f4[x]f5[x](uf'')[x] \\
&-0.00994569f4[x]f5[x]uf[x]^2(uf'')[x]
\end{aligned} \tag{1.36}
$$

$$
\begin{aligned}
u[8] \;=\;& 0.08f5[x] - 0.015345f2[x]f5[x] + 0.002891f5[x]^2 + 0.8uf[x] \\
&+0.0445896f2[x]uf[x] - 0.00871597f2[x]^2uf[x] + 0.0723048f5[x]uf[x] \\
&-0.0117866f2[x]f5[x]uf[x] + 0.00260131f5[x]^2uf[x] \\
&-0.00564143f5[x]^2uf[x]^2 - 0.0619403f5[x]uf[x]^3 \\
&+0.00783579f2[x]f5[x]uf[x]^3 - 0.00745182f5[x]^2uf[x]^3 \\
&+0.0445896f3[x](uf')[x] - 0.0174319f2[x]f3[x](uf')[x] \\
&-0.0117866f3[x]f5[x](uf')[x] + 0.0297437f3[x]f5[x]uf[x]^2(uf')[x] \\
&-0.00871597f3[x]^2(uf'')[x] + 0.0220765f4[x](uf'')[x] \\
&-0.00363515f2[x]f4[x](uf'')[x] + 0.00171483f4[x]f5[x](uf'')[x] \\
&-0.00389297f4[x]f5[x]uf[x]^2(uf'')[x].
\end{aligned} \tag{1.37}
$$

1.6 Conclusion

In this article, we present an advance concerning the computation of a solution for a partial differential equation through the generalized method of lines. In particular, in its previous versions, we used to truncate the series in d^2 however, we have realized the results are much better by taking line solutions in series for $u_f[x]$ and its derivatives, as it is indicated in the present software.

This is a little different from the previous procedure, but with a great result in improvement as the parameter $\varepsilon > 0$ is small.

Indeed, with a sufficiently large N (number of lines), we may obtain very good results even as $\varepsilon > 0$ is very small.

Chapter 2

An Approximate Proximal Numerical Procedure Concerning the Generalized Method of Lines

2.1 Introduction

This chapter develops an approximate proximal approach for the generalized method of lines. We recall that for the generalized method of lines, the domain of the partial differential equation in question is discretized in lines (or in curves), and the concerning solution is developed on these lines, as functions of the boundary conditions and the domain boundary shape. Considering such a context, along the text we develop an approximate numerical procedure of proximal nature applicable to a large class of models in physics and engineering. Finally, in the last sections, we present numerical examples and results related to a Ginzburg-Landau type equation.

Remark 2.1.1 *This chapter has been published in a similar article format by the MDPI Journal Mathematics, reference:*
Fabio Silva Botelho, Mathematics 2022, 10(16), 2950;
https://doi.org/10.3390/math10162950 - 16 Aug 2022.

We start by recalling that the generalized method of lines was originally introduced in the book entitled "Topics on Functional Analysis, Calculus of Variations and Duality" [22], published in 2011.

Indeed, the present results are extensions and applications of previous ones which have been published since 2011, in books and articles such as [22, 17, 12, 13]. About the Sobolev spaces involved, we would mention [1, 2]. Concerning the applications, related models in physics are addressed in [4, 52].

We also emphasize that, in such a method, the domain of the partial differential equation in question is discretized in lines (or more generally, in curves), and the concerning solution is written on these lines as functions of boundary conditions and the domain boundary shape.

In fact, in its previous format, this method consists of an application of a kind of partial finite differences procedure combined with the Banach fixed point theorem to obtain the relation between two adjacent lines (or curves).

In the present article, we propose an approximate approach and a related iterative procedure of proximal nature. We highlight this as a proximal method inspired by some models concerning duality principles in D.C optimization in the calculus of variations, such as those found in Toland [81].

In the following lines and sections, we develop in detail such a numerical procedure.

With such statements in mind, let $\Omega \subset \mathbb{R}^2$ be an open, bounded and connected set where

$$\Omega = \{(x,y) \in \mathbb{R}^2 \ : \ y_1(x) \leq y \leq y_2(x), \ a \leq x \leq b\}.$$

Here, we assume, $y_1, y_2 : [a,b] \to \mathbb{R}$ are continuous functions.
Consider the Ginzburg-Landau type equation, defined by

$$\begin{cases} -\varepsilon \nabla^2 u + \alpha u^3 - \beta u = f, & \text{in } \Omega, \\ u = 0, & \text{on } \partial\Omega, \end{cases} \tag{2.1}$$

where $\varepsilon > 0$, $\alpha > 0$, $\beta > 0$ and $f \in L^2(\Omega)$.

Also, $u \in W_0^{1,2}(\Omega)$ and the equation in question must be considered in a distributional sense.

In the next section, we address the problem of approximately solving this concerning equation. We highlight the methods and the ideas exposed are applicable to a large class of similar models in physics and engineering.

2.2 The numerical method

We discretize the interval $[a,b]$ into N same measure sub-intervals, through a partition

$$P = \{x_0 = a, x_1, \cdots, x_N = b\},$$

where $x_n = a + nd$, $\forall n \in \{1, \cdots, N-1\}$. Here

$$d = \frac{(b-a)}{N}.$$

Through such a procedure, we generate N vertical lines parallel to the Cartesian axis $0y$, so that for each line n based on the point x_n we are going to compute an approximate solution $u_n(y)$ corresponding to values of u on such a line.

Considering this procedure, the equation system obtained in partial finite differences (please see [71], for concerning models in finite differences) is given by

$$-\varepsilon \left(\frac{u_{n+1} - 2u_n + u_{n-1}}{d^2} + \frac{\partial^2 u_n}{\partial y^2} \right) + \alpha u_n^3 - \beta u_n = f_n,$$

$\forall n \in \{1, \cdots, N-1\}$, with the boundary conditions

$$u_0 = 0,$$

and

$$u_N = 0.$$

Let $K > 0$ be an appropriate constant to be specified.
In a proximal approach, considering an initial solution

$$\{(u_0)_n\}$$

we redefine the system of equations in question as below indicated.

$$-\varepsilon \left(\frac{u_{n+1} - 2u_n + u_{n-1}}{d^2} + \frac{\partial^2 u_n}{\partial y^2} \right) + \alpha u_n^3 - \beta u_n + Ku_n - K(u_0)_n = f_n,$$

$\forall n \in \{1, \cdots, N-1\}$, with the boundary conditions

$$u_0 = 0,$$

and

$$u_N = 0.$$

Hence, we may denote

$$u_{n+1} - \left(2 + K\frac{d^2}{\varepsilon} \right) u_n + u_{n-1} + T(u_n) + \tilde{f}_n \frac{d^2}{\varepsilon} = 0,$$

where

$$T(u_n) = \left(-\alpha u_n^3 + \beta u_n \right) \frac{d^2}{\varepsilon} + \frac{\partial^2 u_n}{\partial y^2} d^2,$$

and $\tilde{f}_n = K(u_0)_n + f_n$, $\forall n \in \{1, \cdots, N-1\}$.

In particular, for $n = 1$, we get

$$u_2 - \left(2 + K\frac{d^2}{\varepsilon}\right)u_1 + T(u_1) + \tilde{f}_1\frac{d^2}{\varepsilon} = 0,$$

so that

$$u_1 = a_1 u_2 + b_1 T(u_2) + c_1 + E_1,$$

where

$$a_1 = \frac{1}{2 + K\frac{d^2}{\varepsilon}},$$

$$b_1 = a_1,$$

and

$$c_1 = a_1\tilde{f}_1\frac{d^2}{\varepsilon}$$

and the error E_1, proportional to $1/K$, is given by

$$E_1 = b_1(T(u_1) - T(u_2)).$$

Now, reasoning inductively, having

$$u_{n-1} = a_{n-1}u_n + b_{n-1}T(u_n) + c_{n-1} + E_{n-1}$$

for the line n, we have

$$u_{n+1} - \left(2 + K\frac{d^2}{\varepsilon}\right)u_n + a_{n-1}u_n + b_{n-1}T(u_n) + c_{n-1} + E_{n-1}$$

$$+ T(u_n) + \tilde{f}_n\frac{d^2}{\varepsilon} = 0, \tag{2.2}$$

so that

$$u_n = a_n u_{n+1} + b_n T(u_{n+1}) + c_n + E_n,$$

where

$$a_n = \frac{1}{2 + K\frac{d^2}{\varepsilon} - a_{n-1}},$$

$$b_n = a_n(b_{n-1} + 1),$$

and

$$c_n = a_n\left(c_{n-1} + \tilde{f}_n\frac{d^2}{\varepsilon}\right)$$

and the error E_n, is given by

$$E_n = a_n(E_{n-1} + b_n(T(u_n) - T(u_{n+1}))),$$

$\forall n \in \{1, \cdots, N-1\}$.

In particular, for $n = N - 1$, we have $u_N = 0$ so that,

$$
\begin{aligned}
u_{N-1} &\approx a_{N-1}u_N + b_{N-1}T(u_N) + c_{N-1} \\
&\approx a_{N-1}u_N + b_{N-1}\frac{\partial^2 u_{N-1}}{\partial y^2}d^2 + b_{N-1}(-\alpha u_N^3 + \beta u_N)\frac{d^2}{\varepsilon} + c_{N-1} \\
&= b_{N-1}\frac{\partial^2 u_{N-1}}{\partial y^2}d^2 + c_{N-1}.
\end{aligned} \tag{2.3}
$$

This last equation is an ODE from which we may easily obtain u_{N-1} with the boundary conditions

$$
u_{N-1}(y_1(x_{N-1})) = u_{N-1}(y_2(x_{N-1})) = 0.
$$

Having u_{N-1}, we may obtain u_{N-2} though the equation

$$
\begin{aligned}
u_{N-2} &\approx a_{N-2}u_{N-1} + b_{N-2}T(u_{N-1}) + c_{N-2} \\
&\approx a_{N-2}u_{N-1} + b_{N-2}\frac{\partial^2 u_{N-2}}{\partial y^2}d^2 + b_{N-2}(-\alpha u_{N-1}^3 + \beta u_{N-1})\frac{d^2}{\varepsilon} + c_{N-2}, \tag{2.4}
\end{aligned}
$$

with the boundary conditions

$$
u_{N-2}(y_1(x_{N-2})) = u_{N-2}(y_2(x_{N-2})) = 0.
$$

An so on, up to finding u_1.

The next step is to replace $\{(u_0)_n\}$ by $\{u_n\}$ and then to repeat the process until an appropriate convergence criterion is satisfied.

The problem is then approximately solved.

2.3 A numerical example

We present numerical results for $\Omega = [0,1] \times [0,1]$, $\alpha = \beta = 1$, $f \equiv 1$ in Ω, $N = 100$, $K = 50$ and for

$$
\varepsilon = 0.1, \ 0.01 \ \text{and} \ 0.001.
$$

For such values of ε, please see Figures 2.1, 2.2 and 2.3, respectively.

Remark 2.3.1 *We observe that as $\varepsilon > 0$ decreases to the value 0.001 the solution approaches the constant value 1.3247 along the domain, up to the satisfaction of boundary conditions. This is expected, since this value is an approximate solution of equation $u^3 - u - 1 = 0$.*

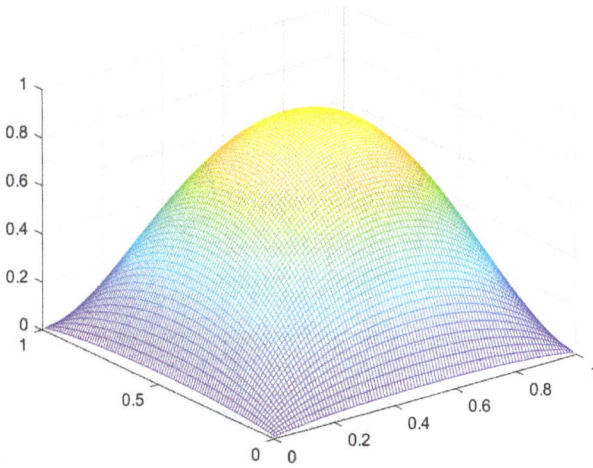

Figure 2.1: Solution u for $\varepsilon = 0.1$.

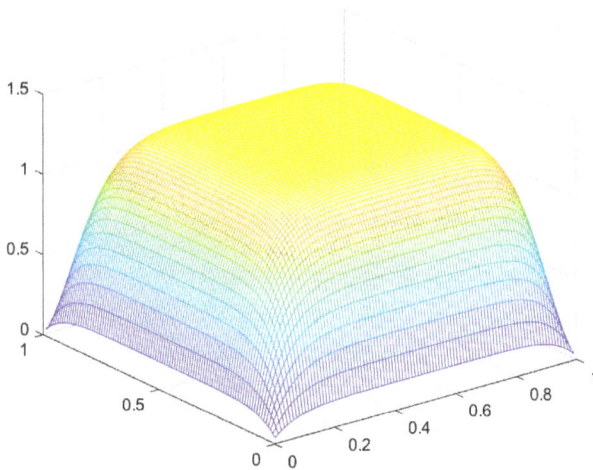

Figure 2.2: Solution u for $\varepsilon = 0.01$.

2.4 A general proximal explicit approach

Based on the algorithm presented in the last section, we develop a software in MATHEMATICA to approximately solve the following equation.

$$\begin{cases} -\varepsilon \left(\frac{\partial^2 u}{\partial r^2} + \frac{1}{r}\frac{\partial u}{\partial r} + \frac{1}{r^2}\frac{\partial^2 u}{\partial \theta^2} \right) + \alpha u^3 - \beta u = f, & \text{in } \Omega, \\ u = 0, & \text{on } \partial\Omega_1, \\ u = u_f(\theta), & \text{on } \partial\Gamma_2. \end{cases} \quad (2.5)$$

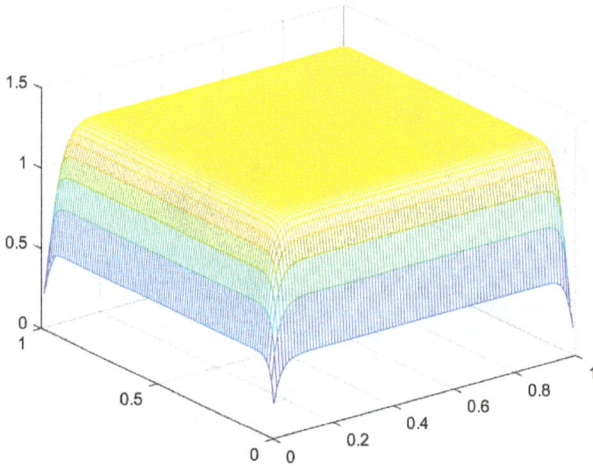

Figure 2.3: Solution u for $\varepsilon = 0.001$.

Here

$$\Omega = \{(r,\theta) \in \mathbb{R}^2 \ : \ 1 \le r \le 2, \, 0 \le \theta \le 2\pi\},$$

$$\partial\Omega_1 = \{(1,\theta) \in \mathbb{R}^2 \ : \ 0 \le \theta \le 2\pi\},$$

$$\partial\Omega_2 = \{(2,\theta) \in \mathbb{R}^2 \ : \ 0 \le \theta \le 2\pi\},$$

$\alpha = \beta = 1, K = 10, N = 100$ and $f \equiv 1$, on Ω.

At this point we present such a software in MATHEMATICA.

****

1. $m8 = 100$;

2. $d = 1.0/m8$;

3. $K = 10.0$;

4. $e1 = 0.01$;

5. $A = 1.0$;

6. $B = 1.0$;

7. For[i = 1, i < m8 + 1, i++,
   uo[i] = 0.0];

8. For[k = 1, k < 150, k++,

    Print[k];

    a[1] = 1/(2.0 + $K*d^2/e1$);

    b[1] = a[1];

    c[1] = a[1]*(K*uo[1] + 1.0)*$d^2/e1$;

    For[i = 2, i < m8, i++,

    a[i] = 1/(2.0 + $K*d^2/e1 - a[i-1]$);

    b[i] = a[i]*(b[i - 1] + 1);

    c[i] = a[i]*(c[i - 1] + $(K*uo[i] + 1.0)*d^2/e1)$];

9. u[m8] = uf[x]; d1 = 1.0;

10. For[i = 1, i < m8, i++,

    t[m8 - i] = 1 + (m8 - i)*d;

    A1 = (a[m8 - i]*u[m8 - i + 1] +

    b[m8 - i]*(-A*$u[m8-i+1]^3$ + B*u[m8 - i + 1])*$d^2/e1*d1^2$ +

    c[m8 - i] +

    d^2*d1^2*b[m8 - i]*(D[u[m8 - i + 1], x, 2]/$t[m8-i]^2$) +

    $d1^2*1/t[m8-i]$*b[m8 - i]* d^2 (uo[m8 - i + 1] - uo[m8 - i])/d)/(1.0);

    A1 = Expand[A1];

    A1 = Series[A1, {uf[x], 0, 3}, {uf'[x], 0, 1}, {uf"[x], 0, 1}, {uf'"[x], 0, 0}, {uf""[x], 0, 0}];

    A1 = Normal[A1];

    u[m8 - i] = Expand[A1]];

    For[i = 1, i < m8 + 1, i++,

    uo[i] = u[i]]; d1 = 1.0;

    Print[Expand[u[m8/2]]]]

****

For such a general approach, for $\varepsilon = 0.1$, we have obtained the following lines (here x stands for θ).

$$u[10](x) = 0.4780 + 0.0122\, u_f[x] - 0.0115\, u_f[x]^2 + 0.0083\, u_f[x]^3 + 0.00069 u_f''[x]$$
$$- 0.0014\, u_f[x](u_f'')[x] + 0.0016\, uf[x]^2 u_f''[x] - 0.00092\, u_f[x]^3 u_f''[x] \quad (2.6)$$

$$u[20](x) = 0.7919 + 0.0241\,u_f[x] - 0.0225\,u_f[x]^2 + 0.0163\,u_f[x]^3 + 0.0012\,u_f''[x]$$
$$- 0.0025\,u_f[x](u_f'')[x] + 0.0030\,u_f[x]^2(u_f'')[x] - 0.0018\,u_f[x]^3(u_f'')[x]. \quad (2.7)$$

$$u[30](x) = 0.9823 + 0.0404\,u_f[x] - 0.0375\,u_f[x]^2 + 0.0266\,u_f[x]^3 + 0.00180\,(u_f'')[x]$$
$$- 0.00362\,u_f[x](u_f'')[x] + 0.0043\,u_f[x]^2(uf'')[x] - 0.0028\,u_f[x]^3(uf'')[x] \quad (2.8)$$

$$u[40](x) = 1.0888 + 0.0698\,u_f[x] - 0.0632\,u_f[x]^2 + 0.0433\,u_f[x]^3 + 0.0026(u_f'')[x]$$
$$- 0.0051\,u_f[x](u_f'')[x] + 0.0061\,u_f[x]^2(u_f'')[x] - 0.0043\,u_f[x]^3(u_f'')[x] \quad (2.9)$$

$$u[50](x) = 1.1316 + 0.1277\,u_f[x] - 0.1101\,u_f[x]^2 + 0.0695\,u_f[x]^3 + 0.0037\,(u_f'')[x]$$
$$- 0.0073\,u_f[x](u_f'')[x] + 0.0084\,u_f[x]^2(u_f'')[x] - 0.0062\,u_f[x]^3(u_f'')[x] \quad (2.10)$$

$$u[60](x) = 1.1104 + 0.2389\,u_f[x] - 0.1866\,u_f[x]^2 + 0.0988\,u_f[x]^3 + 0.0053(u_f'')[x]$$
$$- 0.0099\,u_f[x](u_f'')[x] + 0.0105\,u_f[x]^2(u_f'')[x] - 0.0075\,u_f[x]^3(u_f'')[x] \quad (2.11)$$

$$u[70](x) = 1.0050 + 0.4298\,u_f[x] - 0.273813\,u_f[x]^2 + 0.0949\,u_f[x]^3 + 0.0070\,(u_f'')[x]$$
$$- 0.0116\,u_f[x](u_f'')[x] + 0.0102\,u_f[x]^2(u_f'')[x] - 0.0061\,u_f[x]^3(u_f'')[x] \quad (2.12)$$

$$u[80](x) = 0.7838 + 0.6855\,u_f[x] - 0.2892\,u_f[x]^2 + 0.0161\,u_f[x]^3 + 0.0075\,(u_f'')[x]$$
$$- 0.0098 u_f[x](u_f'')[x] + 0.0063084 uf[x]^2(uf'')[x] - 0.0027 u_f[x]^3(u_f'')[x] \quad (2.13)$$

$$u[90](x) = 0.4359 + 0.9077\,u_f[x] - 0.1621 u_f[x]^2 - 0.0563\,u_f[x]^3 + 0.0051\,(u_f'')[x]$$
$$- 0.0047\,u_f[x](u_f'')[x] + 0.0023\,u_f[x]^2(u_f'')[x] - 0.00098\,u_f[x]^3(u_f'')[x] \quad (2.14)$$

For $\varepsilon = 0.01$, we have obtained the following line expressions.

$$u[10](x) = 1.0057 + 2.07 * 10^{-11}\,u_f[x] - 1.85 * 10^{-11}\,u_f[x]^2 + 1.13 * 10^{-11}\,u_f[x]^3$$
$$+ 4.70 * 10^{-13}\,(u_f'')[x] - 8.44 * 10^{-13} u_f[x](u_f'')[x]$$
$$+ 7.85 * 10^{-13}\,u_f[x]^2(u_f'')[x] - 6.96 * 10^{-14}\,u_f[x]^3(u_f'')[x] \quad (2.15)$$

$$u[20](x) = 1.2512 + 2.13 * 10^{-10}\,u_f[x] - 1.90 * 10^{-10}\,u_f[x]^2 + 1.16 * 10^{-10}\,u_f[x]^3$$
$$+ 3.94 * 10^{-12}\,(u_f'')[x] - 7.09 * 10^{-12} u_f[x](uf'')[x] + 6.61 * 10^{-12}$$
$$u_f[x]^2(u_f'')[x] - 7.17 * 10^{-13}\,u_f[x]^3(u_f'')[x] \quad (2.16)$$

$$
\begin{aligned}
u[30](x) = \ & 1.3078 + 3.80 * 10^{-9} uf[x] - 3.39 * 10^{-9}\, u_f[x]^2 + 2.07 * 10^{-9}\, u_f[x]^3 \\
& + 5.65 * 10^{-11} (u_f'')[x] - 1.018 * 10^{-10}\, u_f[x](u_f'')[x] \\
& + 9.52 * 10^{-11}\, u_f[x]^2 (u_f'')[x] - 1.27 * 10^{-11}\, uf[x]^3 (u_f'')[x] \quad (2.17)
\end{aligned}
$$

$$
\begin{aligned}
u[40](x) = \ & 1.3208 + 7.82 * 10^{-8} u_f[x] - 6.98 * 10^{-8}\, u_f[x]^2 + 4.27 * 10^{-8}\, u_f[x]^3 \\
& + 9.27 * 10^{-10} (u_f'')[x] - 1.67 * 10^{-9} u_f[x](u_f'')[x] \\
& + 1.57 * 10^{-9}\, u_f[x]^2 (u_f'')[x] - 2.62 * 10^{-10}\, u_f[x]^3 (u_f'')[x] \quad (2.18)
\end{aligned}
$$

$$
\begin{aligned}
u[50](x) = \ & 1.3238 + 1.67 * 10^{-6}\, u_f[x] - 1.49 * 10^{-6}\, u_f[x]^2 + 9.15 * 10^{-7}\, uf[x]^3 \\
& + 1.54 * 10^{-8} (u_f'')[x] - 2.79 * 10^{-8}\, u_f[x](u_f'')[x] \\
& + 2.64 * 10^{-8}\, u_f[x]^2 (u_f'')[x] - 5.62 * 10^{-9}\, u_f[x]^3 (u_f'')[x] \quad (2.19)
\end{aligned}
$$

$$
\begin{aligned}
u[60](x) = \ & 1.32449 + 0.000036\, u_f[x] - 0.000032\, u_f[x]^2 + 0.000019\, u_f[x]^3 \\
& + 2.51 * 10^{-7} (u_f'')[x] - 4.57 * 10^{-7}\, u_f[x](u_f'')[x] \\
& + 4.36 * 10^{-7}\, u_f[x]^2 (u_f'')[x] - 1.21 * 10^{-7}\, u_f[x]^3 (u_f'')[x] \quad (2.20)
\end{aligned}
$$

$$
\begin{aligned}
u[70](x) = \ & 1.32425 + 0.00079\, u_f[x] - 0.00070\, u_f[x]^2 + 0.00043\, u_f[x]^3 \\
& + 3.89 * 10^{-6} (u_f'')[x] - 7.12 * 10^{-6}\, u_f[x](u_f'')[x] \\
& + 6.89 * 10^{-6}\, u_f[x]^2 (u_f'')[x] - 2.64 * 10^{-6}\, u_f[x]^3 (u_f'')[x] \quad (2.21)
\end{aligned}
$$

$$
\begin{aligned}
u[80](x) = \ & 1.31561 + 0.017\, u_f[x] - 0.015\, u_f[x]^2 + 0.009\, u_f[x]^3 \\
& + 0.000053 (u_f'')[x] - 0.000098\, u_f[x](u_f'')[x] \\
& + 0.000095\, u_f[x]^2 (u_f'')[x] - 0.000051\, u_f[x]^3 (u_f'')[x] \quad (2.22)
\end{aligned}
$$

$$
\begin{aligned}
u[90](x) = \ & 1.14766 + 0.296 u_f[x] - 0.1991\, u_f[x]^2 + 0.0638\, u_f[x]^3 \\
& + 0.00044 (u_f'')[x] - 0.00067\, u_f[x](u_f'')[x] \\
& + 0.00046\, u_f[x]^2 (u_f'')[x] - 0.00018\, u_f[x]^3 (u_f'')[x] \quad (2.23)
\end{aligned}
$$

Remark 2.4.1 *We observe that as $\varepsilon > 0$ decreases to the value 0.01 the solution approaches the constant value 1.3247 along the domain, up to the satisfaction of boundary conditions. This is expected, since this value is an approximate solution of equation $u^3 - u - 1 = 0$.*

2.5   Conclusion

In this article, we develop an approximate numerical procedure related to the generalized method of lines. Such a procedure is proximal and involves a parameter $K > 0$ which minimizes the concerning numerical error of the initial approximation.

We have presented numerical results concerning a Ginzburg-Landau type equation. The results obtained are consistent with those expected for such a mathematical model.

In the last section, we present a software in MATHEMATICA for approximately solving a large class of similar systems of partial differential equations.

Finally, for future research, we intend to extend the results for the Navier-Stokes system in fluid mechanics and related time dependent models in physics and engineering.

Chapter 3

Approximate Numerical Procedures for the Navier-Stokes System through the Generalized Method of Lines

3.1 Introduction

In this chapter, we develop approximate solutions for the time independent incompressible Navier-Stokes system through the generalized method of lines. We recall again, for such a method, the domain of the partial differential equation in question is discretized in lines, and the concerning solution is written on these lines as functions of the boundary conditions and boundary shape. We emphasize the first article part concerns the application and extension of an approximate proximal approach published in [18]. We develop an analogous algorithm as those presented in [18] but now for a Navier-Stokes system, which is more complex than the systems previously addressed. In this first step, we present an algorithm and respective software in MATLAB®.

Furthermore, we have developed and presented related software in MATHE-MATICA for a simpler type of domain but also concerning the mentioned proximal approach. Finally, in the last section, we present a software and related line expressions through the original conception of the generalized method of lines, so that in such related numerical examples, the main results are established through applications of the Banach fixed point theorem.

Remark 3.1.1 *We also highlight that the following two paragraphs in this article (a relatively small part) overlap with Chapter 28, starting on page 526, in the book by F.S. Botelho, [13], published in 2020, by CRC Taylor and Francis. However, we emphasize the present article includes substantial new parts, including a concerning software not included in the previous version of 2020. Another novelty in the present version is the establishment of appropriate boundary conditions for an elliptic system equivalent to original Navier-Stokes one. Such new boundary conditions and concerning results are indicated in Section 3.2.*

At this point we describe the system in question.

Consider $\Omega \subset \mathbb{R}^2$ an open, bounded and connected set with a regular (Lipschitzian) internal boundary denoted Γ_0, and a regular external one denoted by Γ_1. For a two-dimensional motion of a fluid on Ω, we denote by $u : \Omega \to \mathbb{R}$ the velocity field in the direction x of the Cartesian system (x, y), by $v : \Omega \to \mathbb{R}$, the velocity field in the direction y and by $p : \Omega \to \mathbb{R}$, the pressure one. Moreover, ρ denotes the fluid density, μ is the viscosity coefficient and g denotes the gravity field. Under such notation and statements, the time-independent incompressible Navier-Stokes system of partial differential equations stands for,

$$\begin{cases} \mu \nabla^2 u - \rho\, u\, u_x - \rho\, v\, u_y - p_x + \rho\, g_x = 0, & \text{in } \Omega, \\ \mu \nabla^2 v - \rho\, u\, v_x - \rho\, v\, v_y - p_y + \rho\, g_y = 0, & \text{in } \Omega, \\ u_x + v_y = 0, & \text{in } \Omega, \end{cases} \quad (3.1)$$

$$\begin{cases} u = v = 0, & \text{on } \Gamma_0, \\ u = u_\infty,\ v = 0,\ p = p_\infty, & \text{on } \Gamma_1 \end{cases} \quad (3.2)$$

At first we look for solutions $(u, v, P) \in W^{2,2}(\Omega) \times W^{2,2}(\Omega) \times W^{1,2}(\Omega)$. We emphasize that details about such Sobolev spaces may be found in [1].

About the references, we emphasize that related existence, numerical and theoretical results for similar systems may be found in [39, 40, 41, 42] and [73], respectively. In particular [73] addresses extensively both theoretical and numerical methods and an interesting interplay between them. Moreover, related finite difference schemes are addressed in [71].

3.2   Details about an equivalent elliptic system

Defining now $P = p/\rho$ and $\nu = \mu/\rho$, consider again the Navier-Stokes system in the following format

$$
\begin{cases}
\nu\nabla^2 u - u\partial_x u - v\partial_y u - \partial_x P + g_x = 0, & \text{in } \Omega, \\
\nu\nabla^2 v - u\partial_x v - v\partial_y v - \partial_y P + g_y = 0, & \text{in } \Omega, \\
\partial_x u + \partial_y v = 0, & \text{in } \Omega,
\end{cases}
\tag{3.3}
$$

$$
\begin{cases}
u = v = 0, & \text{on } \Gamma_0, \\
u = u_\infty, \; v = 0, \; P = P_\infty, & \text{on } \Gamma_1
\end{cases}
\tag{3.4}
$$

As previously mentioned, at first we look for solutions $(u, v, P) \in W^{2,2}(\Omega) \times W^{2,2}(\Omega) \times W^{1,2}(\Omega)$.

We are going to obtain an equivalent Elliptic system with appropriate boundary conditions.

Our main result is summarized by the following theorem.

Theorem 3.2.1 *Let $\Omega \subset \mathbb{R}^2$ be an open, bounded, connected set with a regular (Lipschitzian) boundary.*

Assume $u, v, P \in W^{2,2}(\Omega)$ are such that

$$
\begin{cases}
\nu\nabla^2 u - u\, u_x - v\, u_y - P_x + g_x = 0, & \text{in } \Omega, \\
\nu\nabla^2 v - u\, v_x - v\, v_y - P_y + g_y = 0, & \text{in } \Omega, \\
\nabla^2 P + u_x^2 + v_y^2 + 2u_y v_x - \operatorname{div} g = 0, & \text{in } \Omega,
\end{cases}
\tag{3.5}
$$

$$
\begin{cases}
u = u_0, \; v = v_0, & \text{on } \partial\Omega, \\
u_x + v_y = 0, & \text{on } \partial\Omega.
\end{cases}
\tag{3.6}
$$

Suppose also the unique solution of equation in w

$$
\nu\nabla^2 w - u\, w_x - v\, w_y = 0, \text{ in } \Omega
$$

with the boundary conditions

$$
w = 0 \text{ on } \partial\Omega,
$$

is

$$
w = 0, \text{ in } \Omega.
$$

Under such hypotheses, u, v, P solve the following Navier-Stokes system

$$\begin{cases} v\nabla^2 u - u\, u_x - v\, u_y - P_x + g_x = 0, & in\ \Omega, \\ v\nabla^2 v - u\, v_x - v\, v_y - P_y + g_y = 0, & in\ \Omega, \\ u_x + v_y = 0, & in\ \Omega, \end{cases} \tag{3.7}$$

$$\begin{cases} u = u_0,\ v = v_0, & on\ \partial\Omega, \\ u_x + v_y = 0, & on\ \partial\Omega. \end{cases} \tag{3.8}$$

Proof 3.1   In (3.5), taking the derivative in x of the first equation and adding with the derivative in y of the second equation, we obtain

$$v\nabla^2(u_x + v_y) - u(u_x + v_y)_x - v(u_x + v_y)_y$$
$$-\nabla^2 P - u_x^2 - v_y^2 - 2u_y v_x + \operatorname{div} g = 0,\ in\ \Omega \tag{3.9}$$

From the hypotheses, u, v, P are such that

$$\nabla^2 P + u_x^2 + v_y^2 + 2u_y v_x - \operatorname{div} g = 0,\ in\ \Omega,$$

From this and (13.14), we get

$$v\nabla^2(u_x + v_y) - u(u_x + v_y)_x - v(u_x + v_y)_y = 0,\ in\ \Omega. \tag{3.10}$$

Denoting $w = u_x + v_y$, from this last equation we obtain

$$v\nabla^2 w - uw_x - vw_y = 0,\ in\ \Omega.$$

From the hypothesis, the unique solution of this last equation with the boundary conditions $w = 0$, on $\partial\Omega$, is $w = 0$.

From this and (3.10) we have

$$u_x + v_y = 0,$$

in Ω with the boundary conditions

$$u_x + v_y = 0,\ on\ \partial\Omega.$$

The proof is complete.

Remark 3.2.2 *The process of obtaining such a system with a Laplace operator in P in the third equation is a standard and well known one.*

The novelty here is the identification of the corrected related boundary conditions obtained through an appropriate solution of equation (3.10).

3.3 An approximate proximal approach

In this section, we develop an approximate proximal numerical procedure for the model in question.

Such results are extensions of previous ones published in F.S. Botelho [18] now for the Navier-Stokes system context.

More specifically, neglecting the gravity field, we solve the system of equations

$$
\begin{cases}
v\nabla^2 u - u\partial_x u - v\partial_y u - \partial_x P = 0, & \text{in } \Omega, \\[2mm]
v\nabla^2 v - u\partial_x v - v\partial_y v - \partial_y P = 0, & \text{in } \Omega, \\[2mm]
\nabla^2 P + (\partial_x u)^2 + (\partial_y v)^2 + 2(\partial_y u)(\partial_x v) = 0, & \text{in } \Omega.
\end{cases}
\tag{3.11}
$$

We present a software similar to those presented in [18], with $v = 0.0177$, and with

$$
\Omega = [0,1] \times [0,1]
$$

with the boundary conditions

$$
u = u_0 = 0.65y(1-y), \; v = v_0 = 0, \; P = p_0 = 0.15 \text{ on } [0,y], \forall y \in [0,1],
$$

$$
u = v = P_y = 0, \text{ on } [x,0] \text{ and } [x,1], \; \forall x \in [0,1],
$$

$$
u_x = v_x = 0, \text{ and } P_x = 0 \text{ on } [1,y], \; \forall y \in [0,1].
$$

The equation (3.11), in partial finite differences, stands for

$$
v\left(\frac{u_{n+1} - 2u_n + u_{n-1}}{d^2} + \frac{\partial^2 u_n}{\partial y^2}\right) - u_n \frac{(u_n - u_{n-1})}{d} - v_n \frac{\partial u_n}{\partial y}
$$
$$
- \frac{P_n - P_{n-1}}{d} = 0,
\tag{3.12}
$$

$$
v\left(\frac{v_{n+1} - 2v_n + v_{n-1}}{d^2} + \frac{\partial^2 v_n}{\partial y^2}\right) - u_n \frac{(v_n - v_{n-1})}{d} - v_n \frac{\partial v_n}{\partial y}
$$
$$
- \frac{\partial P_n}{\partial y} = 0,
\tag{3.13}
$$

$$
\left(\frac{P_{n+1} - 2P_n + P_{n-1}}{d^2} + \frac{\partial^2 P_n}{\partial y^2}\right) + (u_n - u_{n-1})\frac{(u_n - u_{n-1})}{d^2} + \left(\frac{\partial v_n}{\partial y}\right)^2
$$
$$
+ 2\frac{\partial u_n}{\partial y}\left(\frac{v_n - v_{n-1}}{d}\right) = 0.
\tag{3.14}
$$

After linearizing such a system about U_0, V_0, P_0 and introducing the proximal formulation, for an appropriate non-negative real constant K, we get

$$\nu \left(\frac{u_{n+1} - 2u_n + u_{n-1}}{d^2} + \frac{\partial^2 u_n}{\partial y^2} \right) - (U_0)_n \frac{(u_n - (U_0)_{n-1})}{d} - (V_0)_n \frac{\partial u_n}{\partial y}$$

$$- \frac{(P_0)_n - (P_0)_{n-1}}{d} - K u_n + K (U_0)_n = 0, \tag{3.15}$$

$$\nu \left(\frac{v_{n+1} - 2v_n + v_{n-1}}{d^2} + \frac{\partial^2 v_n}{\partial y^2} \right) - (U_0)_n \frac{(v_n - (V_0)_{n-1})}{d} - (V_0)_n \frac{\partial v_n}{\partial y}$$

$$- \frac{\partial (P_0)_n}{\partial y} - K v_n + K (V_0)_n = 0, \tag{3.16}$$

$$\left(\frac{P_{n+1} - 2P_n + P_{n-1}}{d^2} + \frac{\partial^2 P_n}{\partial y^2} \right) + (u_{n+1} - (U_0)_n) \frac{((u)_{n+1} - (U_0)_n)}{d^2} + \left(\frac{\partial (V_0)_n}{\partial y} \right)^2$$

$$+ 2 \frac{\partial (U_0)_n}{\partial y} \left(\frac{v_{n+1} - (V_0)_n}{d} \right) - K P_n + K (P_0)_n = 0. \tag{3.17}$$

At this point denoting $\nu = e_1$, we define

$$(T_1)_n = \left(-(U_0)_n \frac{(u_n - (U_0)_{n-1})}{d} - (V_0)_n \frac{\partial u_n}{\partial y} - \frac{(P_0)_n - (P_0)_{n-1}}{d} \right) \frac{d^2}{e_1} + \frac{\partial^2 u_n}{\partial y^2} d^2,$$

$$(T_2)_n = \left(-(U_0)_n \frac{(v_n - (V_0)_{n-1})}{d} - (V_0)_n \frac{\partial v_n}{\partial y} - \frac{\partial (P_0)_n}{\partial y} \right) \frac{d^2}{e_1} + \frac{\partial^2 v_n}{\partial y^2} d^2,$$

and

$$(T_3)_n = (u_{n+1} - (U_0)_n) \frac{((u)_{n+1} - (U_0)_n)}{d^2} d^2 + \left(\frac{\partial (V_0)_n}{\partial y} \right)^2 d^2$$

$$+ 2 \frac{\partial (U_0)_n}{\partial y} \left(\frac{v_{n+1} - (V_0)_n}{d} \right) d^2 + \frac{\partial^2 P_n}{\partial y^2} d^2. \tag{3.18}$$

Therefore, we may write

$$u_{n+1} - 2u_n + u_{n-1} - K u_n \frac{d^2}{e_1} + (T_1)_n + (f_1)_n = 0,$$

where

$$(f_1)_n = K(U_0)_n \frac{d^2}{e_1},$$

$\forall n \in \{1, \cdots, N-1\}$.

In particular for $n = 1$, we obtain

$$u_2 - 2u_1 + u_0 - K u_1 \frac{d^2}{e_1} + (T_1)_1 + (f_1)_1 = 0,$$

so that

$$u_1 = a_1 u_2 + b_1 u_0 + c_1 (T_1)_1 + (h_1)_1 + (E_r)_1,$$

where

$$a_1 = \left(2 + K\frac{d^2}{e_1}\right)^{-1},$$

$$b_1 = a_1$$

$$c_1 = a_1$$

$$(h_1)_1 = a_1 (f_1)_1,$$

$$(E_r)_1 = 0.$$

Similarly, for $n = 2$ we get

$$u_3 - 2u_2 + u_1 - K u_2 \frac{d^2}{e_2} + (T_1)_2 + (f_1)_2 = 0,$$

so that

$$u_2 = a_2 u_3 + b_2 u_0 + c_2 (T_1)_2 + (h_1)_2 + (E_r)_2,$$

where

$$a_2 = \left(2 + K\frac{d^2}{e_1} - a_1\right)^{-1},$$

$$b_2 = a_2 b_1$$

$$c_2 = a_2 (c_1 + 1)$$

$$(h_1)_2 = a_2 ((h_1)_1 + (f_1)_2),$$

$$(E_1)_2 = a_2 (c_1 ((T_1)_1 - (T_1)_2)).$$

Reasoning inductively, having

$$u_{n-1} = a_{n-1} u_n + b_{n-1} u_0 + c_{n-1} (T_1)_{n-1} + (h_1)_{n-1} + (E_r)_{n-1},$$

we obtain

$$u_n = a_n u_{n+1} + b_n u_0 + c_n (T_1)_n + (h_1)_n + (E_r)_n, \tag{3.19}$$

where

$$a_n = \left(2 + K\frac{d^2}{e_1} - a_{n-1}\right)^{-1},$$

$$b_n = a_n b_{n-1}$$

$$c_n = a_n(c_{n-1} + 1)$$

$$(h_1)_n = a_n((h_1)_{n-1} + (f_1)_n),$$

$$(E_r) = a_n((E_r)_{n-1} + c_{n-1}((T_1)_{n-1} - (T_1)_n)),$$

$\forall n \in \{1, \cdots, N-1\}$.

Observe now that $n = N - 1$ we have $u_{N-1} = u_N$, so that

$$
\begin{aligned}
u_{N-1} \approx\ & a_{N-1}u_{N-1} + b_{N-1}u_0 + c_{N-1}(T_1)_{N-1} + (h_1)_{N-1} \\
& a_{N-1}u_{N-1} + b_{N-1}u_0 \\
& + c_{N-1}\frac{\partial^2 u_{N-1}}{\partial y^2}d^2 \\
& + c_{N-1}\left(-(U_0)_n\frac{(u_n - (U_0)_{n-1})}{d} - (V_0)_n\frac{\partial u_n}{\partial y} - \frac{(P_0)_n - (P_0)_{n-1}}{d}\right)\frac{d^2}{e_1} \\
& + (h_1)_{N-1}
\end{aligned}
\tag{3.20}
$$

This last equation is a second order ODE in u_{N-1} which must be solved with the boundary conditions

$$u_{N-1}(0) = u_{N-1}(1) = 0.$$

Summarizing we have obtained u_{N-1}.

Similarly, we may obtain v_{N-1} and P_{N-1}.

Having u_{N-1} we may obtain u_{N-2} with $n = N - 2$ in equation (3.19) (neglecting $(E_r)_{N-2}$.)

Similarly, we may obtain v_{N-2} and P_{N-2}.

Having u_{N-2} we may obtain u_{N-3} with $n = N - 3$ in equation (3.19) (neglecting $(E_r)_{N-3}$.)

Similarly, we may obtain v_{N-3} and P_{N-3}.

And so on up to obtaining u_1, v_1 and P_1.

The next step is to replace $\{(U_0)_n, (V_0)_n, (P_0)_n\}$ by $\{u_n, v_n, P_n\}$ and repeat the process until an appropriate convergence criterion is satisfied.

Here, we present a concerning software in MATLAB based in this last algorithm (with small changes and differences where we have set $K = 155$ and $v = 0.047$).

```
******************************
clearall
    m8 = 500;
    d = 1/m8;
    m9 = 140;
    d1 = 1/m9;
    e1 = 0.05;
    K = 155.0;
    m2 = zeros(m9 − 1, m9 − 1);
    for i = 2 : m9 − 2
    m2(i, i) = −2.0;
    m2(i, i + 1) = 1.0;
    m2(i, i − 1) = 1.0;
    end;
    m2(1, 1) = −1.0;
    m2(1, 2) = 1.0;
    m2(m9 − 1, m9 − 1) = −1.0;
    m2(m9 − 1, m9 − 2) = 1.0;
    m22 = zeros(m9 − 1, m9 − 1);
    for i = 2 : m9 − 2
    m22(i, i) = −2.0;
    m22(i, i + 1) = 1.0;
    m22(i, i − 1) = 1.0;
    end;
    m22(1, 1) = −2.0;
    m22(1, 2) = 1.0;
    m22(m9 − 1, m9 − 1) = −2.0;
    m22(m9 − 1, m9 − 2) = 1.0;
    m1a = zeros(m9 − 1, m9 − 1);
    m1b = zeros(m9 − 1, m9 − 1);
    for i = 1 : m9 − 2
    m1a(i, i) = −1.0;
    m1a(i, i + 1) = 1.0;
    end;
    m1a(m9 − 1, m9 − 1) = −1.0;
    for i = 2 : m9 − 1
    m1b(i, i) = 1.0;
    m1b(i, i − 1) = −1.0;
    end;
    m1b(1, 1) = 1.0;
    m1 = (m1a + m1b)/2;
    Id = eye(m9 − 1);
    a(1) = 1/(2 + K * d²/e1);
```

$b(1) = 1/(2 + K * d^2/e1);$
$c(1) = 1/(2 + K * d^2/e1);$
$for\ i = 2 : m8 - 1$
$a(i) = 1/(2 - a(i-1) + K * d^2/e1);$
$b(i) = a(i) * b(i-1);$
$c(i) = (c(i-1) + 1) * a(i);$
$end;$
$for\ i = 1 : m9 - 1$
$u5(i,1) = 0.55 * i * d1 * (1 - i * d1);$
$end;$
$uo = u5;$
$vo = zeros(m9 - 1, 1);$
$po = 0.15 * ones(m9 - 1, 1);$
$for\ i = 1 : m8 - 1$
$Uo(:,i) = 0.25 * ones(m9 - 1, 1);$
$Vo(:,i) = 0.05 * ones(m9 - 1, 1);$
$Po(:,i) = 0.05 * ones(m9 - 1, 1);$
$end;$
$for\ k7 = 1 : 1$
$e1 = e1 * .94;$
$b14 = 1.0;$
$k1 = 1;$
$k1max = 1000;$
$\ while\ (b14 > 10^{-4.0})\ and\ (k1 < 1000)$
$k1 = k1 + 1;$
$a(1) = 1/(2 + K * d^2/e1);$
$b(1) = a(1);$
$c1(:,1) = a(1) * K * Uo(:,1) * d^2/e1;$
$c2(:,1) = a(1) * K * Vo(:,1) * d^2/e1;$
$c3(:,1) = a(1) * (K * Po(:,1) * d^2/e1);$
$for\ i = 2 : m8 - 1$
$a(i) = 1/(2 + K * d^2/e1 - a(i-1));$
$b(i) = a(i) * (b(i-1));$
$c1(:,i) = a(i) * (c1(:,i-1) + K * Uo(:,i) * d^2/e1);$
$c2(:,i) = a(i) * (c2(:,i-1) + K * Vo(:,i) * d^2/e1);$
$c3(:,i) = a(i) * (c3(:,i-1) + K * Po(:,i) * d^2/e1);$
$end;$
$i = 1;$
$M50 = (Id - a(m8 - 1) * Id - c(m8 - 1) * m22/d1^2 * d^2);$
$z1 = b(m8 - 1) * uo + c1(:,m8 - i)$
$z1 = z1 + c(m8 - 1) * (-Vo(:,m8 - i).*(m1 * Uo(:,m8 - i)))/d1 * d^2/e1;$
$M60 = (Id - a(m8 - 1) * Id - c(m8 - 1) * m22/d1^2 * d^2);$
$z2 = b(m8 - 1) * vo + c2(:,m8 - i)$

$z2 = z2 + c(m8 - 1) * (-Vo(:,m8 - i)). * (m1 * Vo(:,m8 - i))/d1 * d^2/e1;$
$M70 = (Id - a(m8 - 1) * Id - c(m8 - 1) * m2/d1^2 * d^2);$
$z3 = b(m8 - 1) * po;$
$z3 = z3 + c(m8 - 1) * ((m1/d1 * Vo(:,m8 - i)). * (m1/d1 * Vo(:,m8 - i)) * d^2);$
$z3 = z3 + c3(:,m8 - i);$
$U(:,m8 - 1) = inv(M50) * z1;$
$V(:,m8 - 1) = inv(M60) * z2;$
$P(:,m8 - 1) = inv(M70) * z3;$
$for\ i = 2 : m8 - 1$
$M50 = (Id - c(m8 - i) * m22/d1^2 * d^2);$
$z1 = b(m8 - i) * uo + a(m8 - i) * U(:,m8 - i + 1)$
$z1 = z1 + c(m8 - i) * (-U(:,m8 - i + 1). * (Uo(:,m8 - i + 1) - Uo(:,m8 - i)) * d/e1)$
$z1 = z1 - c(m8 - i)(Po(:,m8 - i + 1) - Po(:,m8 - i)) * d/e1;$
$z1 = z1 + c1(:,m8 - i) + c(m8 - i) * (-V(:,m8 - i + 1). * (m1 * Uo(:,m8 - i))/d1 * d^2)/e1;$
$M60 = (Id - c(m8 - i) * m22/d1^2 * d^2);$
$z2 = b(m8 - i) * vo + a(m8 - i) * Vo(:,m8 - i + 1)$
$z2 = z2 + c(m8 - i) * (-U(:,m8 - i + 1). * (Vo(:,m8 - i + 1) - Vo(:,m8 - i)) * d/e1)$
$z2 = z2 - c(m8 - i) * V(:,m8 - i + 1). * (m1 * Vo(:,m8 - i)/d1 * d^2)/e1;$
$z2 = z2 - c(m8 - i) * (m1 * Po(:,m8 - i)/d1 * d^2)/e1;$
$z2 = z2 + c2(:,m8 - i);$
$M70 = (Id - c(m8 - i) * m2/d1^2 * d^2);$
$z3 = b(m8 - i) * po + a(m8 - i) * P(:,m8 - i + 1)$
$z3 = z3$
$+ c(m8 - i) * ((Uo(:,m8 - i + 1) - Uo(:,m8 - i)). * (Uo(:,m8 - i) - Uo(:,m8 - i)));$
$z3 = z3 + c(m8 - i)(m1/d1 * Vo(:,m8 - i)). * (m1/d1 * Vo(:,m8 - i)) * d^2;$
$z3 = z3$
$+ 2 * (m1/d1 * Uo(:,m8 - i)). * (Vo(:,m8 - i + 1) - Vo(:,m8 - i)) * d + c3(:,m8 - i);$
$U(:,m8 - i) = inv(M50) * z1;$
$V(:,m8 - i) = inv(M60) * z2;$
$P(:,m8 - i) = inv(M70) * z3;$
$end;$
$b14 = max(max(abs(U - Uo)));$
$b14$
$Uo = U;$
$Vo = V;$
$Po = P;$
$k1$
$U(m9/2, 10)$

end;
k7
end;
for i = 1 : *m9* − 1
y(*i*) = *i* ∗ *d*1;
end;
for i = 1 : *m8* − 1
x(*i*) = *i* ∗ *d*;
end;
mesh(*x*, *y*, *U*);

****

For the field of velocities U, V and the pressure field P, please see Figures 3.1, 3.2 and 3.3, respectively.

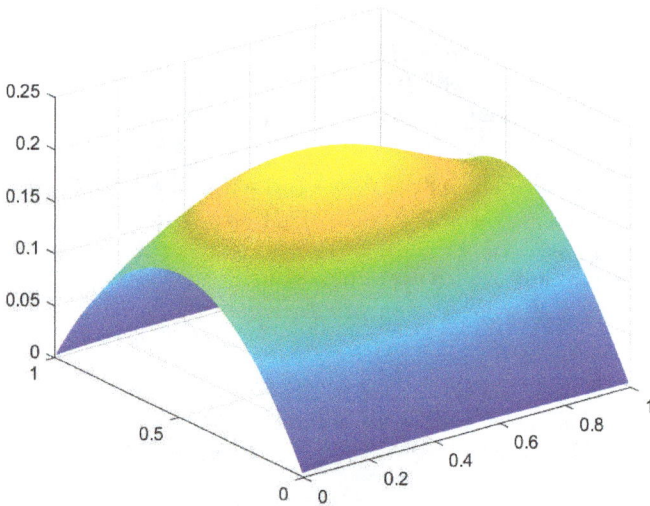

Figure 3.1: Solution $U(x,y)$ for the case $v = 0.047$.

3.4 A software in MATHEMATICA related to the previous algorithm

In this section, we develop the solution for the Navier-Stokes system through the generalized method of lines, similar to the results presented in [18], but now in a Navier-Stokes system context.

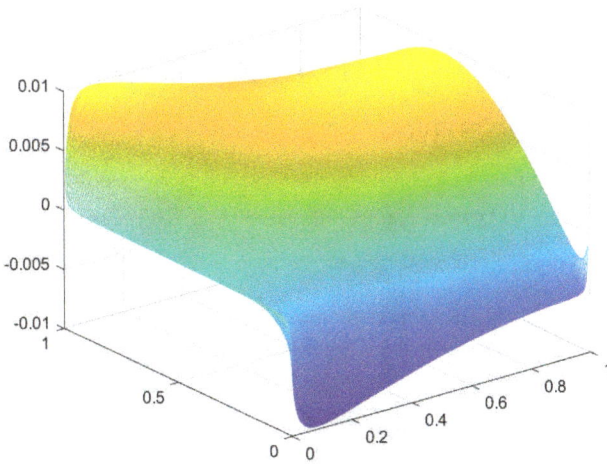

Figure 3.2: Solution $V(x,y)$ for the case $v = 0.047$.

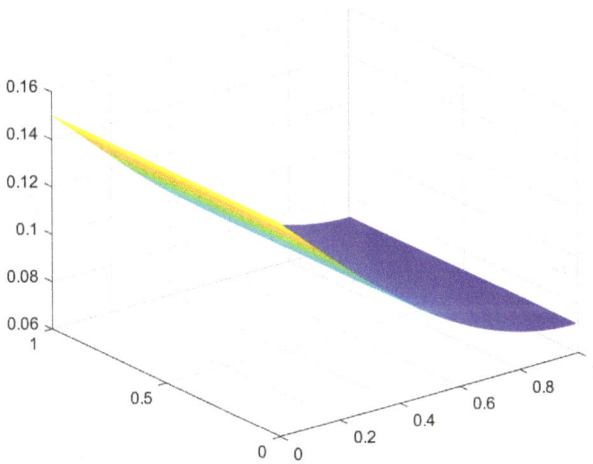

Figure 3.3: Solution $P(x,y)$ for the case $v = 0.047$.

We present a software in MATHEMATICA for $N = 10$ lines for the case in which

$$\begin{cases} v\nabla^2 u - u\partial_x u - v\partial_y u - \partial_x P = 0, & \text{in } \Omega, \\ v\nabla^2 v - u\partial_x v - v\partial_y v - \partial_y P = 0, & \text{in } \Omega, \quad (3.21) \\ \nabla^2 P + (\partial_x u)^2 + (\partial_y v)^2 + 2(\partial_y u)(\partial_x v) = 0, & \text{in } \Omega, \end{cases}$$

We consider it in polar coordinates, with $v = e_1 = 0.1$, and with

$$\Omega = \{(r,\theta) \in \mathbb{R}^2 : 1 \leq r \leq 2, 0 \leq \theta \leq 2\pi\},$$

$$\partial\Omega_1 = \{(1,\theta) \in \mathbb{R}^2 : 0 \leq \theta \leq 2\pi\},$$

and

$$\partial\Omega_2 = \{(2,\theta) \in \mathbb{R}^2 : 0 \leq \theta \leq 2\pi\}.$$

The boundary conditions are

$$u = v = 0, \ P = 0.15 \text{ on } \partial\Omega_1,$$

$$u = u_f[x], \ v = 0, \ P = 0.12 \text{ on } \partial\Omega_2.$$

From now and on, x stands for θ.

We remark some changes have been made, concerning the original conception, to make it suitable through the software MATHEMATICA for such a Navier-Stokes system.

We highlight, as $K > 0$ is larger, the related approximation is of a better quality. However if $K > 0$ is too much large, the converging process gets slower.

Here the concerning software.

****

1. $m8 = 10$;
   $Clear[t3,t4]$;
   $d = 1.0/m8$;
   $K = 4.0$;
   $e1 = 0.1$;
   $Uoo[x_] = 0.0$;
   $Voo[x_] = 0.0$;
   $Poo[x_] = 0.15$;

2. $For[i = 1, i < m8 + 1, i++,$
   $uo[i] = 0.05$;
   $vo[i] = 0.05$;
   $Po[i] = 0.05]$;

3. $For[k = 1, k < 80, k++,$ (here we have fixed the number of iterations)
   $Print[k]$;
   $a[1] = 1/(2.0 + K * d^2/e1)$;
   $b[1] = a[1]$;

$b11[1] = a[1];$

$c1[1] = a[1] * (K * uo[1]) * d^2/e1;$

$c2[1] = a[1] * (K * vo[1]) * d^2/e1;$

$c3[1] = a[1] * (K * Po[1] + P1) * d^2/e1;$

4. *For*$[i = 2, i < m8, i++,$

$a[i] = 1/(2.0 + K * d^2/e1 - a[i-1]);$

$b[i] = a[i] * (b[i-1] + 1);$

$b11[i] = a[i] * b11[i-1];$

$c1[i] = a[i] * (c1[i-1] + (K * uo[i]) * d^2/e1);$

$c2[i] = a[i] * (c2[i-1] + (K * vo[i]) * d^2/e1);$

$c3[i] = a[i] * (c3[i-1] + (K * Po[i]) * d^2/e1)];$

$u[m8] = uf[x] * t3; v[m8] = vf[x] * t3; P[m8] = 0.12; d1 = 1.0;$

5. *For*$[i = 1, i < m8, i++,$

Print$[i];$

$t[m8-i] = 1.0 + (m8-i) * d;$

$Dxu = (uo[m8-i+1] - uo[m8-i])/d * f1[x] * t4 - D[uo[m8-i], x] * f2[x]/t[m8-i] * t4;$

$Dyu = (uo[m8-i+1] - uo[m8-i])/d * f2[x] * t4 + D[uo[m8-i], x] * f1[x]/t[m8-i] * t4;$

$Dxv = (vo[m8-i+1] - vo[m8-i])/d * f1[x] * t4 - D[vo[m8-i], x] * f2[x]/t[m8-i] * t4;$

$Dyv = (vo[m8-i+1] - vo[m8-i])/d * f2[x] * t4 + D[vo[m8-i], x] * f1[x]/t[m8-i] * t4;$

$DxP = (Po[m8-i+1] - Po[m8-i])/d * f1[x] * t4 - D[Po[m8-i], x] * f2[x]/t[m8-i] * t4;$

$DyP = (Po[m8-i+1] - Po[m8-i])/d * f2[x] * t4 + D[Po[m8-i], x] * f1[x]/t[m8-i] * t4;$

$T1 = -(u[m8-i+1] * Dxu + v[m8-i+1] * Dyu + DxP) * d^2/e1$
$+ (uo[m8-i+1] - uo[m8-i])/d/t[m8-i] * d^2 + D[uo[m8-i+1], \{x, 2\}]/t[m8-i]^2 * d^2;$

$T2 = -(u[m8-i+1] * Dxv + v[m8-i+1] * Dyv + DyP) * d^2/e1 + (vo[m8-i+1] - vo[m8-i])/d/t[m8-i] * d^2$
$+ D[vo[m8-i+1], \{x, 2\}]/t[m8-i]^2 * d^2;$

$T3 = (Dxu^2 + Dyv^2 + 2 * Dyu * Dxv) * d^2$
$+ (Po[m8-i+1] - Po[m8-i])/d/t[m8-i] * d^2 + D[Po[m8-i+1], \{x, 2\}]/t[m8-i]^2 * d^2;$

6. $A1 = a[m8-i]*u[m8-i+1]+b[m8-i]*T1+c1[m8-i];$

   $A1 = Expand[A1];$

   $A1 = Series[A1,\{t4,0,1\},\{t3,0,2\},\{uf[x],0,2\},\{uf'[x],0,1\},$
   $\{uf''[x],0,1\},$
   $\{uf'''[x],0,0\},\{uf''''[x],0,0\},\{vf[x],0,1\},\{vf'[x],0,1\},\{vf''[x],0,1\},$
   $\{vf'''[x],0,0\},\{vf''''[x],0,0\},$
   $\{f1[x],0,1\},\{f2[x],0,1\},\{f1'[x],0,0\},\{f2'[x],0,0\},\{f1''[x],0,0\},$
   $\{f2''[x],0,0\}];$

   $A1 = Normal[A1];$

   $u[m8-i] = Expand[A1];$

7. $A2 = a[m8-i]*v[m8-i+1]+b[m8-i]*T2+c2[m8-i];$

   $A2 = Expand[A2];$

   $A2 = Series[A2,\{t4,0,1\},\{t3,0,2\},\{uf[x],0,1\},\{uf'[x],0,1\},$
   $\{uf''[x],0,1\},$
   $\{uf'''[x],0,0\},\{uf''''[x],0,0\},\{vf[x],0,2\},\{vf'[x],0,1\},\{vf''[x],0,1\},$
   $\{vf'''[x],0,0\},\{vf''''[x],0,0\},$
   $\{f1[x],0,1\},\{f2[x],0,1\},\{f1'[x],0,0\},\{f2'[x],0,0\},$
   $\{f1''[x],0,0\},\{f2''[x],0,0\}];$

   $A2 = Normal[A2];$

   $v[m8-i] = Expand[A2];$

8. $A3 = a[m8-i]*P[m8-i+1]+b[m8-i]*T3+c3[m8-i]+b11[m8-i]*$
   $Poo[x];$

   $A3 = Expand[A3];$

   $A3 = Series[A3,\{t4,0,2\},\{t3,0,2\},\{uf[x],0,1\},\{uf'[x],0,1\},$
   $\{uf''[x],0,0\},\{uf'''[x],0,0\},$
   $\{uf''''[x],0,0\},\{vf[x],0,1\},\{vf'[x],0,1\},\{vf''[x],0,0\},\{vf'''[x],0,0\},$
   $\{vf''''[x],0,0\},\{f1[x],0,1\},\{f2[x],0,1\},$
   $\{f1'[x],0,0\},\{f2'[x],0,0\},\{f1''[x],0,0\},\{f2''[x],0,0\}];$

   $A3 = Normal[A3];$

   $P[m8-i] = Expand[A3]];$

9. $For[i=1, i<m8+1, i++,$

   $uo[i] = u[i];$

   $vo[i] = v[i];$

   $Po[i] = P[i]];d1 = 1.0;$

10. *Print*[*Expand*[*U*[*m*8/2]]]]

11. *For*[*i* = 1, *i* < *m*₈, *i* ++,

$\qquad$ *Print*["u[", *i*, "] = ", *u*[*i*][*x*]]]

Here we present the related line expressions obtained for the lines $n = 1$, $n = 5$ and n=9 of a total of $N = 10$ lines.

1. *Line n* = 1

$$
\begin{aligned}
u[1] \quad = \quad & 1.58658 * 10^{-9} + 0.019216 f_1[x] + 3.68259 * 10^{-11} f_2[x] + 0.132814 u_f[x] \\
& + 2.28545 * 10^{-8} f_1[x] u_f[x] - 1.69037 * 10^{-8} f_2[x] u_f[x] - 0.263288 f_1[x] u_f[x]^2 \\
& - 6.02845 * 10^{-9} f_1[x](u'_f)[x] + 6.02845 * 10^{-9} f_2[x](u'_f)[x] \\
& + 0.104284 f_2[x] u_f[x](u'_f])[x] + 0.0127544(u''_f)[x] \\
& + 9.39885 * 10^{-9} f_1[x](u''_f)[x] - 5.26303 * 10^{-9} f_2[x](u''_f)[x] \\
& - 0.0340276 f_1[x] u_f[x](u''_f)[x] + 0.0239544 f_2[x](u'_f)[x](u''_f)[x] \qquad (3.22)
\end{aligned}
$$

2. *Line n* = 5

$$
\begin{aligned}
u[5] \quad = \quad & 4.25933 * 10^{-9} + 0.0436523 f_1[x] + 9.88625 * 10^{-11} f_2[x] + 0.572969 u_f[x] \\
& + 6.87985 * 10^{-8} f_1[x] u_f[x] - 4.40534 * 10^{-8} f_2[x] uf[x] - 0.765222 f_1[x] u_f[x]^2 \\
& - 1.61319 * 10^{-8} f_1[x](u'_f)[x] + 1.61319 * 10^{-8} f_2[x](u'_f)[x] \\
& + 0.363471 f_2[x] u_f[x](u'_f)[x] + 0.0333685(u''_f)[x] \\
& + 2.39576 * 10^{-8} f_1[x](u''_f)[x] - 1.27491 * 10^{-8} f_2[x](u''_f)[x] \\
& - 0.0342544 f_1[x] u_f[x](u''_f)[x] + 0.0509889 f_2[x](u'_f)[x](u''_f)[x] \qquad (3.23)
\end{aligned}
$$

3. *Line n* = 9

$$
\begin{aligned}
u[9] \quad = \quad & 1.15848 * 10^{-9} + 0.0136828 f_1[x] + 2.68892 * 10^{-11} f_2[x] + 0.922534 u_f[x] \\
& + 2.16498 * 10^{-8} f_1[x] u_f[x] - 1.16065 * 10^{-8} f_2[x] u_f[x] - 0.278966 f_1[x] u_f[x]^2 \\
& - 4.25263 * 10^{-9} f_1[x](uf')[x] + 4.25263 * 10^{-9} f_2[x](u'_f)[x] \\
& + 0.154642 f_2[x] u_f[x](u'_f)[x] + 0.0110114(u''_f)[x] \\
& + 6.13523 * 10^{-9} f_1[x](u''_f)[x] - 3.23081 * 10^{-9} f_2[x](u''_f)[x] \\
& + 0.0146222 f_1[x] u_f[x](u''_f)[x] + 0.0090088 f_2[x](u'_f)[x](u''_f)[x] \qquad (3.24)
\end{aligned}
$$

3.5 The software and numerical results for a more specific example

In this section, we present numerical results for the same Navier-Stokes system and domain as in the previous one, but now with different boundary conditions.

In this example, we set $v = 0.1$ and the boundary conditions are

$$u = v = 0, \; P = 0.15 \text{ on } \partial\Omega_1,$$

$$u = -1.0\sin[x], \; v = 1.0\cos[x], \; P = 0.12 \text{ on } \partial\Omega_2.$$

Here the concerning software:

****

1. $m8 = 10$;
   $Clear[t3, t4]$;
   $d = 1.0/m8$;
   $K = 4.0$;
   $e1 = 0.1$;
   $Uoo[x_-] = 0.0$;
   $Voo[x_-] = 0.0$;
   $Poo[x_-] = 0.15$;

2. $For[i = 1, \; i < m8 + 1, \; i++,$
   $uo[i] = 0.05$;
   $vo[i] = 0.05$;
   $Po[i] = 0.05]$;
   $f1[x_-] = Cos[x]$;
   $f2[x_-] = Sin[x]$;

3. $For[k = 1, \; k < 80, \; k++,$ (here we have fixed the number of iterations)
   $Print[k]$;
   $a[1] = 1/(2.0 + K * d^2/e1)$;
   $b[1] = a[1]$;
   $b11[1] = a[1]$;
   $c1[1] = a[1] * (K * uo[1]) * d^2/e1$;
   $c2[1] = a[1] * (K * vo[1]) * d^2/e1$;
   $c3[1] = a[1] * (K * Po[1] + P1) * d^2/e1$;

4. $For[i = 2, i < m8, i++,$

$a[i] = 1/(2.0 + K * d^2/e1 - a[i-1]);$

$b[i] = a[i] * (b[i-1] + 1);$

$b11[i] = a[i] * b11[i-1];$

$c1[i] = a[i] * (c1[i-1] + (K * uo[i]) * d^2/e1);$

$c2[i] = a[i] * (c2[i-1] + (K * vo[i]) * d^2/e1);$

$c3[i] = a[i] * (c3[i-1] + (K * Po[i]) * d^2/e1)];$

$uf[x\_] = -1.0 * Sin[x];$

$vf[x] = 1.0 * Cos[x];$

$u[m8] = uf[x] * t3;$

$v[m8] = vf[x] * t3;$

$P[m8] = 0.12;$

5. $For[i = 1, i < m8, i++,$

$Print[i];$

$t[m8 - i] = 1.0 + (m8 - i) * d;$

$Dxu = (uo[m8 - i + 1] - uo[m8 - i])/d * f1[x] * t4 - D[uo[m8 - i], x] * f2[x]/t[m8 - i] * t4;$

$Dyu = (uo[m8 - i + 1] - uo[m8 - i])/d * f2[x] * t4 + D[uo[m8 - i], x] * f1[x]/t[m8 - i] * t4;$

$Dxv = (vo[m8 - i + 1] - vo[m8 - i])/d * f1[x] * t4 - D[vo[m8 - i], x] * f2[x]/t[m8 - i] * t4;$

$Dyv = (vo[m8 - i + 1] - vo[m8 - i])/d * f2[x] * t4 + D[vo[m8 - i], x] * f1[x]/t[m8 - i] * t4;$

$DxP = (Po[m8 - i + 1] - Po[m8 - i])/d * f1[x] * t4 - D[Po[m8 - i], x] * f2[x]/t[m8 - i] * t4;$

$DyP = (Po[m8 - i + 1] - Po[m8 - i])/d * f2[x] * t4 + D[Po[m8 - i], x] * f1[x]/t[m8 - i] * t4;$

$T1 = -(u[m8 - i + 1] * Dxu + v[m8 - i + 1] * Dyu + DxP) * d^2/e1$

$+(uo[m8 - i + 1] - uo[m8 - i])/d/t[m8 - i] * d^2$

$+D[uo[m8 - i + 1], \{x, 2\}]/t[m8 - i]^2 * d^2;$

$T2 = -(u[m8 - i + 1] * Dxv + v[m8 - i + 1] * Dyv + DyP) * d^2/e1$

$+(vo[m8 - i + 1] - vo[m8 - i])/d/t[m8 - i] * d^2 + D[vo[m8 - i + 1], \{x, 2\}]$

$/t[m8 - i]^2 * d^2;$

$T3 = (Dxu^2 + Dyv^2 + 2 * Dyu * Dxv) * d^2$

$+(Po[m8-i+1]-Po[m8-i])/d/t[m8-i]*d^2+D[Po[m8-i+1],\{x,2\}]$
$/t[m8-i]^2*d^2;$

$A1=a[m8-i]*u[m8-i+1]+b[m8-i]*T1+c1[m8-i];$

$A1=Expand[A1];$

$A1=Series[A1,\{t4,0,1\},\{t3,0,2\},\{Sin[x],0,2\},\{Cos[x],0,2\}];$

$A1=Normal[A1];$

$u[m8-i]=Expand[A1];$

$A2=a[m8-i]*v[m8-i+1]+b[m8-i]*T2+c2[m8-i];$

$A2=Expand[A2];$

$A2=Series[A2,\{t4,0,1\},\{t3,0,2\},\{Sin[x],0,2\},\{Cos[x],0,2\}];$

$A2=Normal[A2];$

$v[m8-i]=Expand[A2];$

$A3=a[m8-i]*P[m8-i+1]+b[m8-i]*T3+c3[m8-i]+b11[m8-i]*$
$Poo[x];$

$A3=Expand[A3]; A3=Series[A3,\{t4,0,2\},\{t3,0,2\},\{Sin[x],0,2\},$
$\{Cos[x],0,2\}];$

$A3=Normal[A3];$

$P[m8-i]=Expand[A3]];$

6. $For[i=1, i<m8+1, i++,$

$uo[i]=u[i];$

$vo[i]=v[i];$

$Po[i]=P[i]];$

$Print[Expand[P[m8/2]]]]$

7. $For[i=1,i<m_8,i++,$

$Print["u[",i,"]=",u[i][x]]]$

Here the corresponding line expressions for $N=10$ lines

1. *Line n = 1*

$$
\begin{aligned}
u[1] = \quad & 1.445*10^{-10}+0.0183921Cos[x]+1.01021*10^{-9}Cos[x]^2-0.120676Sin[x] \\
& +3.62358*10^{-10}Cos[x]Sin[x]+1.37257*10^{-9}Sin[x]^2 \\
& +0.0620534Cos[x]Sin[x]^2 \quad\quad\quad\quad\quad\quad\quad\quad\quad\quad\quad\quad (3.25)
\end{aligned}
$$

2. *Line n = 2*

$$
\begin{aligned}
u[2] = \quad & 2.60007*10^{-10}+0.0307976Cos[x]+1.81088*10^{-9}Cos[x]^2-0.233061Sin[x] \\
& +6.53242*10^{-10}Cos[x]Sin[x]+2.46412*10^{-9}Sin[x]^2 \\
& +0.123121Cos[x]Sin[x]^2 \quad\quad\quad\quad\quad\quad\quad\quad\quad\quad\quad\quad (3.26)
\end{aligned}
$$

3. *Line n = 3*

$$u[3] = 3.40796 * 10^{-10} + 0.0384482Cos[x] + 2.36167 * 10^{-9}Cos[x]^2 - 0.339657Sin[x]$$
$$+ 8.5759 * 10^{-10}Cos[x]Sin[x] + 3.21926 * 10^{-9}Sin[x]^2$$
$$+ 0.180891Cos[x]Sin[x]^2 \tag{3.27}$$

4. *Line n = 4*

$$u[4] = 3.83612 * 10^{-10} + 0.0420843Cos[x] + 2.64262 * 10^{-9}Cos[x]^2 - 0.441913Sin[x]$$
$$+ 9.66336 * 10^{-10}Cos[x]Sin[x] + 3.60895 * 10^{-9}Sin[x]^2$$
$$+ 0.230559Cos[x]Sin[x]^2 \tag{3.28}$$

5. *Line n = 5*

$$u[5] = 3.87923 * 10^{-10} + 0.0421948Cos[x] + 2.65457 * 10^{-9}Cos[x]^2 - 0.540729Sin[x]$$
$$+ 9.77606 * 10^{-10}Cos[x]Sin[x] + 3.63217 * 10^{-9}Sin[x]^2$$
$$+ 0.266239Cos[x]Sin[x]^2 \tag{3.29}$$

6. *Line n = 6*

$$u[6] = 3.56064 * 10^{-10} + 0.0391334Cos[x] + 2.419 * 10^{-9}Cos[x]^2 - 0.636718Sin[x]$$
$$+ 8.97185 * 10^{-10}Cos[x]Sin[x] + 3.31618 * 10^{-9}Sin[x]^2$$
$$+ 0.281514Cos[x]Sin[x]^2 \tag{3.30}$$

7. *Line n = 7*

$$u[7] = 2.93128 * 10^{-10} + 0.033175Cos[x] + 1.97614 * 10^{-9}Cos[x]^2 - 0.730328Sin[x]$$
$$+ 7.38127 * 10^{-10}Cos[x]Sin[x] + 2.71426 * 10^{-9}Sin[x]^2$$
$$+ 0.269642Cos[x]Sin[x]^2 \tag{3.31}$$

8. *Line n = 8*

$$u[8] = 2.0656 * 10^{-10} + 0.0245445Cos[x] + 1.38127 * 10^{-9}Cos[x]^2 - 0.821902Sin[x]$$
$$+ 5.19585 * 10^{-10}Cos[x]Sin[x] + 1.90085 * 10^{-9}Sin[x]^2$$
$$+ 0.22362Cos[x]Sin[x]^2 \tag{3.32}$$

9. *Line n = 9*

$$u[9] = 1.05509 * 10^{-10} + 0.0134316Cos[x] + 6.99554 * 10^{-10}Cos[x]^2 - 0.911718Sin[x]$$
$$+ 2.65019 * 10^{-10}Cos[x]Sin[x] + 9.64573 * 10^{-10}Sin[x]^2$$
$$+ 0.13622Cos[x]Sin[x]^2 \tag{3.33}$$

Here we present the related plots for the Lines $n = 2$, $n = 4$, $n = 6$ and $n = 8$ of a total of $N = 10$ lines.

For each line we set $N = 500$ nodes on the interval $[0, 2\pi]$, so that the units in x are $2\pi/500$, where again x stands for θ.

For such lines, please see Figures 3.4, 3.5, 3.6 and 3.7, respectively.

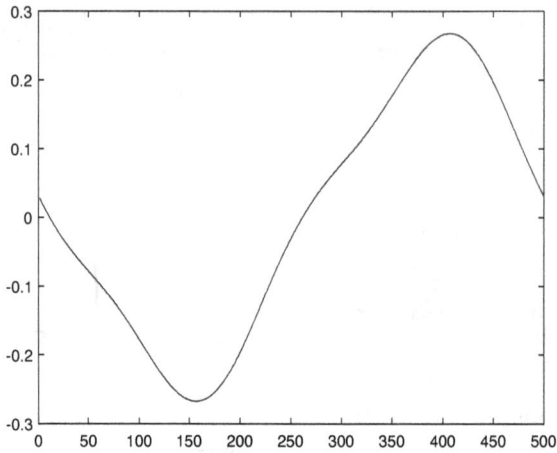

Figure 3.4: Solution $u_2(x)$ for the line $n = 2$, for the case $v = 0.1$.

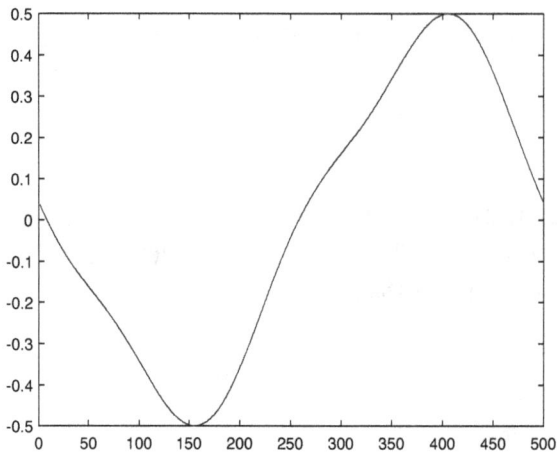

Figure 3.5: Solution $u_4(x)$ for the line $n = 4$, for the case $v = 0.1$.

3.6 Numerical results through the original conception of the generalized method of lines for the Navier-Stokes system

In this section, we develop the solution for the Navier-Stokes system through the generalized method of lines, as originally introduced in [22], with further developments in [17].

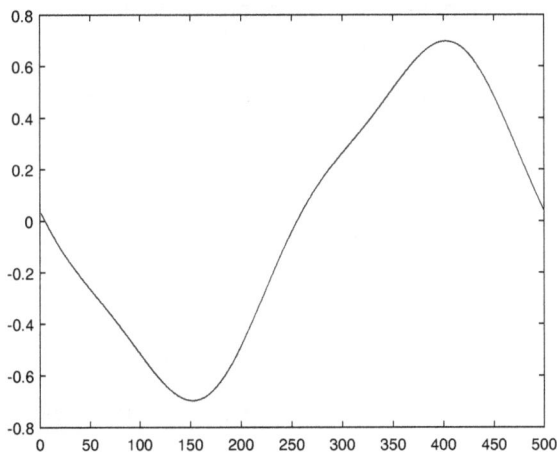

Figure 3.6: Solution $u_6(x)$ for the line $n = 6$, for the case $v = 0.1$.

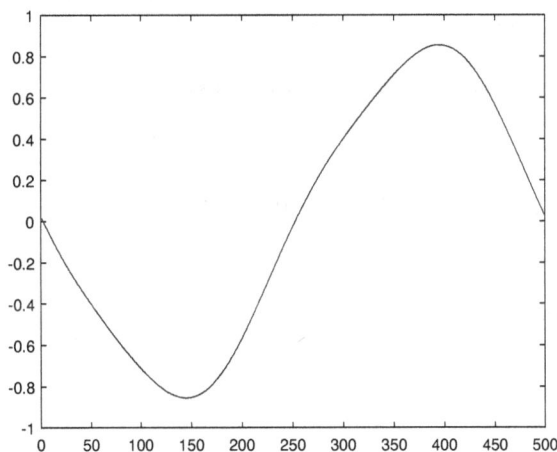

Figure 3.7: Solution $u_8(x)$ for the line $n = 8$, for the case $v = 0.1$.

We present a software in MATHEMATICA for $N = 10$ lines for the case in which

$$
\begin{cases}
v\nabla^2 u - u\partial_x u - v\partial_y u - \partial_x P = 0, & \text{in } \Omega, \\[2mm]
v\nabla^2 v - u\partial_x v - v\partial_y v - \partial_y P = 0, & \text{in } \Omega, \\[2mm]
\nabla^2 P + (\partial_x u)^2 + (\partial_y v)^2 + 2(\partial_y u)(\partial_x v) = 0, & \text{in } \Omega.
\end{cases} \tag{3.34}
$$

Such a software refers to an algorithm presented in Chapter 27, in [13], in polar coordinates, with $v = 1.0$, and with

$$\Omega = \{(r, \theta) \in \mathbb{R}^2 : 1 \leq r \leq 2, 0 \leq \theta \leq 2\pi\},$$

$$\partial\Omega_1 = \{(1, \theta) \in \mathbb{R}^2 : 0 \leq \theta \leq 2\pi\},$$

and

$$\partial\Omega_2 = \{(2, \theta) \in \mathbb{R}^2 : 0 \leq \theta \leq 2\pi\}.$$

The boundary conditions are

$$u = v = 0, \ P = 0.15 \text{ on } \partial\Omega_1,$$

$$u = -1.0\sin(\theta), \ v = 1.0\cos(\theta), \ P = 0.10 \text{ on } \partial\Omega_2.$$

We remark some changes have been made, concerning the original conception, to make it suitable through the software MATHEMATICA for such a Navier-Stokes system.

We highlight the nature of this approximation is qualitative.

Here the concerning software in MATHEMATICA.

****

1. $m8 = 10$;

2. $Clear[z_1, z_2, z_3, u_1, u_2, P, b_1, b_2, b_3, a_1, a_2, a_3]$;

3. $Clear[P_f, t, a_{11}, a_{12}, a_{13}, b_{11}, b_{12}, b_{13}, t_3]$;

4. $d = 1.0/m8$;

5. $e_1 = 1.0$;

6. $a_1 = 0.0$;

7. $a_2 = 0.0$;

8. $a_3 = 0.15$;

9. $For[i = 1, i < m8, i++,$
   $Print[i]$;
   $Clear[b_1, b_2, b_3, u_1, u_2, P]$;
   $b_1[x\_] = u_1[i+1][x]$;
   $b_2[x\_] = u_2[i+1][x]$;
   $b_3[x\_] = P[i+1][x]$;
   $t[i] = 1 + i*d$;

$du1x = Cos[x] * (b_1[x] - a_1)/d * t_3 - 1/t[i] * Sin[x] * D[b_1[x],x] * t_3;$

$du1y = Sin[x] * (b_1[x] - a_1)/d * t_3 + 1/t[i] * Cos[x] * D[b_1[x],x] * t_3;$

$du2x = Cos[x] * (b_2[x] - a_2)/d * t_3 - 1/t[i] * Sin[x] * D[b_2[x],x] * t_3;$

$du2y = Sin[x] * (b_2[x] - a_2)/d * t_3 + 1/t[i] * Cos[x] * D[b_2[x],x] * t_3;$

$dPx = Cos[x] * (b_3[x] - a_3)/d * t3 - 1/t[i] * Sin[x] * D[b_3[x],x] * t_3;$

$dPy = Sin[x] * (b_3[x] - a_3)/d * t3 + 1/t[i] * Cos[x] * D[b_3[x],x] * t_3;$

10. *For* $k = 1, k < 6, k + +$, (in this example, we have fixed a relatively small number of iterations)

   Print $[k];$

   $z_1 = (u_1[i + 1][x] + b_1[x] + a_1 + 1/t[i] * (b_1[x] - a_1) * d + 1/t[i]^2 * D[b_1[x],x,2] * d^2$
   $-(b_1[x] * du1x + b_2[x] * du1y) * d^2/e1 - dPx * d^2/e_1)/3.0;$

   $z_2 = (u_2[i + 1][x] + b_2[x] + a_2 + 1/t[i] * (b_2[x] - a_2) * d + 1/t[i]^2 * D[b_2[x],x,2] * d^2$
   $-(b_1[x] * du2x + b_2[x] * du2y) * d^2/e1 - dPy * d^2/e_1)/3.0;$

   $z_3 = (P[i + 1][x] + b_3[x] + a_3 + 1/t[i] * (b_3[x] - a_3) * d + 1/t[i]^2 * D[b_3[x],x,2] * d^2$
   $+(du1x * du1x + du2y * du2y + 2.0 * du1y * du2x) * d^2)/3.0;$

11. $z_1 = Series[z_1, \{u_1[i+1][x],0,2\}, \{u_1[i+1]'[x],0,1\}, \{u_1[i+1]''[x],0,1\},$
   $\{u_1[i+1]'''[x],0,0\}, \{u_1[i+1]''''[x],0,0\}, \{u_2[i+1][x],0,1\},$
   $\{u_2[i+1]'[x],0,0\}, \{u_2[i+1]''[x],0,0\}, \{u_2[i+1]'''[x],0,0\},$
   $\{u_2[i+1]''''[x],0,0\}, \{P[i+1][x],0,1\}, \{P[i+1]'[x],0,0\},$
   $\{P[i+1]''[x],0,0\}, \{P[i+1]'''[x],0,0\}, \{P[i+1]''''[x],0,0\},$
   $\{Sin[x],0,1\}, \{Cos[x],0,1\}];$

   $z_1 = Normal[z_1];$
   $z_1 = Expand[z_1];$
   Print $[z_1];$

12. $z_2 = Series[z_2, \{u_1[i+1][x],0,1\}, \{u_1[i+1]'[x],0,1\}, \{u_1[i+1]''[x],0,1\},$
   $\{u_1[i+1]'''[x],0,0\}, \{u_1[i+1]''''[x],0,0\}, \{u_2[i+1][x],0,2\},$
   $\{u_2[i+1]'[x],0,0\}, \{u_2[i+1]''[x],0,0\}, \{u_2[i+1]'''[x],0,0\},$
   $\{u_2[i+1]''''[x],0,0\}, \{P[i+1][x],0,1\}, \{P[i+1]'[x],0,0\},$
   $\{P[i+1]''[x],0,0\}, \{P[i+1]'''[x],0,0\}, \{P[i+1]''''[x],0,0\},$
   $\{Sin[x],0,1\}, \{Cos[x],0,1\}];$

```
z2 = Normal[z2];
z2 = Expand[z2];
Print[z2];
```

13.
```
z3 = Series[z3, {u1[i + 1][x], 0, 2}, {u1[i + 1]'[x], 0, 1}, {u1[i + 1]"[x], 0, 1},
    {u1[i + 1]'''[x], 0, 0}, {u1[i + 1]''''[x], 0, 0}, {u2[i + 1][x], 0, 2},
    {u2[i + 1]'[x], 0, 1}, {u2[i + 1]"[x], 0, 0}, {u2[i + 1]'''[x], 0, 0},
    {u2[i + 1]''''[x], 0, 0}, {P[i + 1][x], 0, 1}, {P[i + 1]'[x], 0, 1},
    {P[i + 1]"[x], 0, 0}, {P[i + 1]'''[x], 0, 0}, {P[i + 1]''''[x], 0, 0},
    {Sin[x], 0, 1}, {Cos[x], 0, 1}];
```

```
z3 = Normal[z3];
z3 = Expand[z3];
Print[z3];
```

14.
```
b1[x_] = z1;
b2[x_] = z2;
b3[x_] = z3;
b11 = z1;
b12 = z2;
b13 = z3;];
```

15.
```
a11[i] = b11;
a12[i] = b12;
a13[i] = b13;
Print[a11[i]];
Clear[b1, b2, b3];
u1[i + 1][x_] = b1[x];
u2[i + 1][x_] = b2[x];
P[i + 1][x_] = b3[x];
a1 = Series[b11, {t3, 0, 1}; {b1[x], 0, 1}, {b2[x], 0, 1}, {b3[x], 0, 1},
    {b1'[x], 0, 0}, {b2'[x], 0, 0}, {b3'[x], 0, 0}, {b1"[x], 0, 0}, {b2"[x], 0, 0},
    {b3"[x], 0, 0}];
a1 = Normal[a1];
a1 = Expand[a1];
a2 = Series[b12, {t3, 0, 1}; {b1[x], 0, 1}, {b2[x], 0, 1}, {b3[x], 0, 1},
```

$\{b_1'[x], 0, 0\}, \{b_2'[x], 0, 0\}, \{b_3'[x], 0, 0\}, \{b_1''[x], 0, 0\}, \{b_2''[x], 0, 0\},$
$\{b_3''[x], 0, 0\}];$

$a_2 = Normal[a_2];$

$a_2 = Expand[a_2];$

$a_3 = Series[b_{13}, \{t_3, 0, 1\}; \{b_1[x], 0, 1\}, \{b_2[x], 0, 1\}, \{b_3[x], 0, 1\},$
$\{b_1'[x], 0, 0\}, \{b_2'[x], 0, 0\}, \{b_3'[x], 0, 0\}, \{b_1''[x], 0, 0\}, \{b_2''[x], 0, 0\},$
$\{b_3''[x], 0, 0\}];$

$a_3 = Normal[a_3];$

$a_3 = Expand[a_3];$

16. $b_1[x] = -1.0 * Sin[x];$
    $b_2[x] = 1.0 * Cos[x];$
    $b_3[x] = 0.10;$

17. $For[i = 1, i < m8, i++,$
    $A_{11} = a_{11}[m8 - i];$
    $A_{11} = Series[A_{11}, \{Sin[x], 0, 2\}, \{Cos[x], 0, 2\}];$
    $A_{11} = Normal[A_{11}];$
    $A_{11} = Expand[A_{11}];$
    $A_{12} = a_{12}[m8 - i];$
    $A_{12} = Series[A_{12}, \{Sin[x], 0, 2\}, \{Cos[x], 0, 2\}];$
    $A_{12} = Normal[A_{12}];$
    $A_{12} = Expand[A_{12}];$
    $A_{13} = a_{13}[m8 - i];$
    $A_{13} = Series[A_{13}, \{Sin[x], 0, 2\}, \{Cos[x], 0, 2\}];$
    $A_{13} = Normal[A_{13}];$
    $A_{13} = Expand[A_{13}];$

18. $u_1[m8 - i][x_-] = A_{11};$
    $u_2[m8 - i][x_-] = A_{12};$
    $P[m8 - i][x_-] = Expand[A_{13}];$

19. $t_3 = 1.0;$

20. $Print["u_1[", m8 - i, "] = ", A_{11}];$
    $Clear[t_3];$
    $b_1[x] = A_{11};$
    $b_2[x] = A_{12};$
    $b_3[x] = A_{13};];$

21. $t_3 = 1.0;$

22. $For[i = 1, i < m_8, i++,$

   $Print["u_1[", i, "] = ", u_1[i][x]]]$

****

Here the line expressions for the field of velocity $u = \{u_1[n](x)\}$, where again we emphasize $N = 10$ lines and $v = e_1 = 1.0$:

1.

$$
\begin{aligned}
u_1[1](x) = {} & 0.0044548 Cos[x] - 0.174091 Sin[x] + 0.00041254 Cos[x]^2 Sin[x] \\
& + 0.0260471 Cos[x] Sin[x]^2 - 0.000188598 Cos[x]^2 Sin[x]^2 \quad (3.35)
\end{aligned}
$$

2.

$$
\begin{aligned}
u_1[2](x) = {} & 0.00680614 Cos[x] - 0.331937 Sin[x] + 0.000676383 Cos[x]^2 Sin[x] \\
& + 0.0501544 Cos[x] Sin[x]^2 - 0.000176433 Cos[x]^2 Sin[x]^2 \quad (3.36)
\end{aligned}
$$

3.

$$
\begin{aligned}
u_1[3](x) = {} & 0.00775103 Cos[x] - 0.470361 Sin[x] + 0.000863068 Cos[x]^2 Sin[x] \\
& + 0.0682792 Cos[x] Sin[x]^2 - 0.000121656 Cos[x]^2 Sin[x]^2 \quad (3.37)
\end{aligned}
$$

4.

$$
\begin{aligned}
u_1[4](x) = {} & 0.00771379 Cos[x] - 0.589227 Sin[x] + 0.000994973 Cos[x]^2 Sin[x] \\
& + 0.0781784 Cos[x] Sin[x]^2 - 0.00006958 Cos[x]^2 Sin[x]^2 \quad (3.38)
\end{aligned}
$$

5.

$$
\begin{aligned}
u_1[5](x) = {} & 0.00701567 Cos[x] - 0.690152 Sin[x] + 0.00106158 Cos[x]^2 Sin[x] \\
& + 0.0796091 Cos[x] Sin[x]^2 - 0.0000330485 Cos[x]^2 Sin[x]^2 \quad (3.39)
\end{aligned}
$$

6.

$$
\begin{aligned}
u_1[6](x) = {} & 0.00589597 Cos[x] - 0.775316 Sin[x] + 0.00104499 Cos[x]^2 Sin[x] \\
& + 0.0734277 Cos[x] Sin[x]^2 - 0.0000121648 Cos[x]^2 Sin[x]^2 \quad (3.40)
\end{aligned}
$$

7.

$$
\begin{aligned}
u_1[7](x) = {} & 0.00452865 Cos[x] - 0.846947 Sin[x] + 0.000931782 Cos[x]^2 Sin[x] \\
& + 0.0609739 Cos[x] Sin[x]^2 - 2.74137 * 10^{-6} Cos[x]^2 Sin[x]^2 \quad (3.41)
\end{aligned}
$$

8.

$$u_1[8](x) = 0.00303746Cos[x] - 0.907103Sin[x] + 0.000716865Cos[x]^2Sin[x]$$
$$+0.0437018Cos[x]Sin[x]^2 \qquad (3.42)$$

9.

$$u_1[9](x) = 0.00150848Cos[x] - 0.957599Sin[x] + 0.000403216Cos[x]^2Sin[x]$$
$$+0.0229802Cos[x]Sin[x]^2 \qquad (3.43)$$

3.7 Conclusion

In this chapter, we develop solutions for examples concerning the two-dimensional, time-independent, and incompressible Navier-Stokes system through the generalized method of lines. We also obtain the appropriate boundary conditions for an equivalent elliptic system. Finally, the extension of such results to $\mathbb{R}^3$, compressible and time dependent cases is planned for a future work.

Chapter 4

An Approximate Numerical Method for Ordinary Differential Equation Systems with Applications to a Flight Mechanics Model

4.1 Introduction

This short communication develops a new numerical procedure suitable for a large class of ordinary differential equation systems found in models in physics and engineering. The main numerical procedure is analogous to those concerning the generalized method of lines, originally published in the referenced books of 2011 and 2014 [22, 12], respectively. Finally, in the last section, we apply the method to a model in flight mechanics.

Consider the first order system of ordinary differential equations given by

$$\frac{du_j}{dt} = f_j(\{u_l\}), \text{ on } [0,t_f] \ \forall j \in \{1,\cdots,4\},$$

with the boundary conditions

$$u_1(0) = 0, \ u_2(0), \ u_4(0) = 0, \ u_4(t_f) = u_f.$$

Here $\mathbf{u} = \{u_l\} \in W^{1,2}([0,t_f],\mathbb{R}^4)$ and f_j are functions at least of C^1 class, $\forall j \in \{1,2,3,4\}$.

Our proposed method is iterative so that we choose an starting solution denoted by $\tilde{u}$.

At this point, we define the number of nodes on $[0,t_f]$ by N and set $d = t_f/N$.

Similarly to a proximal approach, we propose the following algorithm (in a similar fashion as those found in [22, 12, 13]).

1. Choose $\tilde{u} \in W^{1,2}([0,t_f],\mathbb{R}^4)$.

2. Solve the equation system

$$K\frac{du_j}{dt} - (K-1)\frac{d\tilde{u}_j}{dt} = f_j(\{u_l\}), \text{ on } [0,t_f] \ \forall j \in \{1,\cdots,4\}, \qquad (4.1)$$

   with the boundary conditions

$$u_1(0) = 0, \ u_2(0), \ u_4(0) = 0, \ u_4(t_f) = u_f.$$

3. Replace $\tilde{u}$ by u and go to item (2) up to the satisfaction of an appropriate convergence criterion.

In finite differences, observe that the system may be approximated by

$$(u_j)_n - (u_j)_{n-1} = \frac{K-1}{K}((\tilde{u}_j)_n - (\tilde{u}_j)_{n-1}) + f_j(\{(u_l)_{n-1}\})\frac{d}{K}$$

In particular, for $n = 1$, we get

$$(u_j)_1 = (u_j)_0 + f_j(\{(u_l)_1\})\frac{d}{K} + \frac{K-1}{K}((\tilde{u}_j)_1 - (\tilde{u}_j)_0) + (E_j)_1,$$

where

$$(E_j)_1 = [f_j(\{(u_l)_0\}) - f_j(\{(u_l)_1\})]\frac{d}{K} \approx \mathcal{O}\left(\frac{d^2}{K}\right).$$

Similarly, for $n = 2$ we have

$$
\begin{aligned}
(u_j)_2 - (u_j)_0 &= [(u_j)_2 - (u_j)_1] + [(u_j)_1 - (u_j)_0] \\
&= f_j(\{(u_l)_1\})\frac{d}{K} + \frac{K-1}{K}((\tilde{u}_j)_2 - (\tilde{u}_j)_1) \\
&\quad + f_j(\{(u_l)_1\})\frac{d}{K} + \frac{K-1}{K}((\tilde{u}_j)_1 - (\tilde{u}_j)_0) + (E_j)_1 \\
&= 2f_j(\{(u_l)_1\})\frac{d}{K} - 2f_j(\{(u_l)_2\})\frac{d}{K} + 2f_j(\{(u_l)_2\})\frac{d}{K} \\
&\quad + \frac{K-1}{K}((\tilde{u}_j)_2 - (\tilde{u}_j)_0) + (E_j)_1,
\end{aligned} \tag{4.2}
$$

Summarizing

$$
(u_j)_2 = (u_j)_0 + 2f_j(\{(u_l)_2\})\frac{d}{K} + \frac{K-1}{K}((\tilde{u}_j)_2 - (\tilde{u}_j)_0) + (E_j)_2,
$$

where

$$
(E_j)_2 = (E_j)_1 + (f_j(\{(u_l)_1\}) - f_j(\{(u_l)_2\}))\frac{2d}{K}.
$$

Reasoning inductively, for all $1 \le k \le N$, we obtain

$$
(u_j)_k = (u_j)_0 + f_j(\{(u_l)_k\})\frac{kd}{K} + \frac{K-1}{K}((\tilde{u}_j)_k - (\tilde{u}_j)_0) + (E_j)_k,
$$

where,

$$
\begin{aligned}
(E_j)_k &= (E_j)_{k-1} + k(f_j(\{u_l\}_{k-1}) - f_j(\{u_l\}_k)\frac{d}{K} \\
&= \sum_{m=1}^{k} m[f_j(\{(u_l)_{m-1}\}) - f_j(\{(u_l)_m\})]\frac{d}{K} \approx \mathcal{O}\left(\frac{(k^2+k)d^2}{2K}\right),
\end{aligned} \tag{4.3}
$$

$\forall j \in \{1, \cdots, 4\}$.

In particular, for $k = N$, for $K > 0$ sufficiently big, we obtain

$$
\begin{aligned}
(u_j)_N &= (u_j)_0 + f_j(\{(u_l)_N\})\frac{Nd}{K} + \frac{K-1}{K}((\tilde{u}_j)_N - (\tilde{u}_j)_0) + (E_j)_N \\
&\approx (u_j)_0 + f_j(\{(u_l)_N\})\frac{Nd}{K} + \frac{K-1}{K}((\tilde{u}_j)_N - (\tilde{u}_j)_0), \forall j \in \{1,2,3,4\}. \tag{4.4}
\end{aligned}
$$

In such a system we have 4 equations suitable to find the unknown variables $(u_0)_3, (u_N)_1, (u_N)_2, (u_N)_3$, considering that $(u_N)_4 = (u_4)_f$ is known.

Through the system (4.4), we obtain u_N through the Newton's Method for a system of only the 4 indicated variables.

Having u_N, we obtain u_{N-1} through the equations

$$(u_j)_N - (u_j)_{N-1} \approx f_j(\{(u_l)_{N-1}\})\frac{d}{K} + \frac{K-1}{K}((\tilde{u}_j)_N) - (\tilde{u}_j)_{N-1}),$$

also through the Newton's Method.

Similarly, having u_{N-1} we obtain u_{N-2} through the system

$$(u_j)_{N-1} - (u_j)_{N-2} \approx f_j(\{(u_l)_{N-2}\})\frac{d}{K} + \frac{K-1}{K}((\tilde{u}_j)_{N-1}) - (\tilde{u}_j)_{N-2}),$$

and so on up to finding u_1.

Having calculated u, we replace $\tilde{u}$ by u and repeat the process up to an appropriate convergence criterion is satisfied.

The problem is then approximately solved.

4.2   Applications to a flight mechanics model

We present numerical results for the following system of equations, which models the in plane climbing motion of an airplane (please, see more details in [83]).

$$
\begin{cases}
\dot{h} = V\sin\gamma, \\
\dot{\gamma} = \frac{1}{m_f V}(F\sin[a+a_F]+L) - \frac{g}{V}\cos\gamma, \\
\dot{V} = \frac{1}{m_f}(F\cos[a+a_F]) - D) - g\sin\gamma \\
\dot{x} = V\cos\gamma,
\end{cases}
\tag{4.5}
$$

with the boundary conditions,

$$
\begin{cases}
h(0) = h_0, \\
V(0) = V_0 \\
x(0) = x_0 \\
h(t_f) = h_f,
\end{cases}
\tag{4.6}
$$

where $t_f = 505s$, h is the airplane altitude, V is its speed, γ is the angle between its velocity and the horizontal axis, and finally x denotes the horizontal coordinate position.

For numerical purposes, we assume (Air bus 320)
$m_f = 120,000Kg$, $S_f = 260m^2$, $a = 0.138$ rad, $g = 9.8m/s^2$,

$$\rho(h) = 1.225(1 - 0.0065h/288.15)^{4.225}Kg/m^3,$$

$$a_F = 0.0175,$$

$$(C_L)_a = 5$$

$$(C_D)_0 = 0.0175,$$

$$K_1 = 0.0,$$

$$K_2 = 0.06$$

$$C_D = (C_D)_0 + K_1 1 C_L + K_2 C_L^2,$$

$$C_L = (C_L)_0 + (C_L)_a a,$$

$$L = \frac{1}{2} \rho(h) V^2 C_L S_f,$$

$$D = \frac{1}{2} \rho(h) V^2 C_D S_f,$$

$$F = 240000$$

and where units refer to the International System.

To simplify the analysis, we redefine the variables as below indicated:

$$\begin{cases} h = u_1, \\ \gamma = u_2 \\ V = u_3 \\ x = u_4. \end{cases} \tag{4.7}$$

Thus, denoting $\mathbf{u} = (u_1, u_2, u_3, u_4) \in U = W^{1,2}([0, t_f]; \mathbb{R}^4)$, the system above indicated may be expressed by

$$\begin{cases} \dot{u}_1 = f_1(\mathbf{u}) \\ \dot{u}_2 = f_2(\mathbf{u}) \\ \dot{u}_3 = f_3(\mathbf{u}) \\ \dot{u}_4 = f_4(\mathbf{u}), \end{cases} \tag{4.8}$$

where,

$$\begin{cases} f_1(\mathbf{u}) = u_3 \sin(u_2), \\ f_2(\mathbf{u}) = \frac{1}{m_f u_3} (F \sin[a + a_F] + L(\mathbf{u})) - \frac{g}{u_3} \cos(u_2), \\ f_3(\mathbf{u}) = \frac{1}{m_f} (F \cos[a + a_F] - D(\mathbf{u})) - g \sin(u_2) \\ f_4(\mathbf{u}) = u_3 \cos(u_2). \end{cases} \tag{4.9}$$

We solve this last system for the following boundary conditions:

$$\begin{cases} h(0) = 0 \, m, \\ V(0) = 120 m/s, \\ x(0) = 0 \, m, \\ h(t_f) = 11000 \, m. \end{cases} \tag{4.10}$$

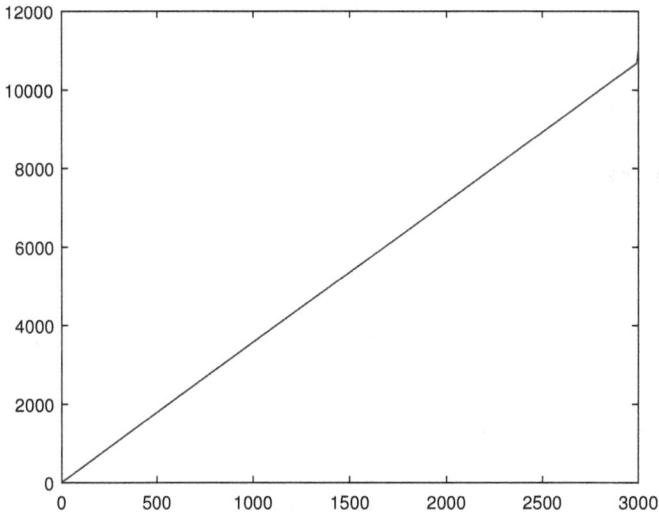

Figure 4.1: The solution h (in m) for $t_f = 505s$.

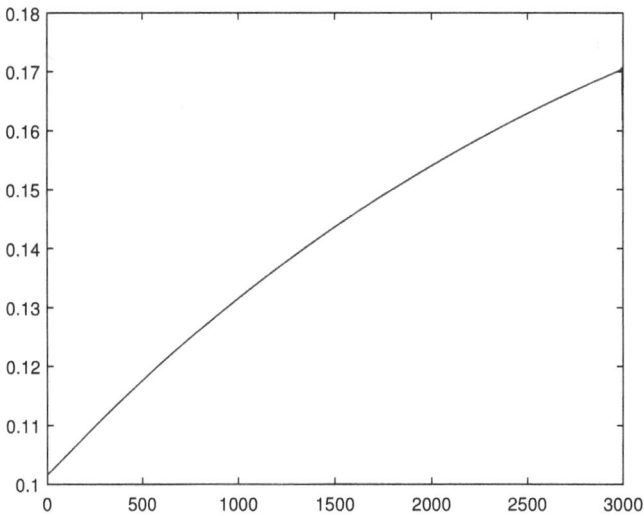

Figure 4.2: The solution γ (in rad) for $t_f = 505s$.

We have obtained the following solutions for h, γ, V and x. Please see Figures 4.1, 4.2, 4.3 and 4.4, respectively. We have set $N = 3000$ nodes and $K = 100$.

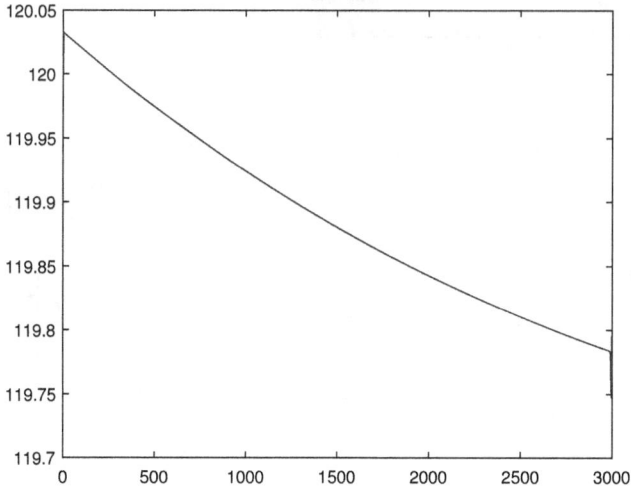

Figure 4.3: The solution V (in m/s) for $t_f = 505s$.

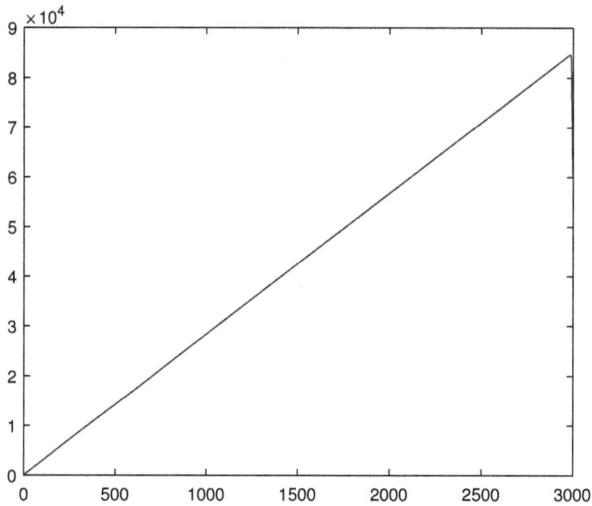

Figure 4.4: The solution x (in m) for $t_f = 505s$.

4.3 Acknowledgements

The author is grateful to Professor Pedro Paglione of Technological Institute of Aeronautics, ITA, SP-Brazil, for his valuable suggestions and comments, which help me a lot to improve some important parts of this text, in particular on the part concerning the model in flight mechanics addressed.

CALCULUS OF VARIATIONS, CONVEX ANALYSIS AND RESTRICTED OPTIMIZATION

II

Chapter 5

Basic Topics on the Calculus of Variations

5.1 Banach spaces

We start by recalling the norm definition.

Definition 5.1.1 *Let V be a vectorial space. A norm in V is a function denoted by $\| \cdot \|_V : V \to \mathbb{R}^+ = [0, +\infty)$, for which the following properties hold:*

1.
$$\|u\|_V > 0, \forall u \in V \text{ such that } u \neq \mathbf{0}$$

   and

$$\|u\|_V = 0, \text{ if, and only if } u = \mathbf{0}.$$

2. *Triangular inequality, that is*

$$\|u + v\|_V \leq \|u\|_V + \|v\|_V, \ \forall u, v \in V$$

3.
$$\|\alpha u\|_V = |\alpha| \|u\|, \ \forall u \in V, \ \alpha \in \mathbb{R}.$$

In such a case we say that the space V is a normed one.

Definition 5.1.2 (Convergent sequence) *Let V be a normed space and let $\{u_n\} \subset V$ be a sequence. We say that $\{u_n\}$ converges to $u_0 \in V$, if for each $\varepsilon > 0$, there exists $n_0 \in \mathbb{N}$ such that if $n > n_0$, then*

$$\|u_n - u_0\|_V < \varepsilon.$$

In such a case, we write

$$\lim_{n \to \infty} u_n = u_0, \text{ in norm.}$$

Definition 5.1.3 (Cauchy sequence in norm) *Let V be a normed space and let $\{u_n\} \subset V$ be a sequence.*
We say that $\{u_n\}$ is a Cauchy one, as for each $\varepsilon > 0$, there exists $n_0 \in \mathbb{N}$ such that if $m, n > n_0$, then

$$\|u_n - u_m\|_V < \varepsilon.$$

At this point we recall the definition of Banach space.

Definition 5.1.4 (Banach space) *A normed space V is said to be a Banach space as it is complete, that is, for each Cauchy sequence de Cauchy $\{u_n\} \subset V$ there exists a $u_0 \in V$ such that*

$$\|u_n - u_0\|_V \to 0, \text{ as } n \to \infty.$$

Example 5.1.5 *Examples of Banach spaces:*
Consider $V = C([a,b])$, the space of continuous functions on $[a,b]$. We shall prove that such a space is a Banach one with the norm,

$$\|f\|_V = \max\{|f(x)| \; : \; x \in [a,b]\}.$$

Exercise 5.1.6 *Prove that*

$$\|f\|_V = \max\{|f(x)| \; : \; x \in [a,b]\}$$

is a norm for $V = C([a,b])$.

Solution:

1. Clearly
$$\|f\|_V \geq 0, \forall f \in V$$

   and
   $$\|f\|_V = 0 \text{ if, and only if } f(x) = 0, \; \forall x \in [a,b],$$
   $$\text{that is if, and only if } f = \mathbf{0}.$$

2. Let $f, g \in V$.
   Thus,

$$
\begin{aligned}
\|f + g\|_V &= \max\{|f(x) + g(x)|, \; x \in [a,b]\} \\
&\leq \max\{|f(x)| + |g(x)|, \; x \in [a,b]\} \\
&\leq \max\{|f(x)|, \; x \in [a,b]\} + \max\{|g(x)| \; x \in [a,b]\} \\
&= \|f\|_V + \|g\|_V.
\end{aligned}
\tag{5.1}
$$

3. Finally, let $\alpha \in \mathbb{R}$ and $f \in V$.

Hence,

$$
\begin{aligned}
\|\alpha f\|_V &= \max\{|\alpha f(x)|, \ x \in [a,b]\} \\
&= \max\{|\alpha||f(x)|, \ x \in [a,b]\} \\
&= |\alpha| \max\{|f(x)|, \ x \in [a,b]\} \\
&= |\alpha| \|f\|.
\end{aligned}
\tag{5.2}
$$

From this we may infer that $\|\cdot\|_V$ is a norm for V.

The solution is complete.

Theorem 5.1.7 $V = C([a,b])$ *is a Banach space with the norm*

$$
\|f\|_V = \max\{|f(x)| \ : \ x \in [a,b]\}, \ \forall f \in V.
$$

Proof 5.1 The proof that $C([a,b])$ is a vector space is left as an exercise.

From the last exercise, $\|\cdot\|_V$ is a norm for V.

Let $\{f_n\} \subset V$ be a Cauchy sequence.

We shall prove that there exists $f \in V$ such that

$$
\|f_n - f\|_V \to 0, \ \text{as } n \to \infty.
$$

Let $\varepsilon > 0$.

Thus, there exists $n_0 \in \mathbb{N}$ such that if $m, n > n_0$, then

$$
\|f_n - f_m\|_V < \varepsilon.
$$

Hence

$$
\max\{|f_n(x) - f_m(x)| \ : \ x \in [a,b]\} < \varepsilon,
$$

that is,

$$
|f_n(x) - f_m(x)| < \varepsilon, \ \forall x \in [a,b], \ m, n > n_0.
\tag{5.3}
$$

Let $x \in [a,b]$.

From (5.3), $\{f_n(x)\}$ is a real Cauchy sequence, therefore it is convergent.

So, define

$$
f(x) = \lim_{n \to \infty} f_n(x), \ \forall x \in [a,b].
$$

Also from (5.3), we have that

$$
\lim_{m \to \infty} |f_n(x) - f_m(x)| = |f_n(x) - f(x)| \le \varepsilon, \ \forall n > n_0.
$$

From this we may infer that

$$
\|f_n - f\|_V \to 0, \ \text{as } n \to \infty.
$$

We shall prove now that f is continuous on $[a,b]$.

From the exposed above,

$$f_n \to f$$

uniformly on $[a,b]$ as $n \to \infty$.

Thus, there exists $n_1 \in \mathbb{N}$ such that if $n > n_1$, then

$$|f_n(x) - f(x)| < \frac{\varepsilon}{3}, \ \forall x \in [a,b].$$

Choose $n_2 > n_1$.
Let $x \in [a,b]$. From

$$\lim_{y \to x} f_{n_2}(y) = f_{n_2}(x),$$

there exists $\delta > 0$ such that if $y \in [a,b]$ and $|y - x| < \delta$, then

$$|f_{n_2}(y) - f_{n_2}(x)| < \frac{\varepsilon}{3}.$$

Thus, if $y \in [a,b]$ e $|y - x| < \delta$, then

$$
\begin{aligned}
|f(y) - f(x)| &= |f(y) - f_{n_2}(y) + f_{n_2}(y) - f_{n_2}(x) + f_{n_2}(x) - f(x)| \\
&\leq |f(y) - f_{n_2}(y)| + |f_{n_2}(y) - f_{n_2}(x)| + |f_{n_2}(x) - f(x)| \\
&< \frac{\varepsilon}{3} + \frac{\varepsilon}{3} + \frac{\varepsilon}{3} \\
&= \varepsilon.
\end{aligned}
\tag{5.4}
$$

So, we may infer that f is continuous at x, $\forall x \in [a,b]$, that is $f \in V$.
The proof is complete.

Exercise 5.1.8 *Let $V = C^1([a,b])$ be the space of functions $f : [a,b] \to \mathbb{R}$ which the the first derivative is continuous on $[a,b]$.*
Define s function (in fact a functional) $\| \cdot \|_V : V \to \mathbb{R}^+$ by

$$\|f\|_V = \max\{|f(x)| + |f'(x)| \ : \ x \in [a,b]\}.$$

1. Prove that $\| \cdot \|_V$ is a norm.

2. Prove that V is a Banach space with such a norm.

Solution: The proof of item 1 is left as an exercise.
Now, we shall prove that V is complete.
Let $\{f_n\} \subset V$ be a Cauchy sequence.
Let $\varepsilon > 0$. Thus, there exists $n_0 \in \mathbb{N}$ such that if $m,n > n_0$, then

$$\|f_n - f_m\|_V < \varepsilon/2.$$

Therefore,

$$|f_n(x) - f_m(x)| + |f'_n(x) - f'_m(x)| < \varepsilon/2, \ \forall x \in [a,b], \ m,n > n_0. \tag{5.5}$$

Let $x \in [a,b]$. hence, $\{f_n(x)\}$ are $\{f'_n(x)\}$ real Cauchy sequences, and therefore, they are convergent.

Denote

$$f(x) = \lim_{n \to \infty} f_n(x)$$

and

$$g(x) = \lim_{n \to \infty} f'_n(x).$$

From this and (5.5), we obtain

$$
\begin{aligned}
|f_n(x) - f(x)| + |f'_n(x) - g(x)| &= \lim_{m \to \infty} |f_n(x) - f_m(x)| + |f'_n(x) - f'_m(x)| \\
&\leq \varepsilon/2, \ \forall x \in [a,b], \ n > n_0.
\end{aligned}
\tag{5.6}
$$

Similarly to the last example, we may obtain that f and g are continuous, therefore uniformly continuous on the compact set $[a,b]$.

Thus, there exists $\delta > 0$ such that if $x, y \in [a,b]$ and $|y - x| < \delta$, then

$$|g(y) - g(x)| < \varepsilon/2. \tag{5.7}$$

Choose $n_1 > n_0$. Let $x \in (a,b)$.

Hence, if $0 < |h| < \delta$, then from (5.6) and (5.7) we have

$$
\begin{aligned}
&\left| \frac{f_{n_1}(x+h) - f_{n_1}(x)}{h} - g(x) \right| \\
&= |f'_{n_1}(x+th) - g(x+th) + g(x+th) - g(x)| \\
&\leq |f'_{n_1}(x+th) - g(x+th)| + |g(x+th) - g(x)| \\
&< \varepsilon/2 + \varepsilon/2 \\
&= \varepsilon,
\end{aligned}
\tag{5.8}
$$

where from mean value theorem, $t \in (0,1)$ (it depends on h). Therefore, letting $n_1 \to \infty$, we get

$$
\begin{aligned}
&\left| \frac{f_{n_1}(x+h) - f_{n_1}(x)}{h} - g(x) \right| \\
&\to \left| \frac{f(x+h) - f(x)}{h} - g(x) \right| \\
&\leq \varepsilon, \forall 0 < |h| < \delta.
\end{aligned}
\tag{5.9}
$$

From this we may infer that

$$f'(x) = \lim_{h \to 0} \frac{f(x+h) - f(x)}{h} = g(x), \ \forall x \in (a,b).$$

The cases in which $x = a$ or $x = b$ may be dealt similarly with one-sided limits.

From this and (5.6), we have

$$\|f_n - f\|_V \to 0, \text{ as } n \to \infty$$

e

$$f \in C^1([a,b]).$$

The solution is complete.

Definition 5.1.9 (Functional) *Let V be a Banach space. A functional F defined on V is a function whose the co-domain is $\mathbb{R}$ ($F : V \to \mathbb{R}$).*

Example 5.1.10 *Let $V = C([a,b])$ and $F : V \to \mathbb{R}$ where*

$$F(y) = \int_a^b (\text{sen}^3 x + y(x)^2) \, dx, \ \forall y \in V.$$

Example 5.1.11 *Let $V = C^1([a,b])$ and let $J : V \to \mathbb{R}$ where*

$$J(y) = \int_a^b \sqrt{1 + y'(x)^2} \, dx, \ \forall y \in C^1([a,b]).$$

In our first frame-work we consider functionals defined as

$$F(y) = \int_a^b f(x, y(x), y'(x)) \, dx,$$

where we shall assume

$$f \in C([a,b] \times \mathbb{R} \times \mathbb{R})$$

and $V = C^1([a,b])$.

Thus, for $F : D \subset V \to \mathbb{R}$ where

$$F(y) = \int_a^b f(x, y(x), y'(x)) \, dx,$$

we assume

$$V = C^1([a,b]),$$

and

$$D = \{y \in V : y(a) = A \text{ and } y(b) = B\},$$

where $A, B \in \mathbb{R}$.

Observe that if $y \in D$, then $y + v \in D$ if, and only if, $v \in V$ and

$$v(a) = v(b) = 0.$$

Indeed, in such a case

$$y + v \in V$$

e

$$y(a) + v(a) = y(a) = A,$$

and
$$y(b) + v(b) = y(b) = B.$$

Thus, we define the space of admissible directions for F, denoted by V_a, as,
$$V_a = \{v \in V : v(a) = v(b) = 0\}.$$

Definition 5.1.12 (Global minimum) *Let V be a Banach space and let $F : D \subset V \to \mathbb{R}$ be a functional. We say that $y_0 \in D$ is a point of global minimum for F, if*
$$F(y_0) \leq F(y), \ \forall y \in D.$$

Observe that denoting $y = y_0 + v$ where $v \in V_a$, we have
$$F(y_0) \leq F(y_0 + v), \ \forall v \in V_a.$$

Example 5.1.13 *Consider $J : D \subset V \to \mathbb{R}$ where $V = C^1([a,b])$,*
$$D = \{y \in V : y(a) = 0 \ e \ y(b) = 1\}$$

and
$$J(y) = \int_a^b (y'(x))^2 \, dx.$$

Thus,
$$V_a = \{v \in V : v(a) = v(b) = 0\}.$$

Let $y_0 \in D$ be a candidate to global minimum for F and let $v \in V_a$ be admissible direction.

Hence, we must have
$$J(y_0 + v) - J(y_0) \geq 0, \tag{5.10}$$

where

$$
\begin{aligned}
J(y_0 + v) - J(y_0) &= \int_a^b (y_0'(x) + v'(x))^2 \, dx - \int_a^b y_0'(x)^2 \, dx \\
&= 2\int_a^b y_0'(x)v'(x) \, dx + \int_a^b v'(x)^2 \, dx \\
&\geq 2\int_a^b y_0'(x)v'(x) \, dx. \tag{5.11}
\end{aligned}
$$

Observe that if $y_0'(x) = c$ em $[a,b]$, we have (5.10) satisfied, since in such a case,

$$
\begin{aligned}
J(y_0 + v) - J(y_0) &\geq 2\int_a^b y_0'(x)v'(x) \, dx \\
&= 2c\int_a^b v'(x) \, dx \\
&= 2c[v(x)]_a^b \\
&= 2c(v(b) - v(a)) \\
&= 0. \tag{5.12}
\end{aligned}
$$

Summarizing, if $y_0'(x) = con$ $[a,b]$, then

$$J(y_0 + v) \geq J(y_0), \ \forall v \in V_a.$$

Observe that in such a case,

$$y_0(x) = cx + d,$$

for some $d \in \mathbb{R}$.
  However, from
$$y(a) = 0, \ we \ get \ ca + d = 0.$$

From $y_0(b) = 1$, we have $cb + d = 1$.
Solving this last system in c and d we obtain,

$$c = \frac{1}{b-a},$$

and

$$d = \frac{-a}{b-a}.$$

From this, we have,

$$y_0(x) = \frac{x-a}{b-a}.$$

Observe that the graph of y_0 corresponds to the straight line connecting the points $(a,0)$ and $(b,1)$.

5.2 The Gâteaux variation

Definition 5.2.1 *Let V be a Banach space and let $J : D \subset V \to \mathbb{R}$ be a functional. Let $y \in D$ and $v \in V_a$.*
  We define the Gâteaux variation of J at y in the direction v, denoted by $\delta J(y;v)$, by

$$\delta J(y;v) = \lim_{\varepsilon \to 0} \frac{J(y + \varepsilon v) - J(y)}{\varepsilon},$$

if such a limit exists. Equivalently,

$$\delta J(y;v) = \frac{\partial J(y + \varepsilon v)}{\partial \varepsilon}\Big|_{\varepsilon=0}.$$

Example 5.2.2 *Let $V = C^1([a,b])$ and $J : V \to \mathbb{R}$ where*

$$F(y) = \int_a^b (sen^3 x + y(x)^2) \, dx.$$

Let $y, v \in V$. Let us calculate
$$\delta J(y;v).$$

Observe that,

$$
\begin{aligned}
\delta J(y;v) &= \lim_{\varepsilon \to 0} \frac{J(y+\varepsilon v) - J(y)}{\varepsilon} \\
&= \lim_{\varepsilon \to 0} \frac{\int_a^b \left(sen^3 x + (y(x) + \varepsilon v(x))^2 \right) dx - \int_a^b \left(sen^3 x + y(x)^2 \right) dx}{\varepsilon} \\
&= \lim_{\varepsilon \to 0} \frac{\int_a^b (2\varepsilon y(x)v(x) + \varepsilon^2 v(x)) \, dx}{\varepsilon} \\
&= \lim_{\varepsilon \to 0} \left(\int_a^b 2y(x)v(x) \, dx + \varepsilon \int_a^b v(x)^2 \, dx \right) \\
&= \int_a^b 2y(x)v(x) \, dx.
\end{aligned} \tag{5.13}
$$

Example 5.2.3 *Let $V = C^1([a,b])$ and let $J : V \to \mathbb{R}$ where*

$$
J(y) = \int_a^b \rho(x)\sqrt{1 + y'(x)^2} \, dx,
$$

and where $\rho : [a,b] \to (0,+\infty)$ is a fixed function.
    Let $y, v \in V$.
    Thus,

$$
\delta J(y;v) = \frac{\partial J(y+\varepsilon v)}{\partial \varepsilon}\Big|_{\varepsilon=0}, \tag{5.14}
$$

where

$$
J(y+\varepsilon v) = \int_a^b \rho(x)\sqrt{1 + (y'(x) + \varepsilon v'(x))^2} \, dx.
$$

Hence,

$$
\begin{aligned}
\frac{\partial J(y+\varepsilon v)}{\partial \varepsilon}\Big|_{\varepsilon=0} &= \frac{\partial}{\partial \varepsilon}\left(\int_a^b \rho(x)\sqrt{1 + (y'(x) + \varepsilon v'(x))^2} \, dx \right) \\
&=^{(*)} \int_a^b \rho(x) \frac{\partial}{\partial \varepsilon}\left(\sqrt{1 + (y'(x) + \varepsilon v'(x))^2} \right) dx \\
&= \int_a^b \frac{\rho(x)}{2} \frac{2(y'(x) + \varepsilon v'(x))v'(x)}{\sqrt{1 + (y'(x) + \varepsilon v'(x))^2}} \, dx.
\end{aligned} \tag{5.15}
$$

(*): We shall prove this step is valid in the subsequent pages.

From this we get,

$$
\begin{aligned}
\delta J(y;v) &= \frac{\partial J(y+\varepsilon v)}{\partial \varepsilon}\Big|_{\varepsilon=0} \\
&= \int_a^b \frac{\rho(x)y'(x)v'(x)}{\sqrt{1 + y'(x)^2}} \, dx.
\end{aligned} \tag{5.16}
$$

The example is complete.

Example 5.2.4 *Let $V = C^1([a,b])$ and $f \in C^1([a,b] \times \mathbb{R} \times \mathbb{R})$. Thus f is a function of three variables, namely, $f(x,y,z)$.*

Consider the functional $F : V \to \mathbb{R}$, defined by

$$F(y) = \int_a^b f(x, y(x), y'(x)) \, dx.$$

Let $y, v \in V$. Thus,

$$\delta F(y; v) = \frac{\partial}{\partial \varepsilon} F(y + \varepsilon v)|_{\varepsilon=0}.$$

Observe que

$$F(y + \varepsilon v) = \int_a^b f(x, y(x) + \varepsilon v(x), y'(x) + \varepsilon v'(x)) \, dx,$$

and therefore

$$
\begin{aligned}
\frac{\partial}{\partial \varepsilon} F(y + \varepsilon v) &= \frac{\partial}{\partial \varepsilon} \left(\int_a^b f(x, y(x) + \varepsilon v(x), y'(x) + \varepsilon v'(x)) \, dx \right) \\
&= \int_a^b \frac{\partial}{\partial \varepsilon} \left(f(x, y(x) + \varepsilon v(x), y'(x) + \varepsilon v'(x)) \right) dx \\
&= \int_a^b \left(\frac{\partial f(x, y(x) + \varepsilon v(x), y'(x) + \varepsilon v'(x))}{\partial y} v(x) \right. \\
&\quad \left. + \frac{\partial f(x, y(x) + \varepsilon v(x), y'(x) + \varepsilon v'(x))}{\partial z} v'(x) \right) dx. \quad (5.17)
\end{aligned}
$$

Thus

$$
\begin{aligned}
\delta F(y; v) &= \frac{\partial F(y + \varepsilon v)}{\partial \varepsilon} |_{\varepsilon=0} \\
&= \int_a^b \left(\frac{\partial f(x, y(x), y'(x))}{\partial y} v(x) + \frac{\partial f(x, y(x), y'(x))}{\partial z} v'(x) \right) dx. \quad (5.18)
\end{aligned}
$$

5.3   Minimization of convex functionals

Definition 5.3.1 (Convex function) *A function $f : \mathbb{R}^n \to \mathbb{R}$ is said to be convex if*

$$f(\lambda x + (1 - \lambda)y) \le \lambda f(x) + (1 - \lambda) f(y), \, \forall x, y \in \mathbb{R}^n, \, \lambda \in [0, 1].$$

Proposition 5.3.2 *Let $f : \mathbb{R}^n \to \mathbb{R}$ be a convex and differentiable function. Under such hypotheses,*

$$f(y) - f(x) \ge \langle f'(x), y - x \rangle_{\mathbb{R}^n}, \, \forall x, y \in \mathbb{R}^n,$$

where $\langle \cdot, \cdot \rangle_{\mathbb{R}^n} : \mathbb{R}^n \times \mathbb{R}^n \to \mathbb{R}$ denotes the usual inner product for $\mathbb{R}^n$, that is,

$$\langle x, y \rangle_{\mathbb{R}^n} = x_1 y_1 + \cdots + x_n y_n,$$

$\forall x = (x_1, \cdots, x_n), \, y = (y_1, \cdots, y_n) \in \mathbb{R}^n$.

Proof 5.2 Choose $x, y \in \mathbb{R}^n$.

From the hypotheses,

$$f((1-\lambda)x + \lambda y) \leq (1-\lambda)f(x) + \lambda f(y), \ \forall \lambda \in (0,1).$$

Thus,

$$\frac{f(x + \lambda(y-x)) - f(x)}{\lambda} \leq f(y) - f(x), \ \forall \lambda \in (0,1).$$

Therefore,

$$\begin{aligned}
\langle f'(x), y-x \rangle_{\mathbb{R}^n} &= \lim_{\lambda \to 0^+} \frac{f(x + \lambda(y-x)) - f(x)}{\lambda} \\
&\leq f(y) - f(x), \ \forall x, y \in \mathbb{R}^n.
\end{aligned} \tag{5.19}$$

The proof is complete.

Proposition 5.3.3 *Let $f : \mathbb{R}^n \to \mathbb{R}$ be a differentiable function on $\mathbb{R}^n$.*
Assume

$$f(y) - f(x) \geq \langle f'(x), y-x \rangle_{\mathbb{R}^n}, \ \forall x, y \in \mathbb{R}^n.$$

Under such hypotheses, f is convex.

Proof 5.3 Define $f^* : \mathbb{R}^n \to \mathbb{R} \cup \{+\infty\}$ by

$$f^*(x^*) = \sup_{x \in \mathbb{R}^n} \{\langle x, x^* \rangle_{\mathbb{R}^n} - f(x)\}.$$

Such a function is said to the polar function for f.

Let $x \in \mathbb{R}^n$. From the hypotheses,

$$\langle f'(x), x \rangle_{\mathbb{R}^n} - f(x) \geq \langle f'(x), y \rangle_{\mathbb{R}^n} - f(y), \ \forall y \in \mathbb{R}^n,$$

that is,

$$\begin{aligned}
f^*(f'(x)) &= \sup_{y \in \mathbb{R}^n} \{\langle f'(x), y \rangle_{\mathbb{R}^n} - f(y)\} \\
&= \langle f'(x), x \rangle_{\mathbb{R}^n} - f(x).
\end{aligned} \tag{5.20}$$

On the other hand

$$f^*(x^*) \geq \langle x, x^* \rangle_{\mathbb{R}^n} - f(x), \ \forall x, x^* \in \mathbb{R}^n,$$

and thus

$$f(x) \geq \langle x, x^* \rangle_{\mathbb{R}^n} - f^*(x^*), \ \forall x^* \in \mathbb{R}^n.$$

Hence,

$$\begin{aligned}
f(x) &\geq \sup_{x^* \in \mathbb{R}^n} \{\langle x, x^* \rangle_{\mathbb{R}^n} - f^*(x^*)\} \\
&\geq \langle f'(x), x \rangle_{\mathbb{R}^n} - f^*(f'(x)).
\end{aligned} \tag{5.21}$$

From this and (5.20), we obtain,

$$
\begin{aligned}
f(x) &= \sup_{x^* \in \mathbb{R}^n} \{\langle x, x^* \rangle_{\mathbb{R}^n} - f^*(x^*)\} \\
&= \langle f'(x), x \rangle_{\mathbb{R}^n} - f^*(f'(x)).
\end{aligned} \tag{5.22}
$$

Summarizing,

$$
f(x) = \sup_{x^* \in \mathbb{R}^n} \{\langle x, x^* \rangle_{\mathbb{R}^n} - f^*(x^*)\}, \ \forall x \in \mathbb{R}^n.
$$

Choose $x, y \in \mathbb{R}^n$ and $\lambda \in [0,1]$.

From the last equation we may write,

$$
\begin{aligned}
f(\lambda x + (1-\lambda)y) &= \sup_{x^* \in \mathbb{R}^n} \{\langle \lambda x + (1-\lambda)y, x^* \rangle_{\mathbb{R}^n} - f^*(x^*)\} \\
&= \sup_{x^* \in \mathbb{R}^n} \{\langle \lambda x + (1-\lambda)y, x^* \rangle_{\mathbb{R}^n} \\
&\qquad - \lambda f^*(x^*) - (1-\lambda)f^*(x^*)\} \\
&= \sup_{x^* \in \mathbb{R}^n} \{\lambda(\langle x, x^* \rangle_{\mathbb{R}^n} - f^*(x^*)) \\
&\qquad + (1-\lambda)(\langle y, x^* \rangle_{\mathbb{R}^n} - f^*(x^*))\} \\
&\leq \lambda \sup_{x^* \in \mathbb{R}^n} \{\langle x, x^* \rangle_{\mathbb{R}^n} - f^*(x^*)\} \\
&\qquad + (1-\lambda) \sup_{x^* \in \mathbb{R}^n} \{\langle y, x^* \rangle_{\mathbb{R}^n} - f^*(x^*)\} \\
&= \lambda f(x) + (1-\lambda)f(y).
\end{aligned} \tag{5.23}
$$

Since $x, y \in \mathbb{R}^n$ and $\lambda \in [0,1]$ are arbitrary, we may infer that f is convex.

This completes the proof.

Definition 5.3.4 (Convex functional) *Let V be a Banach space and let $J : D \subset V \to \mathbb{R}$ be a functional. We say that J is convex if*

$$
J(y+v) - J(y) \geq \delta J(y; v), \ \forall v \in V_a(y),
$$

where

$$
V_a(y) = \{v \in V \ : \ y + v \in D\}.
$$

Theorem 5.3.5 *Let V be a Banach space and let $J : D \subset U$ be a convex functional. Thus, if $y_0 \in D$ is such that*

$$
\delta J(y_0; v) = 0, \ \forall v \in V_a(y_0),
$$

then

$$
J(y_0) \leq J(y), \ \forall y \in D,
$$

that is, y_0 minimizes J on D.

Proof 5.4 Choose $y \in D$. Let $v = y - y_0$. Thus $y = y_0 + v \in D$ so that

$$v \in V_a(y_0).$$

From the hypothesis,
$$\delta J(y_0; v) = 0,$$

and since J is convex, we obtain

$$J(y) - J(y_0) = J(y_0 + v) - J(y_0) \geq \delta J(y_0; v) = 0,$$

that is,
$$J(y_0) \leq J(y), \ \forall y \in D.$$

The proof is complete.

Example 5.3.6 *Let us see this example of convex functional. Let $V = C^1([a,b])$ and let $J : D \subset V \to \mathbb{R}$ be defined by*

$$J(y) = \int_a^b (y'(x))^2 \, dx,$$

where
$$D = \{y \in V \ : \ y(a) = 1 \ e \ y(b) = 5\}.$$

We shall show that J is convex.
Indeed, let $y \in D$ e $v \in V_a$ where

$$V_a = \{v \in V \ : \ v(a) = v(b) = 0\}.$$

Thus,

$$
\begin{aligned}
J(y+v) - J(y) &= \int_a^b (y'(x) + v'(x))^2 \, dx - \int_a^b y'(x)^2 \, dx \\
&= \int_a^b 2y'(x)v'(x) \, dx + \int_a^b v'(x)^2 \, dx \\
&\geq \int_a^b 2y'(x)v'(x) \, dx \\
&= \delta J(y; v).
\end{aligned}
\tag{5.24}
$$

Therefore, J is convex.

5.4 Sufficient conditions of optimality for the convex case

We start this section with a remark.

Remark 5.4.1 *Consider a function* $f : [a,b] \times \mathbb{R} \times \mathbb{R} \to \mathbb{R}$ *where* $f \in C^1([a,b] \times \mathbb{R} \times \mathbb{R})$.

Thus, for $V = C^1([a,b])$, *define* $F : V \to \mathbb{R}$ *by*

$$F(y) = \int_a^b f(x, y(x), y'(x)) \, dx.$$

Let $y, v \in V$. *We have already shown that*

$$\delta F(y; v) = \int_a^b (f_y(x, y(x), y'(x))v(x) + f_z(x, y(x), y'(x))v'(x)) \, dx.$$

Suppose f *is convex in* (y, z) *for all* $x \in [a,b]$, *which we denote by* $f(\underline{x}, y, z)$ *to be convex.*

From the last section, we have that

$$
\begin{aligned}
f(x, y+v, y'+v') - f(x, y, y') &\geq \langle \underline{\nabla} f(x, y, y'), (v, v') \rangle_{\mathbb{R}^2} \\
&= f_y(x, y, y')v + f_z(x, y, y')v', \; \forall x \in [a,b] \quad (5.25)
\end{aligned}
$$

where we denote

$$\underline{\nabla} f(x, y, y') = (f_y(x, y, y'), f_z(x, y, y')).$$

Therefore,

$$
\begin{aligned}
F(y+v) - F(y) &= \int_a^b [f(x, y+v, y'+v') - f(x, y, y')] \, dx \\
&\geq \int_a^b [f_y(x, y, y')v + f_z(x, y, y')v'] \, dx \\
&= \delta J(y; v). \quad (5.26)
\end{aligned}
$$

Thus, F is convex.

Theorem 5.4.2 *Let* $V = C^1([a,b])$. *Let* $f \in C^2([a,b] \times \mathbb{R} \times \mathbb{R})$ *where* $f(\underline{x}, y, z)$ *is convex. Define*

$$D = \{y \in V \ : \ y(a) = a_1 \ e \ y(b) = b_1\},$$

where $a_1, b_1 \in \mathbb{R}$.

Define also $F : D \to \mathbb{R}$ *by*

$$F(y) = \int_a^b f(x, y(x), y'(x)) \, dx.$$

Under such hypotheses, F is convex and if $y_0 \in D$ is such that

$$\frac{d}{dx}[f_z(x,y_0(x),y_0'(x))] = f_y(x,y_0(x),y_0'(x)), \ \forall x \in [a,b],$$

then y_0 minimizes F on D, that is,

$$F(y_0) \leq F(y), \ \forall y \in D.$$

Proof 5.5   From the last remark, F is convex. Suppose now that $y_0 \in D$ is such that

$$\frac{d}{dx}[f_z(x,y_0(x),y_0'(x))] = f_y(x,y_0(x),y_0'(x)), \ \forall x \in [a,b].$$

Let $v \in V_a = \{v \in V : v(a) = v(b) = 0\}$. Thus,

$$
\begin{aligned}
\delta F(y_0;v) &= \int_a^b (f_y(x,y_0(x),y_0'(x))v(x) + f_z(x,y_0(x),y_0'(x))v'(x)) \, dx \\
&= \int_a^b \left(\frac{d}{dx}(f_z(x,y_0(x),y_0'(x))v(x)) + f_z(x,y_0(x),y_0'(x))v'(x) \right) dx \\
&= \int_a^b \left(\frac{d}{dx}[f_z(x,y_0(x),y_0'(x))v(x)] \right) dx \\
&= [f_z(x,y_0(x),y_0'(x))v(x)]_a^b \\
&= f_z(b,y_0(b),y_0'(b))v(b) - f_z(a,y_0(a),y_0'(a))v(a) \\
&= 0, \ \forall v \in V_a.
\end{aligned}
\tag{5.27}
$$

Since F is convex, from this and Theorem 5.3.5, we may infer that y_0 minimizes J on D.

Example 5.4.3 *Let $V = C^1([a,b])$ and*

$$D = \{y \in V : y(0) = 0 \ and \ y(1) = 1\}.$$

Define $F : D \rightarrow \mathbb{R}$ by

$$F(y) = \int_0^1 [y'(x)^2 + 5y(x)] \, dx, \ \forall y \in D.$$

Observe that

$$F(y) = \int_0^1 f(x,y,y') \, dx$$

where

$$f(x,y,z) = z^2 + 5y,$$

that is, $f(\underline{x},y,z)$ is convex.

Thus, from the last theorem F is convex and if $y_0 \in D$ is such that

$$\frac{d}{dx} f_z(x, y_0(x), y_0'(x)) = f_y(x, y_0(x), y_0'(x)), \ \forall x \in [a,b],$$

then y_0 minimizes F on D.

Considering that $f_z(x,y,z) = 2z$ and $f_y(x,y,z) = 5$, from this last equation we get,

$$\frac{d}{dx}(2y_0'(x)) = 5,$$

that is,

$$y_0''(x) = \frac{5}{2}, \ \forall x \in [0,1].$$

Thus,

$$y_0'(x) = \frac{5}{2}x + c,$$

and

$$y_0(x) = \frac{5}{4}x^2 + cx + d.$$

From this and $y_0(0) = 0$, we obtain $d = 0$.

From this and $y_0(1) = 1$, we have

$$\frac{5}{4} + c = 1,$$

de modo que

$$c = -1/4$$

Therefore

$$y_0(x) = \frac{5x^2}{4} - \frac{x}{4}$$

minimizes F on D.

The example is complete.

5.5 Natural conditions, problems with free extremals

We start this section with the following theorem.

Theorem 5.5.1 *Let $V = C^1([a,b])$. Let $f \in C^2([a,b] \times \mathbb{R} \times \mathbb{R})$ be such that $f(x,y,z)$ is convex. Define*

$$D = \{y \in V : y(a) = a_1\},$$

where $a_1 \in \mathbb{R}$.

Define also $F : D \to \mathbb{R}$ by

$$F(y) = \int_a^b f(x, y(x), y'(x)) \, dx.$$

Under such hypotheses, F is convex and if $y_0 \in D$ is such that

$$\frac{d}{dx}[f_z(x, y_0(x), y_0'(x))] = f_y(x, y_0(x), y_0'(x)), \ \forall x \in [a, b]$$

and

$$f_z(b, y_0(b), y_0'(b)) = 0$$

then y_0 minimizes F on D, that is,

$$F(y_0) \le F(y), \ \forall y \in D.$$

Proof 5.6   Since $f(x, y, z)$ is convex from the last remark F is convex. Suppose now that $y_0 \in D$ is such that

$$\frac{d}{dx}[f_z(x, y_0(x), y_0'(x))] = f_y(x, y_0(x), y_0'(x)), \ \forall x \in [a, b]$$

and

$$f_z(b, y_0(b), y_0'(b)) = 0.$$

Let $v \in V_a = \{v \in V \ : \ v(a) = 0\}$. Thus,

$$
\begin{aligned}
\delta F(y_0; v) &= \int_a^b (f_y(x, y_0(x), y_0'(x))v(x) + f_z(x, y_0(x), y_0'(x))v'(x)) \, dx \\
&= \int_a^b \left(\frac{d}{dx}(f_z(x, y_0(x), y_0'(x))v(x)) + f_z(x, y_0(x), y_0'(x))v'(x) \right) dx \\
&= \int_a^b \left(\frac{d}{dx}[f_z(x, y_0(x), y_0'(x))v(x)] \right) dx \\
&= [f_z(x, y_0(x), y_0'(x))v(x)]_a^b \\
&= f_z(b, y_0(b), y_0'(b))v(b) - f_z(a, y_0(a), y_0'(a))v(a) \\
&= 0v(b) - f_z(a, y_0(a), y_0'(b))0 \\
&= 0, \ \forall v \in V_a. \hspace{4cm} (5.28)
\end{aligned}
$$

Since F is convex, from this and Theorem 5.3.5, we may infer that y_0 minimizes J on D.

Remark 5.5.2 *About this last theorem $y(a) = a_1$ is said to be an essential boundary condition, whereas $f_z(b, y_0(b), y_0'(b)) = 0$ is said to be a natural boundary condition.*

Theorem 5.5.3 *Let $V = C^1([a, b])$. Let $f \in C^2([a, b] \times \mathbb{R} \times \mathbb{R})$ where $f(x, y, z)$ is convex. Define*

$$D = V$$

and $F : D \to \mathbb{R}$ by

$$F(y) = \int_a^b f(x, y(x), y'(x)) \, dx.$$

Under such hypotheses, F is convex and if $y_0 \in D$ is such that

$$\frac{d}{dx}[f_z(x,y_0(x),y_0'(x))] = f_y(x,y_0(x),y_0'(x)), \ \forall x \in [a,b],$$

$$f_z(a,y_0(a),y_0'(a)) = 0$$

e

$$f_z(b,y_0(b),y_0'(b)) = 0,$$

then y_0 minimizes F on D, that is,

$$F(y_0) \leq F(y), \ \forall y \in D.$$

Proof 5.7 From the last remark F is convex. Suppose that $y_0 \in D$ is such that

$$\frac{d}{dx}[f_z(x,y_0(x),y_0'(x))] = f_y(x,y_0(x),y_0'(x)), \ \forall x \in [a,b]$$

and

$$f_z(a,y_0(a),y_0'(a)) = f_z(b,y_0(b),y_0'(b)) = 0.$$

Let $v \in D = V$. Thus,

$$
\begin{aligned}
\delta F(y_0;v) &= \int_a^b (f_y(x,y_0(x),y_0'(x))v(x) + f_z(x,y_0(x),y_0'(x))v'(x)) \, dx \\
&= \int_a^b \left(\frac{d}{dx}(f_z(x,y_0(x),y_0'(x))v(x)) + f_z(x,y_0(x),y_0'(x))v'(x) \right) dx \\
&= \int_a^b \left(\frac{d}{dx}[f_z(x,y_0(x),y_0'(x))v(x)] \right) dx \\
&= [f_z(x,y_0(x),y_0'(x))v(x)]_a^b \\
&= f_z(b,y_0(b),y_0'(b))v(b) - f_z(a,y_0(a),y_0'(a))v(a) \\
&= 0v(b) - 0v(a) \\
&= 0, \ \forall v \in D. \tag{5.29}
\end{aligned}
$$

Since F is convex, from this and from Theorem 5.3.5, we may conclude that y_0 minimizes J on $D = V$.

The proof is complete.

Remark 5.5.4 *About this last theorem, the conditions $f_z(a,y_0(a),y_0'(a)) = f_z(b,y_0(b),y_0'(b)) = 0$ are said to be natural boundary conditions and the problem in question a free extremal one.*

Exercise 5.5.5 *Show that F is convex and obtain its point of global minimum on D, D_1 and D_2, where*

$$F(y) = \int_1^2 \frac{y'(x)^2}{x} \, dx,$$

and where

1.

$$D = \{y \in C^1([1,2]) : y(1) = 0, \ y(1) = 3\},$$

2.

$$D_1 = \{y \in C^1([1,2]) : y(2) = 3\}.$$

3.

$$D_2 = C^1([1,2]).$$

Solution: Observe that

$$F(y) = \int_1^2 f(x,y(x),y'(x)) \, dx,$$

where $f(x,y,z) = z^2/x$, so that $f(\underline{x},y,z)$ is convex.
Therefore, F is convex.
Let $y, v \in V$, thus,

$$\delta F(y;v) = \int_1^2 [f_y(x,y,y')v + f_z(x,y,y')v'] \, dx,$$

where

$$f_y(x,y,z) = 0$$

e

$$f_z(x,y,z) = 2z/x.$$

Therefore,

$$\delta F(y;v) = \int_1^2 2x^{-1}y'(x)v'(x) \, dx.$$

For D, from Theorem 5.7.1, sufficient conditions of optimality are given by,

$$\begin{cases} \frac{d}{dx}[f_z(x,y_0(x),y_0'(x))] = f_y(x,y_0(x),y_0'(x)) \ em \ [1,2], \\ y_0(1) = 0, \\ y_0(2) = 3. \end{cases} \tag{5.30}$$

Thus, we must have

$$\frac{d}{dx}[2x^{-1}y_0'(x)] = 0,$$

that is,

$$2x^{-1}y_0'(x) = c,$$

so that

$$y_0'(x) = \frac{cx}{2}.$$

Therefore,

$$y_0(x) = \frac{cx^2}{4} + d.$$

On the other hand, we must have also

$$y_0(1) = \frac{c}{4} + d = 0,$$

and

$$y_0(2) = c + d = 3.$$

Thus, $c = 4$ and $d = -1$ so that $y_0(x) = x^2 - 1$ minimizes F on D.

For D_1, from Theorem 5.5.1, sufficient conditions of global optimality are given by

$$\begin{cases} \frac{d}{dx}[f_z(x, y_0(x), y'_0(x))] = f_y(x, y_0(x), y'_0(x)) \ em \ [1,2], \\ y_0(2) = 3, \\ f_z(1, y_0(1), y'_0(1)) = 0. \end{cases} \tag{5.31}$$

Thus, we must have

$$y_0(x) = \frac{cx^2}{4} + d.$$

On the other hand, we must also have

$$y_0(2) = c + d = 3,$$

and

$$f_z(1, y_0(1), y'_0(1)) = 2(1)^{-1}y'_0(1) = 0,$$

that is,

$$y'_0(1) = c/2 = 0,$$

Therefore, $c = 0$ and $d = 3$ so that $y_0(x) = 3$ minimizes F on D_1.

For D_2, from Theorem 5.5.3, sufficient conditions of global optimality are given by

$$\begin{cases} \frac{d}{dx}[f_z(x, y_0(x), y'_0(x))] = f_y(x, y_0(x), y'_0(x)) \ em \ [1,2], \\ f_z(1, y_0(1), y'_0(1)) = 0 \\ f_z(2, y_0(2), y'_0(2)) = 0. \end{cases} \tag{5.32}$$

Thus, we have

$$y_0(x) = \frac{cx^2}{4} + d.$$

On the other hand we have also

$$f_z(1, y_0(1), y'_0(1)) = 2(1)^{-1}y'_0(1) = 0,$$

$$f_z(2, y_0(2), y'_0(2)) = 2(2)^{-1}y'_0(2) = 0,$$

that is,

$$y'_0(1) = y'_0(2) = 0,$$

where $y'_0(x) = cx/2$.

Thus $c = 0$, so that $y_0(x) = d$, $\forall d \in \mathbb{R}$ minimizes F on D_2.

Exercise 5.5.6 *Let $V = C^2([0,1])$ and $J : D \subset V \to \mathbb{R}$ where*

$$J(y) = \frac{EI}{2} \int_0^1 y''(x)^2 \, dx - \int_0^1 P(x)y(x) \, dx,$$

represents the energy of a straight beam with rectangular cross section with inertial moment I. Here $y(x)$ denotes the vertical displacement of the point $x \in [0,1]$ resulting from the action of distributed vertical load $P(x) = \alpha x$, $\forall x \in [0,1]$, where $E > 0$ is the Young modulus and $\alpha > 0$ is a real constant.
And also

$$D = \{y \in V \ : \ y(0) = y(1) = 0\}.$$

Under such hypotheses,

1. *prove that F is convex.*

2. *Prove that if $y_0 \in D$ is such that*

$$\begin{cases} EI\frac{d^4}{dx^4}[y_0(x)] = P(x), \ \forall x \in [0,1], \\[2mm] y_0''(0) = 0, \\[2mm] y_0''(1) = 0, \end{cases} \tag{5.33}$$

then y_0 minimizes F on D.

3. *Find the optimal solution $y_0 \in D$.*

Solution:
Let $y \in D$ and $v \in V_a = \{v \in V \ : \ v(0) = v(1) = 0\}$.
We recall that

$$\begin{aligned} \delta J(y;v) &= \lim_{\varepsilon \to 0} \frac{F(y + \varepsilon v) - F(y)}{\varepsilon} \\[2mm] &= \lim_{\varepsilon \to 0} \frac{(EI/2) \int_0^1 [(y'' + \varepsilon v'')^2 - (y'')^2] \, dx - \int_0^1 (P(y + \varepsilon v) - P) \, dx}{\varepsilon} \\[2mm] &= \lim_{\varepsilon \to 0} \left(\int_0^1 (EIy''v'' - Pv) \, dx + \frac{\varepsilon EI}{2} \int_0^1 (v'')^2 \, dx \right) \\[2mm] &= \int_0^1 (EIy''v'' - Pv) \, dx. \end{aligned} \tag{5.34}$$

On the other hand,

$$
\begin{aligned}
J(y+v) - J(v) &= (EI/2) \int_0^1 [(y'' + v'')^2 - (y'')^2] \, dx - \int_0^1 (P(y+v) - P) \, dx \\
&= \int_0^1 (EIy''v'' - Pv) \, dx + \frac{EI}{2} \int_0^1 (v'')^2 \, dx \\
&\geq \int_0^1 (EIy''v'' - Pv) \, dx \\
&= \delta J(y; v). \tag{5.35}
\end{aligned}
$$

Since $y \in D$ and $v \in V_a$ are arbitrary, we may infer that J is convex. Assume that $y_0 \in D$ is such that

$$
\begin{cases}
EI \frac{d^4}{dx^4} [y_0(x)] = P(x), \ \forall x \in [0,1], \\[2mm]
y_0''(0) = 0, \\[2mm]
y_0''(1) = 0,
\end{cases} \tag{5.36}
$$

Thus,

$$
\begin{aligned}
\delta J(y; v) &= \int_0^1 (EIy''v'' - Pv) \, dx \\
&= \int_0^1 (EIy''v'' - EIy^{(4)}v) \, dx \\
&= \int_0^1 (EIy''v'' + EIy'''v') \, dx - [EIy'''(x)v(x)]_a^b \\
&= \int_0^1 (EIy''v'' + EIy'''v') \, dx \\
&= \int_0^1 (EIy''v'' - EIy''v'') \, dx + [EIy''(x)v'(x)]_a^b \\
&= 0 \tag{5.37}
\end{aligned}
$$

Summarizing

$$
\delta J(y_0; v) = 0, \ \forall v \in V_a.
$$

Therefore, since J is convex, we may conclude that y_0 minimizes J on D. To obtain the solution of the ODE is question, we shall denote

$$
y_0(x) = y_p(x) + y_h(x),
$$

where a particular solution y_p is given by $y_p(x) = \frac{\alpha x^5}{120EI}$, where claerly

$$
EI \frac{d^4}{dx^4} [y_p(x)] = P(x), \forall x \in [0,1].
$$

The homogeneous associated equation

$$EI\frac{d^4}{dx^4}[y_h(x)] = 0,$$

has the following general solution

$$y_h(x) = ax^3 + bx^2 + cx + d,$$

and thus,

$$y_0(x) = y_p(x) + y_h(x) = \frac{\alpha x^5}{120EI} + ax^3 + bx^2 + cx + d.$$

From $y_0(0) = 0$, we obtain $d = 0$.
Observe that $y_0'(x) = \frac{5\alpha}{120EI}x^4 + 3ax^2 + 2bx + c$ e $y_0''(x) = \frac{\alpha}{6EI}x^3 + 6ax + 2b$.
From this and $y_0''(0) = 0$, we get $b = 0$.
De $y_0''(1) = 0$, obtemos,

$$\frac{\alpha}{6EI}1^3 + 6a\,1 = 0,$$

e assim

$$a = -\frac{\alpha}{36EI}.$$

From such results and from $y_0(1) = 0$, we obtain

$$\frac{\alpha}{120EI} + a\,1^3 + c\,1 = \frac{\alpha}{120EI} - \frac{\alpha}{36EI} + c = 0,$$

that is,

$$c = \frac{\alpha}{EI}\left(\frac{1}{36} - \frac{1}{120}\right) = \frac{7\alpha}{360EI}.$$

Finally, we have that

$$y_0(x) = \frac{\alpha x^5}{120EI} - \frac{\alpha x^3}{36EI} + \frac{7\alpha x}{360EI}$$

minimizes J on D.
The solution is complete.

5.6 The du Bois-Reymond lemma

Lemma 5.6.1 (du Bois-Reymond) *Suppose $h \in C([a,b])$ and*

$$\int_a^b h(x)v'(x)\, dx = 0, \ \forall v \in V_a,$$

where

$$V_a = \{v \in C^1([a,b]) \ : \ v(a) = v(b) = 0\}.$$

Under such hypotheses, there exists $c \in \mathbb{R}$ such that

$$h(x) = c, \ \forall x \in [a,b].$$

Proof 5.8 Let

$$c = \frac{\int_a^b h(t) \, dt}{b-a}.$$

Define

$$v(x) = \int_a^x (h(t) - c) \, dt.$$

Thus,

$$v'(x) = h(x) - c, \ \forall x \in [a,b],$$

so that $v \in C^1([a,b])$.

Moreover,

$$v(a) = \int_a^a (h(t) - c) \, dt = 0,$$

and

$$v(b) = \int_a^b (h(t) - c) \, dt = \int_a^b h(t) \, dt - c(b-a) = c(b-a) - c(b-a) = 0,$$

so that $v \in V_a$.

Observe that, from this and the hypotheses,

$$
\begin{aligned}
0 \ &\leq \ \int_a^b (h(t) - c)^2 \, dt \\
&= \ \int_a^b (h(t) - c)(h(t) - c) \, dt \\
&= \ \int_a^b (h(t) - c)v'(t) \, dt \\
&= \ \int_a^b h(t)v'(t) \, dt - c \int_a^b v'(t) \, dt \\
&= \ 0 - c(v(b) - v(a)) \\
&= \ 0.
\end{aligned}
\tag{5.38}
$$

Thus,

$$\int_a^b (h(t) - c)^2 \, dt = 0.$$

Since h is continuous, we may infer that

$$h(x) - c = 0, \ \forall x \in [a,b],$$

that is,

$$h(x) = c, \ \forall x \in [a,b].$$

The proof is complete.

Theorem 5.6.2 *Let $g,h \in C([a,b])$ and suppose*

$$\int_a^b (g(x)v(x) + h(x)v'(x))\, dx = 0,\ \forall v \in V_a,$$

where

$$V_a = \{v \in C^1([a,b])\ :\ v(a) = v(b) = 0\}.$$

Under such hypotheses, $h \in C^1([a,b])$ e

$$h'(x) = g(x),\ \forall x \in [a,b].$$

Proof 5.9   Define

$$G(x) = \int_a^x g(t)\, dt.$$

Thus,

$$G'(x) = g(x),\ \forall x \in [a,b].$$

Let $v \in V_a$.
From the hypotheses,

$$
\begin{aligned}
0 &= \int_a^b [g(x)v(x) + h(x)v'(x)]\, dx \\
&= \int_a^b [-G(x)v'(x) + h(x)v'(x)]\, dx + [G(x)v(x)]_a^b \\
&= \int_a^b [-G(x) + h(x)]v'(x)\, dx,\ \forall v \in V_a.
\end{aligned}
\tag{5.39}
$$

From this and from the du Bois - Reymond lemma, we may conclude that

$$-G(x) + h(x) = c,\ \forall x \in [a,\ b],$$

for some $c \in \mathbb{R}$.
Thus

$$g(x) = G'(x) = h'(x),\ \forall x \in [a,b],$$

so that

$$g \in C^1([a,b]).$$

The proof is complete.

Lemma 5.6.3 (Fundamental lemma of calculus of variation for one dimension)
Let $g \in C([a,b]) = V$.
   Assume

$$\int_a^b g(x)v(x)\, dx = 0,\ \forall v \in V_a,$$

where again,

$$V_a = \{v \in C^1([a,b])\ :\ v(a) = v(b) = 0\}.$$

Under such hypotheses,

$$g(x) = 0,\ \forall x \in [a,b].$$

Proof 5.10   It suffices to apply the last theorem for $h \equiv 0$.

Exercise 5.6.4 *Let $h \in C([a,b])$.*
Suppose

$$\int_a^b h(x)w(x)\,dx = 0,\ \forall w \in D_0,$$

where

$$D_0 = \left\{ w \in C([a,b])\ :\ \int_a^b w(x)\,dx = 0 \right\}.$$

Show that there exists $c \in \mathbb{R}$ such that

$$h(x) = c,\ \forall x \in [a,b].$$

Solution Define, as above indicated,

$$V_a = \{v \in C^1([a,b])\ :\ v(a) = v(b) = 0\}.$$

Let $v \in V_a$.
Let $w \in C([a,b])$ be such that

$$w(x) = v'(x),\ \forall x \in [a,b].$$

Observe that

$$\int_a^b w(x)\,dx = \int_a^b v'(x)\,dx = [v(x)]_a^b = v(b) - v(a) = 0.$$

From this $\int_a^b h(x)w(x)\,dx = 0$, and thus

$$\int_a^b h(x)v'(x)\,dx = 0.$$

Since $v \in V_a$ is arbitrary, from this and the du Bois-Reymond lemma, there exists $c \in \mathbb{R}$ such that

$$h(x) = c,\ \forall x \in [a,b].$$

The solution is complete.

5.7 Calculus of variations, the case of scalar functions on $\mathbb{R}^n$

Let $\Omega \subset \mathbb{R}^n$ be an open, bounded, connected with a regular boundary $\partial\Omega = S$ (Lipschitzian) (which we define as Ω to be of class $\hat{C}^1$). Let $V = C^1(\overline{\Omega})$ and let $F : D \subset V \to \mathbb{R}$, be such that

$$F(y) = \int_\Omega f(x, y(x), \nabla y(x))\,dx,\ \forall y \in V,$$

where we denote

$$dx = dx_1 \cdots dx_n.$$

Assume $f : \overline{\Omega} \times \mathbb{R} \times \mathbb{R}^n \to \mathbb{R}$ is of C^2 class. Suppose also $f(x,y,\mathbf{z})$ is convex in $(y,\mathbf{z})$, $\forall x \in \overline{\Omega}$, which we denote by $f(\underline{x},y,\mathbf{z})$ to be convex.

Observe that for $y \in D$ e $v \in V_a$, where

$$D = \{y \in V \ : \ y = y_1 \text{ on } \partial\Omega\},$$

and

$$V_a = \{v \in V \ : \ v = 0 \text{ on } \partial\Omega\},$$

where

$$y_1 \in C^1(\overline{\Omega}),$$

we have that

$$\delta F(y;v) = \frac{\partial}{\partial \varepsilon} F(y + \varepsilon v)|_{\varepsilon=0},$$

where

$$F(y + \varepsilon v) = \int_{\Omega} f(x, y + \varepsilon v, \nabla y + \varepsilon \nabla v) \, dx.$$

Therefore,

$$
\begin{aligned}
\frac{\partial}{\partial \varepsilon} F(y + \varepsilon v) &= \int_{\Omega} \left(\frac{\partial}{\partial \varepsilon} (f(x, y + \varepsilon v, \nabla y + \varepsilon \nabla v)) \right) dx \\
&= \int_{\Omega} [f_y(x, y + \varepsilon v, \nabla y + \varepsilon \nabla v) v + \sum_{i=1}^{n} f_{z_i}(x, y + \varepsilon v, \nabla y + \varepsilon \nabla v) v_{x_i}] \, dx. \quad (5.40)
\end{aligned}
$$

Thus,

$$
\begin{aligned}
\delta F(y;v) &= \frac{\partial}{\partial \varepsilon} F(y + \varepsilon v)|_{\varepsilon=0} \\
&= \int_{\Omega} [f_y(x, y, \nabla y) v + \sum_{i=1}^{n} f_{z_i}(x, y, \nabla y) v_{x_i}] \, dx. \quad (5.41)
\end{aligned}
$$

On the other hand, since $f(\underline{x}, y, \mathbf{z})$ is convex, we have that

$$
\begin{aligned}
F(y + v) - F(y) &= \int_{\Omega} [f(x, y + v, \nabla y + \nabla v) - f(x, y, \nabla y)] \, dx \\
&\geq \langle \nabla f(x, y, \nabla y), (v, \nabla v) \rangle_{R^{n+1}} \\
&= \int_{\Omega} [f_y(x, y, \nabla y) v + \sum_{i=1}^{n} f_{z_i}(x, y, \nabla y) v_{x_i}] \, dx \\
&= \delta F(y;v). \quad (5.42)
\end{aligned}
$$

Since $y \in D$ and $v \in V_a$ are arbitrary, w e may infer that F is convex. Here we denote,

$$\underline{\nabla} f(x, y, \nabla y) = (f_y(x, y, \nabla y), f_{z_1}(x, y, \nabla y), \cdots, f_{z_n}(x, y, \nabla y)).$$

Theorem 5.7.1 *Let* $\Omega \subset \mathbb{R}^n$ *be a set of* $\hat{C}^1$ *class and let* $V = C^1(\overline{\Omega})$. *Let* $f \in C^2(\overline{\Omega} \times \mathbb{R} \times \mathbb{R})$ *where* $f(\underline{x}, y, \mathbf{z})$ *is convex. Define*

$$D = \{y \in V : y = y_1 \ em \ \partial\Omega\},$$

where $y_1 \in C^1(\overline{\Omega})$
   Define also $F : D \to \mathbb{R}$ *by*

$$F(y) = \int_{\Omega} f(x, y(x), \nabla y(x)) \, dx.$$

From such hypotheses, F is convex and if $y_0 \in D$ is such that

$$\sum_{i=1}^{n} \frac{d}{dx_i}[f_{z_i}(x, y_0(x), \nabla y_0(x))] = f_y(x, y_0(x), \nabla y_0(x)), \ \forall x \in \overline{\Omega},$$

then y_0 minimizes F on D, that is,

$$F(y_0) \leq F(y), \ \forall y \in D.$$

Proof 5.11   From the last remark, F is convex. Suppose now that $y_0 \in D$ is such that

$$\sum_{i=1}^{n} \frac{d}{dx_i}[f_{z_i}(x, y_0(x), \nabla y_0(x))] = f_y(x, y_0(x), \nabla y_0(x)), \ \forall x \in \overline{\Omega},$$

Let $v \in V_a = \{v \in V : v = 0 \ on \ \partial\Omega\}$. Thus,

$$
\begin{aligned}
\delta F(y_0; v) &= \int_{\Omega} (f_y(x, y_0(x), \nabla y_0(x))v(x) + \sum_{i=1}^{n} f_{z_i}(x, y_0(x), \nabla y_0(x))v_{x_i}(x) \, dx \\
&= \int_{\Omega} \left(\sum_{i=1}^{n} \frac{d}{dx_i}(f_{z_i}(x, y_0(x), \nabla y_0(x)))v(x) + \sum_{i=1}^{n} f_{z_i}(x, y_0(x), \nabla y_0(x))v_{x_i}(x) \right) dx \\
&= \int_{\Omega} \left(-\sum_{i=1}^{n} f_{z_i}(x, y_0(x), \nabla y_0(x))v_{x_i}(x) + \sum_{i=1}^{n} f_{z_i}(x, y_0(x), \nabla y_0(x))v_{x_i}(x) \right) dx \\
&\quad + \int_{\partial\Omega} \sum_{i=1}^{n} f_{z_i}(x, y_0(x), \nabla y_0(x)) \, n_i \, v(x) \, dS \\
&= 0, \ \forall v \in V_a, \quad\quad\quad\quad\quad\quad\quad\quad\quad\quad\quad\quad\quad\quad\quad (5.43)
\end{aligned}
$$

where $\mathbf{n} = (n_1, \cdots, n_n)$ denotes the outward normal field to $\partial\Omega = S$. Since F is convex, from this and Theorem 5.3.5, we have that y_0 minimizes F on D.

5.8   The second Gâteaux variation

Definition 5.8.1 *Let V be a Banach space. Let $F : D \subset V \to \mathbb{R}$ be a functional such that $\delta F(y; v)$ exists on $B_r(y_0)$ for $y_0 \in D$, $r > 0$ and for all $v \in V_a$.*

Let $y \in B_r(y_0)$ e $v, w \in V_a$. We define the second Gâteaux variation of F at the point y in the directions v e w, denoted by $\delta^2 F(y; v, w)$, as

$$\delta^2 F(y; v, w) = \lim_{\varepsilon \to 0} \frac{\delta F(y + \varepsilon w; v) - \delta F(y; v)}{\varepsilon},$$

if such a limit exists.

Remark 5.8.2 *Observe that from this last definition, if the limits in question exist, we have*

$$\delta F(y; v) = \frac{\partial}{\partial \varepsilon} F(y + \varepsilon v)|_{\varepsilon = 0},$$

and

$$\delta^2 F(y; v, v) = \frac{\partial^2}{\partial \varepsilon^2} F(y + \varepsilon v)|_{\varepsilon = 0}, \forall v \in V_a.$$

Thus, for example, for $V = C^1(\overline{\Omega})$ where $\Omega \subset \mathbb{R}^n$ is of $\hat{C}^1$ class and $F : V \to \mathbb{R}$ is given by

$$F(y) = \int_{\Omega} f(x, y, \nabla y) \, dx$$

and where

$$f \in C^2(\overline{\Omega} \times \mathbb{R} \times \mathbb{R}^n),$$

para $y, v \in V$, we have

$$\delta^2 F(y; v, v) = \frac{\partial^2}{\partial \varepsilon^2} F(y + \varepsilon v)|_{\varepsilon = 0},$$

where

$$
\begin{aligned}
\frac{\partial^2}{\partial \varepsilon^2} F(y + \varepsilon v) &= \frac{\partial^2}{\partial \varepsilon^2} \left(\int_{\Omega} f(x, y + \varepsilon v, \nabla y + \varepsilon \nabla v) \, dx \right) \\
&= \int_{\Omega} \frac{\partial^2}{\partial \varepsilon^2} [f(x, y + \varepsilon v, \nabla y + \varepsilon \nabla v)] \, dx \\
&= \int_{\Omega} \left[f_{yy}(x, y + \varepsilon v, \nabla y + \varepsilon \nabla v) v^2 + \sum_{i=1}^{n} 2 f_{yz_i}(x, y + \varepsilon v, \nabla y + \varepsilon \nabla v) v v_{x_i} \right. \\
&\qquad \left. + \sum_{i=1}^{n} \sum_{j=1}^{n} f_{z_i z_j}(x, y + \varepsilon v, \nabla y + \varepsilon \nabla v) v_{x_i} v_{x_j} \right] dx
\end{aligned}
$$
(5.44)

so that

$$
\begin{aligned}
\delta^2 F(y; v, v) &= \frac{\partial^2}{\partial \varepsilon^2} F(y + \varepsilon v)|_{\varepsilon = 0} \\
&= \int_{\Omega} \left[f_{yy}(x, y, \nabla y) v^2 + \sum_{i=1}^{n} 2 f_{yz_i}(x, y, \nabla y) v v_{x_i} \right. \\
&\qquad \left. + \sum_{i=1}^{n} \sum_{j=1}^{n} f_{z_i z_j}(x, y, \nabla y) v_{x_i} v_{x_j} \right] dx.
\end{aligned}
$$
(5.45)

5.9 First order necessary conditions for a local minimum

Definition 5.9.1 *Let V be a Banach space. Let $F : D \subset V \to \mathbb{R}$ be a functional. We say that $y_0 \in D$ is a point of local minimum for F on D, if there exists $\delta > 0$ such that*

$$F(y) \geq F(y_0), \ \forall y \in B_\delta(y_0) \cap D.$$

Theorem 5.9.2 *[First order necessary condition] Let V be a Banach space. Let $F : D \subset V \to \mathbb{R}$ be a functional. Suppose that $y_0 \in D$ is a point of local minimum for F on D. Let $v \in V_a$ and assume $\delta F(y_0; v)$ to exist.*
Under such hypotheses,

$$\delta F(y_0; v) = 0.$$

Proof 5.12 Define $\phi(\varepsilon) = F(y_0 + \varepsilon v)$, which from the existence of $\delta F(y_0; v)$ is well defined for all ε sufficiently small.

Also from the hypotheses, $\varepsilon = 0$ is a point of local minimum for the differentiable at 0 function ϕ.

Thus, from the standard condition for one variable calculus, we have

$$\phi'(0) = 0,$$

that is,

$$\phi'(0) = \delta F(y_0; v) = 0.$$

The proof is complete.

Theorem 5.9.3 (Second order sufficient condition) *Let V be a Banach space. Let $F : D \subset V \to \mathbb{R}$ be a functional. Suppose $y_0 \in D$ is such that $\delta F(y_0; v) = 0$ for all $v \in V_a$ and there exists $\delta > 0$ such that*

$$\delta^2 F(y; v, v) \geq 0, \ \forall y \in B_\delta(y_0) \text{ and } v \in V_a.$$

Under such hypotheses $y_0 \in D$ is a point of local minimum for F, that is

$$F(y) \geq F(y_0), \ \forall y \in B_r(y_0) \cap D.$$

Proof 5.13 Let $y \in B_\delta(y_0) \cap D$. Define $v = y - y_0 \in V_a$.
Define also $\phi : [0, 1] \to \mathbb{R}$ by

$$\phi(\varepsilon) = F(y_0 + \varepsilon v).$$

From the Taylor Theorem for one variable, there exists $t_0 \in (0, 1)$ such that

$$\phi(1) = \phi(0) + \frac{\phi'(0)}{1!}(1 - 0) + \frac{1}{2!}\phi''(t_0)(1 - 0)^2,$$

That is,

$$
\begin{aligned}
F(y) &= F(y_0 + v) \\
&= F(y_0) + \delta F(y_0; v) + \frac{1}{2}\delta^2 F(y_0 + t_0 v; v, v) \\
&= F(y_0) + \frac{1}{2}\delta^2 F(y_0 + t_0 v; v, v) \\
&\geq F(y_0), \ \forall y \in B_\delta(y_0) \cap D. \tag{5.46}
\end{aligned}
$$

The proof is complete.

5.10 Continuous functionals

Definition 5.10.1 *Let V be a Banach space. Let $F : D \subset V \to \mathbb{R}$ be a functional and let $y_0 \in D$.*

We say that F is continuous on $y_0 \in D$, if for each $\varepsilon > 0$ there exists $\delta > 0$ such that if $y \in D$ e $\|y - y_0\|_V < \delta$, then

$$
|F(y) - F(y_0)| < \varepsilon.
$$

Example 5.10.2 *Let $V = C^1([a,b])$ and $f \in C([a,b] \times \mathbb{R} \times \mathbb{R})$.*
Consider $F : V \to \mathbb{R}$ where

$$
F(y) = \int_a^b f(x, y(x), y'(x)) \, dx,
$$

and

$$
\|y\|_V = \max\{|y(x)| + |y'(x)| \ : \ x \in [a,b]\}.
$$

Let $y_0 \in V$. We shall prove that F is continuous at y_0.
Let $y \in V$ be such that

$$
\|y - y_0\|_V < 1.
$$

Thus,

$$
\|y\|_V - \|y_0\|_V \leq \|y - y_0\|_V < 1,
$$

that is,

$$
\|y\|_V < 1 + \|y_0\|_V \equiv \alpha.
$$

Observe that f is uniformly continuous on the compact set

$$
[a,b] \times [-\alpha, \alpha] \times [-\alpha, \alpha] \equiv A.
$$

Let $\varepsilon > 0$. Therefore, there exists $\delta_0 > 0$ such that if (x, y_1, z_1) e $(x, y_2, z_2) \in A$ e

$$
|y_1 - y_2| + |z_1 - z_2| < \delta_0,
$$

then

$$
|f(x, y_1, z_1) - f(x, y_2, z_2)| < \frac{\varepsilon}{b - a}. \tag{5.47}
$$

Let $\delta = \min\{\delta_0, 1\}$.
Hence, if

$$\|y - y_0\|_V < \delta,$$

we have

$$\max\{|y(x) - y_0(x)| + |y'(x) - y_0'(x)| \ : \ x \in [a,b]\} < \delta \leq 1,$$

so that from this and (5.47), we obtain

$$|f(x,y(x),y'(x)) - f(x,y_0(x),y_0'(x))| < \frac{\varepsilon}{b-a}, \ \forall x \in [a,b].$$

Thus,

$$
\begin{aligned}
|F(y) - F(y_0)| &= \left| \int_a^b [f(x,y(x),y'(x)) - f(x,y_0(x),y_0'(x))] \, dx \right| \\
&\leq \int_a^b |f(x,y(x),y'(x)) - f(x,y_0(x),y_0'(x))| \, dx \\
&< \frac{\varepsilon(b-a)}{(b-a)} \\
&= \varepsilon.
\end{aligned}
\tag{5.48}
$$

From this we have that F is continuous at y_0, $\forall y_0 \in V$.
The example is complete.

5.11 The Gâteaux variation, the formal proof of its formula

In the previous sections we had obtained the Gâteaux variations formulas with some informality for a relatively large class of functionals.

In this section, we intend to provide a formal proof for such formulas.

Our main result is summarized by the following theorem.

Theorem 5.11.1 *Let $\Omega \subset \mathbb{R}^n$ be sets of $\hat{C}^1$ class and let $V = C^1(\overline{\Omega})$.*
Let $f : \overline{\Omega} \times \mathbb{R} \times \mathbb{R}^n \to \mathbb{R}$ be a function of C^1 class.
Define $F : V \to \mathbb{R}$ by

$$F(y) = \int_\Omega f(x,y(x),\nabla y(x)) \, dx.$$

Let $y,v \in V$. Under such hypotheses

$$\delta F(y;v) = \int_\Omega \left(f_y(x,y(x),\nabla y(x))v(x) + \sum_{i=1}^n f_{z_i}(x,y(x),\nabla y(x))v_{x_i}(x) \right) dx.$$

Proof 5.14   Let $\{\varepsilon_n\} \subset \mathbb{R} \setminus \{0\}$ be a sequence such that

$$\varepsilon_n \to 0, \text{ as } n \to \infty.$$

Define

$$G_n(x) = \frac{f(x, y(x) + \varepsilon_n v(x), \nabla y(x) + \varepsilon_n \nabla v(x)) - f(x, y(x), \nabla y(x))}{\varepsilon_n},$$

$\forall n \in \mathbb{N}, \ x \in \overline{\Omega}$.
Define also

$$G(x) = f_y(x, y(x), \nabla y(x)) v(x) + \sum_{i=1}^{n} f_{z_i}(x, y(x), \nabla y(x)) v_{x_i}(x), \ \forall x \in \overline{\Omega}.$$

Observe that

$$G_n(x) \to G(x), \ \forall x \in \overline{\Omega}.$$

Now, we are going to prove that, for a not relabeled subsequence,

$$\int_{\Omega} G_n(x) \, dx \to \int_{\Omega} G(x) \, dx, \text{ as } n \to \infty$$

Define

$$c_n = \max_{x \in \overline{\Omega}} \{|G_n(x) - G(x)|\}.$$

From the continuity of the functions in question, for each $n \in \mathbb{N}$, there exists $x_n \in \overline{\Omega}$ such that

$$c_n = |G_n(x_n) - G(x_n)|.$$

Observe that $\{x_n\} \subset \overline{\Omega}$ and such a set is compact. Thus, there exist a subsequence $\{x_{n_j}\}$ de $\{x_n\}$ and $x_0 \in \overline{\Omega}$ such that

$$\lim_{j \to \infty} x_{n_j} = x_0.$$

On the other hand, from the mean value theorem, for each $j \in \mathbb{N}$ there exists $t_j \in (0, 1)$ such that

$$
\begin{aligned}
G_n(x_{n_j}) &= \frac{f(x_{n_j}, y(x_{n_j}) + \varepsilon_{n_j} v(x_{n_j}), \nabla y(x_{n_j}) + \varepsilon_{n_j} \nabla v(x_{n_j})) - f(x_{n_j}, y(x_{n_j}), \nabla y(x_{n_j}))}{\varepsilon_{n_j}} \\
&= f_y(x_{n_j}, y(x_{n_j}) + t_j \varepsilon_{n_j} v(x_{n_j}), \nabla y(x_{n_j}) + t_j \varepsilon_{n_j} \nabla v(x_{n_j})) v(x_{n_j}) \\
&\quad + \sum_{i=1}^{n} f_{z_i}(x_{n_j}, y(x_{n_j}) + t_j \varepsilon_n v(x_{n_j}), \nabla y(x_{n_j}) + t_j \varepsilon_{n_j} \nabla v(x_{n_j})) v_{x_i}(x_{n_j}) \\
&\to G(x_0), \text{ as } j \to \infty.
\end{aligned}
$$

(5.49)

Hence,

$$
\begin{aligned}
c_{n_j} &= |G_{n_j}(x_{n_j}) - G(x_{n_j})| \\
&\to |G(x_0) - G(x_0)| \\
&= 0.
\end{aligned}
$$

(5.50)

Let $\varepsilon > 0$. Thus, there exists $j_0 \in \mathbb{N}$ such that if $j > j_0$, then

$$0 \leq c_{n_j} < \frac{\varepsilon}{m(\Omega)},$$

where

$$m(\Omega) = \int_{\Omega} dx.$$

Therefore, if $j > j_0$, then

$$
\begin{aligned}
\left| \int_{\Omega} [G_{n_j}(x) - G(x)] \, dx \right| &\leq \int_{\Omega} |G_{n_j}(x) - G(x)| \, dx \\
&\leq \int_{\Omega} c_{n_j} \, dx \\
&= c_{n_j} m(\Omega) \\
&< \varepsilon.
\end{aligned}
\tag{5.51}
$$

Thus,

$$\lim_{j \to \infty} \int_{\Omega} G_{n_j}(x) \, dx = \int_{\Omega} G(x) \, dx.$$

Suppose now, to obtain contradiction, that we do not have

$$\lim_{\varepsilon \to 0} \int_{\Omega} G_{\varepsilon}(x) \, dx = \int_{\Omega} G(x) \, dx,$$

where

$$G_{\varepsilon}(x) = \frac{f(x, y(x) + \varepsilon v(x), \nabla y(x) + \varepsilon \nabla v(x)) - f(x, y(x), \nabla y(x))}{\varepsilon},$$

$\forall \varepsilon \in \mathbb{R}$ such that $\varepsilon \neq 0$.

Hence, there exists $\varepsilon_0 > 0$ such that for each $n \in \mathbb{N}$ there exists $\tilde{\varepsilon}_n \in \mathbb{R}$ such that

$$0 < |\tilde{\varepsilon}_n| < \frac{1}{n},$$

and

$$\left| \int_{\Omega} \tilde{G}_n(x) \, dx - \int_{\Omega} G(x) \, dx \right| \geq \varepsilon_0,
\tag{5.52}$$

where

$$\tilde{G}_n(x) = \frac{f(x, y(x) + \tilde{\varepsilon}_n v(x), \nabla y(x) + \tilde{\varepsilon}_n \nabla v(x)) - f(x, y(x), \nabla y(x))}{\tilde{\varepsilon}_n},$$

$\forall n \in \mathbb{N}, \, x \in \overline{\Omega}.$

However, as above indicated, we may obtain a subsequence $\{\tilde{\varepsilon}_{n_j}\}$ of $\{\tilde{\varepsilon}_n\}$ such that

$$\lim_{j \to \infty} \int_\Omega \tilde{G}_{n_j}(x)\, dx = \int_\Omega G(x)\, dx,$$

which contradicts (5.52).

Therefore, necessarily we have that

$$\lim_{\varepsilon \to 0} \int_\Omega G_\varepsilon(x)\, dx = \int_\Omega G(x)\, dx,$$

that is,

$$
\begin{aligned}
\delta F(y; v) &= \lim_{\varepsilon \to 0} \frac{F(y + \varepsilon v) - F(y)}{\varepsilon} \\
&= \lim_{\varepsilon \to 0} \int_\Omega G_\varepsilon(x)\, dx \\
&= \int_\Omega G(x)\, dx \\
&= \int_\Omega \left(f_y(x, y(x), \nabla y(x)) v(x) + \sum_{i=1}^n f_{z_i}(x, y(x), \nabla y(x)) v_{x_i}(x) \right) dx. \quad (5.53)
\end{aligned}
$$

The proof is complete.

Chapter 6

More Topics on the Calculus of Variations

6.1 Introductory remarks

We recall that a functional is a function whose the co-domain is the real set. We denote such functionals by $F : U \to \mathbb{R}$, where U is a Banach space. In our work format, we consider the special cases

1. $F(u) = \int_\Omega f(x, u, \nabla u)\, dx$, where $\Omega \subset \mathbb{R}^n$ is an open, bounded, connected set.

2. $F(u) = \int_\Omega f(x, u, \nabla u, D^2 u)\, dx$, here

$$Du = \nabla u = \left\{ \frac{\partial u_i}{\partial x_j} \right\}$$

and

$$D^2 u = \{D^2 u_i\} = \left\{ \frac{\partial^2 u_i}{\partial x_k \partial x_l} \right\},$$

for $i \in \{1, ..., N\}$ and $j, k, l \in \{1, ..., n\}$.

Also, $f : \overline{\Omega} \times \mathbb{R}^N \times \mathbb{R}^{N \times n} \to \mathbb{R}$ is denoted by $f(x, s, \xi)$ and we assume

1.

$$\frac{\partial f(x, s, \xi)}{\partial s}$$

and

2.

$$\frac{\partial f(x,s,\xi)}{\partial \xi}$$

are continuous $\forall (x,s,\xi) \in \overline{\Omega} \times \mathbb{R}^N \times \mathbb{R}^{N \times n}$.

Remark 6.1.1 *We also recall that the notation $\nabla u = Du$ may be used.*

Now we define our general problem, namely problem $\mathscr{P}$ where

$$\text{Problem } \mathscr{P} : \text{ minimize } F(u) \text{ on } U,$$

that is, to find $u_0 \in U$ such that

$$F(u_0) = \min_{u \in U} \{ F(u) \}.$$

At this point, we introduce some essential definitions.

Theorem 6.1.2 *Consider the hypotheses stated at Section 6.1 on $F : U \to \mathbb{R}$. Suppose F attains a local minimum at $u \in C^2(\overline{\Omega}; \mathbb{R}^N)$ and additionally assume that $f \in C^2(\overline{\Omega}, \mathbb{R}^N, \mathbb{R}^{N \times n})$. Then the necessary conditions for a local minimum for F are given by the Euler-Lagrange equations:*

$$\frac{\partial f(x,u,\nabla u)}{\partial s} - div \left(\frac{\partial f(x,u,\nabla u)}{\partial \xi} \right) = \theta, \text{ in } \Omega.$$

Proof 6.1   Observe that the standard first order necessary condition stands for $\delta F(u,\varphi) = 0, \forall \varphi \in \mathscr{V}$. From above this implies, after integration by parts

$$\int_{\Omega} \left(\frac{\partial f(x,u,\nabla u)}{\partial s} - div \left(\frac{\partial f(x,u,\nabla u)}{\partial \xi} \right) \right) \cdot \varphi \, dx = 0,$$

$$\forall \varphi \in C_c^{\infty}(\Omega, \mathbb{R}^N).$$

The result then follows from the fundamental lemma of calculus of variations.

6.2   The Gâteaux variation, a more general case

Theorem 6.2.1 *Consider the functional $F : U \to \mathbb{R}$, where*

$$U = \{ u \in W^{1,2}(\Omega, \mathbb{R}^N) \mid u = u_0 \text{ in } \partial\Omega \}.$$

Suppose

$$F(u) = \int_{\Omega} f(x,u,\nabla u) \, dx,$$

where $f : \Omega \times \mathbb{R}^N \times \mathbb{R}^{N \times n}$ is such that, for each $K > 0$ there exists $K_1 > 0$ such that

$$|f(x,s_1,\xi_1) - f(x,s_2,\xi_2)| < K_1(|s_1 - s_2| + |\xi_1 - \xi_2|)$$
$$\forall s_1, s_2 \in \mathbb{R}^N, \xi_1, \xi_2 \in \mathbb{R}^{N \times n}, \text{ such that } |s_1| < K, |s_2| < K,$$
$$|\xi_1| < K, |\xi_2| < K.$$

Also assume the hypotheses of Section 6.1 except for the continuity of derivatives of f. Under such assumptions, for each $u \in C^1(\overline{\Omega}; \mathbb{R}^N)$ and $\varphi \in C_c^\infty(\Omega; \mathbb{R}^N)$, we have

$$\delta F(u, \varphi) = \int_\Omega \left\{ \frac{\partial f(x,u,\nabla u)}{\partial s} \cdot \varphi + \frac{\partial f(x,u,\nabla u)}{\partial \xi} \cdot \nabla \varphi \right\} dx.$$

Proof 6.2   First we recall that

$$\delta F(u, \varphi) = \lim_{\varepsilon \to 0} \frac{F(u + \varepsilon\varphi) - F(u)}{\varepsilon}.$$

Observe that

$$\lim_{\varepsilon \to 0} \frac{f(x,u+\varepsilon\varphi, \nabla u + \varepsilon\nabla\varphi) - f(x,u,\nabla u)}{\varepsilon}$$
$$= \frac{\partial f(x,u,\nabla u)}{\partial s} \cdot \varphi + \frac{\partial f(x,u,\nabla u)}{\partial \xi} \cdot \nabla\varphi, \text{ a.e in } \Omega.$$

Define

$$G(x,u,\varphi,\varepsilon) = \frac{f(x,u+\varepsilon\varphi, \nabla u + \varepsilon\nabla\varphi) - f(x,u,\nabla u)}{\varepsilon},$$

and

$$\tilde{G}(x,u,\varphi) = \frac{\partial f(x,u,\nabla u)}{\partial s} \cdot \varphi + \frac{\partial f(x,u,\nabla u)}{\partial \xi} \cdot \nabla\varphi.$$

Thus we have

$$\lim_{\varepsilon \to 0} G(x,u,\varphi,\varepsilon) = \tilde{G}(x,u,\varphi), \text{ a.e in } \Omega.$$

Now will show that

$$\lim_{\varepsilon \to 0} \int_\Omega G(x,u,\varphi,\varepsilon) \, dx = \int_\Omega \tilde{G}(x,u,\varphi) \, dx.$$

It suffices to show that (we do not provide details here)

$$\lim_{n \to \infty} \int_\Omega G(x,u,\varphi,1/n) \, dx = \int_\Omega \tilde{G}(x,u,\varphi) \, dx.$$

Observe that, for an appropriate $K > 0$, we have

$$|G(x,u,\varphi,1/n)| \le K(|\varphi| + |\nabla\varphi|), \text{ a.e. in } \Omega. \tag{6.1}$$

By the Lebesgue dominated convergence theorem, we obtain

$$\lim_{n \to +\infty} \int_\Omega G(x,u,\varphi,1/(n)) \, dx = \int_\Omega \tilde{G}(x,u,\varphi) \, dx,$$

that is,

$$\delta F(u, \varphi) = \int_\Omega \left\{ \frac{\partial f(x,u,\nabla u)}{\partial s} \cdot \varphi + \frac{\partial f(x,u,\nabla u)}{\partial \xi} \cdot \nabla\varphi \right\} dx.$$

6.3   Fréchet differentiability

In this section, we introduce a very important definition namely, Fréchet differentiability.

Definition 6.3.1 *Let* U, Y *be Banach spaces and consider a transformation* $T : U \to Y$. *We say that* T *is Fréchet differentiable at* $u \in U$ *if there exists a bounded linear transformation* $T'(u) : U \to Y$ *such that*

$$\lim_{v \to \theta} \frac{\|T(u+v) - T(u) - T'(u)(v)\|_Y}{\|v\|_U} = 0, \, v \neq \theta.$$

In such a case $T'(u)$ *is called the Fréchet derivative of* T *at* $u \in U$.

6.4   The Legendre-Hadamard condition

Theorem 6.4.1 *If* $u \in C^1(\bar{\Omega}; \mathbb{R}^N)$ *is such that*

$$\delta^2 F(u, \varphi) \geq 0, \forall \varphi \in C_c^\infty(\Omega, \mathbb{R}^N),$$

then

$$f_{\xi_\alpha^i \xi_\beta^k}(x, u(x), \nabla u(x)) \rho^i \rho^k \eta_\alpha \eta_\beta \geq 0, \forall x \in \Omega, \rho \in \mathbb{R}^N, \eta \in \mathbb{R}^n.$$

Such a condition is known as the Legendre-Hadamard condition.

Proof 6.3   Suppose

$$\delta^2 F(u, \varphi) \geq 0, \forall \varphi \in C_c^\infty(\Omega; \mathbb{R}^N).$$

We denote $\delta^2 F(u, \varphi)$ by

$$\begin{aligned}
\delta^2 F(u, \varphi) = &\int_\Omega a(x) D\varphi(x) \cdot D\varphi(x) \, dx \\
&+ \int_\Omega b(x) \varphi(x) \cdot D\varphi(x) \, dx + \int_\Omega c(x) \varphi(x) \cdot \varphi(x) \, dx,
\end{aligned} \tag{6.2}$$

where

$$a(x) = f_{\xi\xi}(x, u(x), Du(x)),$$

$$b(x) = 2 f_{s\xi}(x, u(x), Du(x)),$$

and

$$c(x) = f_{ss}(x, u(x), Du(x)).$$

Now consider $v \in C_c^\infty(B_1(0), \mathbb{R}^N)$. Thus given $x_0 \in \Omega$ for λ sufficiently small we have that $\varphi(x) = \lambda v\left(\frac{x - x_0}{\lambda}\right)$ is an admissible direction. Now we introduce the new coordinates $y = (y^1, ..., y^n)$ by setting $y = \lambda^{-1}(x - x_0)$ and multiply (6.2) by λ^{-n} to obtain

$$\int_{B_1(0)} \{a(x_0 + \lambda y)Dv(y) \cdot Dv(y) + 2\lambda b(x_0 + \lambda y)v(y) \cdot Dv(y)$$
$$+ \lambda^2 c(x_0 + \lambda y)v(y) \cdot v(y)\} \, dy > 0,$$

where $a = \{a_{ij}^{\alpha\beta}\}, b = \{b_{jk}^{\beta}\}$ and $c = \{c_{jk}\}$. Since a, b and c are continuous, we have

$$a(x_0 + \lambda y)Dv(y) \cdot Dv(y) \to a(x_0)Dv(y) \cdot Dv(y),$$
$$\lambda b(x_0 + \lambda y)v(y) \cdot Dv(y) \to 0,$$

and

$$\lambda^2 c(x_0 + \lambda y)v(y) \cdot v(y) \to 0,$$

uniformly on $\bar{\Omega}$ as $\lambda \to 0$. Thus this limit give us

$$\int_{B_1(0)} \tilde{f}_{jk}^{\alpha\beta} D_\alpha v^j D_\beta v^k \, dx \geq 0, \forall v \in C_c^\infty(B_1(0); \mathbb{R}^N), \tag{6.3}$$

where

$$\tilde{f}_{jk}^{\alpha\beta} = a_{jk}^{\alpha\beta}(x_0) = f_{\xi_\alpha^j \xi_\beta^k}(x_0, u(x_0), \nabla u(x_0)).$$

Now define $v = (v^1, ..., v^N)$ where

$$v^j = \rho^j \cos((\eta \cdot y)t)\zeta(y)$$
$$\rho = (\rho^1, ..., \rho^N) \in \mathbb{R}^N$$

and

$$\eta = (\eta_1, ..., \eta_n) \in \mathbb{R}^n$$

and $\zeta \in C_c^\infty(B_1(0))$. From (6.3) we obtain

$$0 \leq \tilde{f}_{jk}^{\alpha\beta} \rho^j \rho^k \left\{ \int_{B_1(0)} (\eta_\alpha t(-\sin((\eta \cdot y)t)\zeta + \cos((\eta \cdot y)t)D_\alpha\zeta) \right.$$
$$\left. \cdot (\eta_\beta t(-\sin((\eta \cdot y)t)\zeta + \cos((\eta \cdot y)t)D_\beta\zeta) \, dy \right\} \tag{6.4}$$

By analogy for

$$v^j = \rho^j \sin((\eta \cdot y)t)\zeta(y)$$

we obtain

$$0 \leq \tilde{f}_{jk}^{\alpha\beta} \rho^j \rho^k \left\{ \int_{B_1(0)} (\eta_\alpha t(\cos((\eta \cdot y)t)\zeta + \sin((\eta \cdot y)t)D_\alpha\zeta) \right.$$
$$\left. \cdot (\eta_\beta t(\cos((\eta \cdot y)t)\zeta + \sin((\eta \cdot y)t)D_\beta\zeta) \, dy \right\} \tag{6.5}$$

Summing up these last two equations, dividing the result by t^2 and letting $t \to +\infty$ we obtain

$$0 \leq \tilde{f}_{jk}^{\alpha\beta} \rho^j \rho^k \eta_\alpha \eta_\beta \int_{B_1(0)} \zeta^2 \, dy,$$

for all $\zeta \in C_c^\infty(B_1(0))$, which implies

$$0 \le \tilde{f}_{jk}^{\alpha\beta} \rho^j \rho^k \eta_\alpha \eta_\beta.$$

The proof is complete.

6.5 The Weierstrass condition for $n = 1$

Here we present the Weierstrass condition for the special case $N \ge 1$ and $n = 1$. We start with a definition.

Definition 6.5.1 *We say that* $u \in \hat{C}([a,b];\mathbb{R}^N)$ *if* $u : [a,b] \to \mathbb{R}^N$ *is continuous in* $[a,b]$*, and Du is continuous except on a finite set of points in* $[a,b]$*.*

Theorem 6.5.2 (Weierstrass) *Let* $\Omega = (a,b)$ *and* $f : \bar{\Omega} \times \mathbb{R}^N \times \mathbb{R}^N \to \mathbb{R}$ *be such that* $f_s(x,s,\xi)$ *and* $f_\xi(x,s,\xi)$ *are continuous on* $\bar{\Omega} \times \mathbb{R}^N \times \mathbb{R}^N$*.*
Define $F : U \to \mathbb{R}$ *by*

$$F(u) = \int_a^b f(x,u(x),u'(x)) \, dx,$$

where

$$U = \{u \in \hat{C}^1([a,b];\mathbb{R}^N) \mid u(a) = \alpha, \ u(b) = \beta\}.$$

Suppose $u \in U$ *minimizes locally* F *on* U*, that is, suppose that there exists* $\varepsilon_0 > 0$ *such that*

$$F(u) \le F(v), \forall v \in U, \text{ such that } \|u - v\|_\infty < \varepsilon_0.$$

Under such hypotheses, we have

$$E(x,u(x),u'(x+),w) \ge 0, \forall x \in [a,b], \ w \in \mathbb{R}^N,$$

and

$$E(x,u(x),u'(x-),w) \ge 0, \forall x \in [a,b], \ w \in \mathbb{R}^N,$$

where

$$u'(x+) = \lim_{h \to 0^+} u'(x+h),$$
$$u'(x-) = \lim_{h \to 0^-} u'(x+h),$$

and,

$$E(x,s,\xi,w) = f(x,s,w) - f(x,s,\xi) - f_\xi(x,s,\xi)(w - \xi).$$

Remark 6.5.3 *The function E is known as the Weierstrass Excess Function.*

Proof 6.4   Fix $x_0 \in (a,b)$ and $w \in \mathbb{R}^N$. Choose $0 < \varepsilon < 1$ and $h > 0$ such that $u + v \in U$ and

$$\|v\|_\infty < \varepsilon_0$$

where $v(x)$ is given by

$$v(x) = \begin{cases} (x - x_0)w, & \text{if } 0 \leq x - x_0 \leq \varepsilon h, \\ \tilde{\varepsilon}(h - x + x_0)w, & \text{if } \varepsilon h \leq x - x_0 \leq h, \\ 0, & \text{otherwise,} \end{cases}$$

where

$$\tilde{\varepsilon} = \frac{\varepsilon}{1 - \varepsilon}.$$

From

$$F(u + v) - F(u) \geq 0$$

we obtain

$$\int_{x_0}^{x_0+h} f(x, u(x) + v(x), u'(x) + v'(x)) \, dx$$

$$- \int_{x_0}^{x_0+h} f(x, u(x), u'(x)) \, dx \geq 0. \quad (6.6)$$

Define

$$\tilde{x} = \frac{x - x_0}{h},$$

so that

$$d\tilde{x} = \frac{dx}{h}.$$

From (6.6) we obtain

$$h \int_0^1 f(x_0 + \tilde{x}h, u(x_0 + \tilde{x}h) + v(x_0 + \tilde{x}h), u'(x_0 + \tilde{x}h) + v'(x_0 + \tilde{x}h)) \, d\tilde{x}$$

$$- h \int_0^1 f(x_0 + \tilde{x}h, u(x_0 + \tilde{x}h), u'(x_0 + \tilde{x}h)) \, d\tilde{x} \geq 0. \quad (6.7)$$

where the derivatives are related to x.

Therefore

$$\int_0^\varepsilon f(x_0 + \tilde{x}h, u(x_0 + \tilde{x}h) + v(x_0 + \tilde{x}h), u'(x_0 + \tilde{x}h) + w) \, d\tilde{x}$$

$$- \int_0^\varepsilon f(x_0 + \tilde{x}h, u(x_0 + \tilde{x}h), u'(x_0 + \tilde{x}h)) \, d\tilde{x}$$

$$+ \int_\varepsilon^1 f(x_0 + \tilde{x}h, u(x_0 + \tilde{x}h) + v(x_0 + \tilde{x}h), u'(x_0 + \tilde{x}h) - \tilde{\varepsilon}w) \, d\tilde{x}$$

$$- \int_\varepsilon^1 f(x_0 + \tilde{x}h, u(x_0 + \tilde{x}h), u'(x_0 + \tilde{x}h)) \, d\tilde{x}$$

$$\geq 0. \quad (6.8)$$

Letting $h \to 0$ we obtain

$$\varepsilon(f(x_0, u(x_0), u'(x_0+) + w) - f(x_0, u(x_0), u'(x_0+)))$$
$$+ (1 - \varepsilon)(f(x_0, u(x_0), u'(x_0+) - \tilde{\varepsilon}w) - f(x_0, u(x_0), u'(x_0+))) \geq 0.$$

Hence, by the mean value theorem we get

$$\varepsilon(f(x_0, u(x_0), u'(x_0+) + w) - f(x_0, u(x_0), u'(x_0+)))$$
$$- (1 - \varepsilon)\tilde{\varepsilon}(f_\xi(x_0, u(x_0), u'(x_0+) + \rho(\tilde{\varepsilon})w)) \cdot w \geq 0. \qquad (6.9)$$

Dividing by ε and letting $\varepsilon \to 0$, so that $\tilde{\varepsilon} \to 0$ and $\rho(\tilde{\varepsilon}) \to 0$ we finally obtain

$$f(x_0, u(x_0), u'(x_0+) + w) - f(x_0, u(x_0), u'(x_0+))$$
$$- f_\xi(x_0, u(x_0), u'(x_0+)) \cdot w \geq 0.$$

Similarly we may get

$$f(x_0, u(x_0), u'(x_0-) + w) - f(x_0, u(x_0), u'(x_0-))$$
$$- f_\xi(x_0, u(x_0), u'(x_0-)) \cdot w \geq 0.$$

Since $x_0 \in [a, b]$ and $w \in \mathbb{R}^N$ are arbitrary, the proof is complete.

6.6 The Weierstrass condition, the general case

In this section, we present a proof for the Weierstrass necessary condition for $N \geq 1, n \geq 1$. Such a result may be found in similar form in [43].

Theorem 6.1
Assume $u \in C^1(\overline{\Omega}; \mathbb{R}^N)$ is a point of strong minimum for a Fréchet differentiable functional $F : U \to \mathbb{R}$ that is, in particular, there exists $\varepsilon > 0$ such that

$$F(u + \varphi) \geq F(u),$$

for all $\varphi \in C_c^\infty(\Omega; \mathbb{R}^n)$ such that

$$\|\varphi\|_\infty < \varepsilon.$$

Here

$$F(u) = \int_\Omega f(x, u, Du) \, dx,$$

where we recall to have denoted

$$Du = \nabla u = \left\{ \frac{\partial u_i}{\partial x_j} \right\}.$$

Under such hypotheses, for all $x \in \Omega$ and each rank-one matrix $\eta = \{\rho_i\beta^\alpha\} = \{\rho \otimes \beta\}$, we have that

$$E(x,u(x),Du(x),Du(x)+\rho \otimes \beta) \geq 0,$$

where

$$
\begin{aligned}
&E(x,u(x),Du(x),Du(x)+\rho \otimes \beta)\\
=\ & f(x,u(x),Du(x)+\rho \otimes \beta) - f(x,u(x),Du(x))\\
& -\rho^i\beta_\alpha f_{\xi^i_\alpha}(x,u(x),Du(x)).
\end{aligned}
\tag{6.10}
$$

Proof 6.5   Since u is a point of local minimum for F, we have that

$$\delta F(u;\varphi) = 0, \forall \varphi \in C_c^\infty(\Omega;\mathbb{R}^N),$$

that is

$$\int_\Omega (\varphi \cdot f_s(x,u(x),Du(x)) + D\varphi \cdot f_\xi(x,u(x),Du(x))\, dx = 0,$$

and hence,

$$
\begin{aligned}
&\int_\Omega (f(x,u(x),Du(x)+D\varphi(x)) - f(x,u(x),Du(x))\, dx\\
&-\int_\Omega (\varphi(x) \cdot f_s(x,u(x),Du(x)) - D\varphi(x) \cdot f_\xi(x,u(x),Du(x))\, dx\\
\geq\ & 0,
\end{aligned}
\tag{6.11}
$$

$\forall \varphi \in \mathcal{V}$, where

$$\mathcal{V} = \{\varphi \in C_c^\infty(\Omega;\mathbb{R}^N) \ : \ \|\varphi\|_\infty < \varepsilon\}.$$

Choose a unite vector $e \in \mathbb{R}^n$ and write

$$x = (x \cdot e)e + \bar{x},$$

where

$$\bar{x} \cdot e = 0.$$

Denote $D_e v = Dv \cdot e$, and let $\rho = (\rho_1,....,\rho_N) \in \mathbb{R}^N$.
Also, let x_0 be any point of Ω. Without loss of generality assume $x_0 = 0$.
Choose $\lambda_0 \in (0,1)$ such that $C_{\lambda_0} \subset \Omega$, where,

$$C_{\lambda_0} = \{x \in \mathbb{R}^n \ : \ |x \cdot e| \leq \lambda_0 \text{ and } \|\bar{x}\| \leq \lambda_0\}.$$

Let $\lambda \in (0,\lambda_0)$ and

$$\phi \in C_c((-1,1);\mathbb{R})$$

and choose a sequence

$$\phi_k \in C_c^\infty((-\lambda^2,\lambda);\mathbb{R})$$

which converges uniformly to the Lipschitz function ϕ_λ given by

$$\phi_\lambda = \begin{cases} t + \lambda^2, & \text{if } -\lambda^2 \le t \le 0, \\ \lambda(\lambda - t), & \text{if } 0 < t < \lambda \\ 0, & \text{otherwise} \end{cases} \qquad (6.12)$$

and such that ϕ_k' converges uniformly to ϕ_λ' on each compact subset of

$$A_\lambda = \{t : -\lambda^2 < t < \lambda, \, t \ne 0\}.$$

We emphasize the choice of $\{\phi_k\}$ may be such that for some $K > 0$ we have $\|\phi\|_\infty < K$, $\|\phi_k\|_\infty < K$ and $\|\phi_k'\|_\infty < K, \forall k \in \mathbb{N}$.

Observe that for any sufficiently small $\lambda > 0$ we have that φ_k defined by

$$\varphi_k(x) = \rho \phi_k(x \cdot e) \phi(|\bar{x}|^2 / \lambda^2) \in \mathcal{V}, \forall k \in \mathbb{N}$$

so that letting $k \to \infty$ we obtain that

$$\varphi(x) = \rho \phi_\lambda(x \cdot e) \phi(|\bar{x}|^2 / \lambda^2),$$

is such that (6.11) is satisfied.

Moreover,

$$D_e \varphi(x) = \rho \phi_\lambda'(x \cdot e) \phi(|\bar{x}|^2 / \lambda^2),$$

and

$$\overline{D}\varphi(x) = \rho \phi_\lambda(x \cdot e) \phi'(|\bar{x}|^2 / \lambda^2) 2\lambda^{-2} \bar{x},$$

where $\overline{D}$ denotes the gradient relating the variable $\bar{x}$.

Note that, for such a $\varphi(x)$ the integrand of (6.11) vanishes if $x \notin C_\lambda$, where

$$C_\lambda = \{x \in \mathbb{R}^n : |x \cdot e| \le \lambda \text{ and } \|\bar{x}\| \le \lambda\}.$$

Define C_λ^+ and C_λ^- by

$$C_\lambda^- = \{x \in C_\lambda : x \cdot e \le 0\},$$

and

$$C_\lambda^+ = \{x \in C_\lambda : x \cdot e > 0\}.$$

Hence, denoting

$$\begin{aligned} g_k(x) &= (f(x, u(x), Du(x) + D\varphi_k(x)) - f(x, u(x), Du(x)) \\ &\quad - (\varphi_k(x) \cdot f_s(x, u(x), Du(x) + D\varphi_k(x) \cdot f_\xi(x, u(x), Du(x)) \end{aligned} \qquad (6.13)$$

and

$$\begin{aligned} g(x) &= (f(x, u(x), Du(x) + D\varphi(x)) - f(x, u(x), Du(x)) \\ &\quad - (\varphi(x) \cdot f_s(x, u(x), Du(x) + D\varphi(x) \cdot f_\xi(x, u(x), Du(x)) \end{aligned} \qquad (6.14)$$

letting $k \to \infty$, using the Lebesgue dominated converge theorem we obtain

$$\int_{C_\lambda^-} g_k(x)\,dx + \int_{C_\lambda^+} g_k(x)\,dx$$
$$\to \int_{C_\lambda^-} g(x)\,dx + \int_{C_\lambda^+} g(x)\,dx \geq 0, \tag{6.15}$$

Now define

$$y = y^e e + \bar{y},$$

where

$$y^e = \frac{x \cdot e}{\lambda^2},$$

and

$$\bar{y} = \frac{\bar{x}}{\lambda}.$$

The sets C_λ^- and C_λ^+ correspond, concerning the new variables, to the sets B_λ^- and B_λ^+, where

$$B_\lambda^- = \{y : \|\bar{y}\| \leq 1, \text{ and } -\lambda^{-1} \leq y^e \leq 0\},$$
$$B_\lambda^+ = \{y : \|\bar{y}\| \leq 1, \text{ and } 0 < y^e \leq \lambda^{-1}\}.$$

Therefore, since $dx = \lambda^{n+1}dy$, multiplying (6.15) by λ^{-n-1}, we obtain

$$\int_{B_1^-} g(x(y))\,dy + \int_{B_\lambda^- \setminus B_1^-} g(x(y))\,dy + \int_{B_\lambda^+} g(x(y))\,dy \geq 0, \tag{6.16}$$

where

$$x = (x \cdot e)e + \bar{x} = \lambda^2 y^e + \lambda \bar{y} \equiv x(y).$$

Observe that

$$D_e\varphi(x) = \begin{cases} \rho\phi(\|\bar{y}\|^2) & \text{if } -1 \leq y^e \leq 0, \\ \rho\phi(\|\bar{y}\|^2)(-\lambda) & \text{if } 0 \leq y^e \leq \lambda^{-1}, \\ 0, & \text{otherwise.} \end{cases} \tag{6.17}$$

Observe also that

$$|g(x(y))| \leq o(\sqrt{|\varphi(x)|^2 + |D\varphi(x)|^2}),$$

so that from the from the expression of $\varphi(x)$ and $D\varphi(x)$ we obtain, for

$$y \in B_\lambda^+, \text{ or } y \in B_\lambda^- \setminus B_1^-,$$

that

$$|g(x(y))| \leq o(\lambda), \text{ as } \lambda \to 0.$$

Since the Lebesgue measures of B_λ^- and B_λ^+ are bounded by

$$2^{n-1}/\lambda$$

the second and third terms in (6.16) are of $o(1)$ where

$$\lim_{\lambda \to 0^+} o(1)/\lambda = 0,$$

so that letting $\lambda \to 0^+$, considering that

$$x(y) \to 0,$$

and on B_1^- (up to the limit set B)

$$
\begin{aligned}
g(x(y)) \quad \to \quad & f(0, u(0), Du(0) + \rho\phi(\|\bar{y}\|^2)e) \\
& -f(0, u(0), Du(0)) - \\
& \rho\phi(\|\bar{y}\|^2)ef_\xi(0, u(0), Du(0))
\end{aligned}
\tag{6.18}
$$

we get,

$$
\begin{aligned}
\int_B [f(0, u(0), Du(0) + \rho\phi(\|\bar{y}\|^2)e) - f(0, u(0), Du(0)) \\
-\rho\phi(\|\bar{y}\|^2)ef_\xi(0, u(0), Du(0))] \, d\bar{y}_2...d\bar{y}_n \\
\geq \quad 0,
\end{aligned}
\tag{6.19}
$$

where B is an appropriate limit set (we do not provide more details here) such that

$$B = \{ y \in \mathbb{R}^n \ : \ y^e = 0 \text{ and } \|\bar{y}\| \leq 1 \}.$$

Here we have used the fact that, on the set in question,

$$D\varphi(x) \to \rho\phi(\|\bar{y}\|^2)e, \text{ as } \lambda \to 0^+.$$

Finally, inequality (6.19) is valid for a sequence $\{\phi_n\}$ (in place of ϕ) such that

$$0 \leq \phi_n \leq 1 \text{ and } \phi_n(t) = 1, \text{ if } |t| < 1 - 1/n,$$

$\forall n \in \mathbb{N}.$

Letting $n \to \infty$, from (6.19) we obtain

$$
\begin{aligned}
f(0, u(0), Du(0) + \rho \otimes e) - f(0, u(0), Du(0)) \\
-\rho \cdot ef_\xi(0, u(0), Du(0)) \geq 0.
\end{aligned}
\tag{6.20}
$$

6.7 The Weierstrass-Erdmann conditions

We start with a definition.

Definition 6.7.1 *Define $I = [a, b]$. A function $u \in \hat{C}([a, b]; \mathbb{R}^N)$ is said to be a weak Lipschitz extremal of*

$$F(u) = \int_a^b f(x, u(x), u'(x)) \, dx,$$

if

$$\int_a^b (f_s(x,u(x),u'(x)) \cdot \varphi + f_\xi(x,u(x),u'(x)) \cdot \varphi'(x))\, dx = 0,$$

$\forall \varphi \in C_c^\infty([a,b];\mathbb{R}^N)$.

Proposition 6.7.2 *For any Lipschitz extremal of*

$$F(u) = \int_a^b f(x,u(x),u'(x))\, dx$$

there exists a constant $c \in \mathbb{R}^N$ such that

$$f_\xi(x,u(x),u'(x)) = c + \int_a^x f_s(t,u(t),u'(t))\, dt, \forall x \in [a,b]. \tag{6.21}$$

Proof 6.6 Fix $\varphi \in C_c^\infty([a,b];\mathbb{R}^N)$. Integration by parts of the extremal condition

$$\delta F(u,\varphi) = 0,$$

implies that

$$\int_a^b f_\xi(x,u(x),u'(x)) \cdot \varphi'(x)\, dx$$

$$- \int_a^b \int_a^x f_s(t,u(t),u'(t))\, dt \cdot \varphi'(x)\, dx = 0.$$

Since φ is arbitrary, from the du Bois-Reymond lemma, there exists $c \in \mathbb{R}^N$ such that

$$f_\xi(x,u(x),u'(x)) - \int_a^x f_s(t,u(t),u'(t))\, dt = c, \forall x \in [a,b].$$

The proof is complete.

Theorem 6.7.3 (Weierstrass-Erdmann Corner Conditions) *Let $I = [a,b]$. Suppose $u \in \hat{C}^1([a,b];\mathbb{R}^N)$ is such that*

$$F(u) \le F(v), \forall v \in \mathscr{C}_r,$$

for some $r > 0$.

where

$$\mathscr{C}_r = \{v \in \hat{C}^1([a,b];\mathbb{R}^N) \mid v(a) = u(a),\ v(b) = u(b),$$

$$\text{and } \|u - v\|_\infty < r\}.$$

Let $x_0 \in (a,b)$ be a corner point of u. Denoting $u_0 = u(x_0)$, $\xi_0^+ = u'(x_0 + 0)$ and $\xi_0^- = u'(x_0 - 0)$, then the following relations are valid:

1. $f_\xi(x_0, u_0, \xi_0^-) = f_\xi(x_0, u_0, \xi_0^+)$,

2.

$$f(x_0, u_0, \xi_0^-) - \xi_0^- f_\xi(x_0, u_0, \xi_0^-)$$
$$= f(x_0, u_0, \xi_0^+) - \xi_0^+ f_\xi(x_0, u_0, \xi_0^+).$$

Remark 6.7.4 *The conditions above are known as the Weierstrass-Erdmann corner conditions.*

Proof 6.7 Condition (1) is just a consequence of equation (6.21). For (2), define

$$\tau_\varepsilon(x) = x + \varepsilon \lambda(x),$$

where $\lambda \in C_c^\infty(I)$. Observe that $\tau_\varepsilon(a) = a$ and $\tau_\varepsilon(b) = b$, $\forall \varepsilon > 0$. Also $\tau_0(x) = x$. Choose $\varepsilon_0 > 0$ sufficiently small such that for each ε satisfying $|\varepsilon| < \varepsilon_0$, we have $\tau_\varepsilon'(x) > 0$ and

$$\tilde{u}_\varepsilon(x) = (u \circ \tau_\varepsilon^{-1})(x) \in \mathscr{C}_r.$$

Define

$$\phi(\varepsilon) = F(x, \tilde{u}_\varepsilon, \tilde{u}_\varepsilon'(x)).$$

Thus ϕ has a local minimum at 0, so that $\phi'(0) = 0$, that is

$$\frac{d(F(x, \tilde{u}_\varepsilon, \tilde{u}_\varepsilon'(x)))}{d\varepsilon}\Big|_{\varepsilon=0} = 0.$$

Observe that

$$\frac{d\tilde{u}_\varepsilon}{dx} = u'(\tau_\varepsilon^{-1}(x))\frac{d\tau_\varepsilon^{-1}(x)}{dx},$$

and

$$\frac{d\tau_\varepsilon^{-1}(x)}{dx} = \frac{1}{1 + \varepsilon\lambda'(\tau_\varepsilon^{-1}(x))}.$$

Thus,

$$F(\tilde{u}_\varepsilon) = \int_a^b f\left(x, u(\tau_\varepsilon^{-1}(x)), u'(\tau_\varepsilon^{-1}(x))\left(\frac{1}{1 + \varepsilon\lambda'(\tau_\varepsilon^{-1}(x))}\right)\right) dx.$$

Defining

$$\bar{x} = \tau_\varepsilon^{-1}(x),$$

we obtain

$$d\bar{x} = \frac{1}{1 + \varepsilon\lambda'(\bar{x})} dx,$$

that is

$$dx = (1 + \varepsilon \lambda'(\bar{x})) \, d\bar{x}.$$

Dropping the bar for the new variable, we may write

$$F(\tilde{u}_\varepsilon) = \int_a^b f\left(x + \varepsilon \lambda(x), u(x), \frac{u'(x)}{1 + \varepsilon \lambda'(x)}\right) (1 + \varepsilon \lambda'(x)) \, dx.$$

From

$$\frac{dF(\tilde{u}_\varepsilon)}{d\varepsilon}\bigg|_{\varepsilon=0},$$

we obtain

$$\int_a^b (\lambda f_x(x, u(x), u'(x)) + \lambda'(x)(f(x, u(x), u'(x))$$
$$- u'(x) f_\xi(x, u(x), u'(x)))) \, dx = 0. \quad (6.22)$$

Since λ is arbitrary, from Proposition 6.7.2, we obtain

$$f(x, u(x), u'(x)) - u'(x) f_\xi(x, u(x), u'(x)) - \int_a^x f_x(t, u(t), u'(t)) \, dt = c_1$$

for some $c_1 \in \mathbb{R}$.

Being $\int_a^x f_x(t, u(t), u'(t)) \, dt + c_1$ a continuous function (in fact absolutely continuous), the proof is complete.

6.8 Natural boundary conditions

Consider the functional $f : U \to \mathbb{R}$, where

$$F(u) \int_\Omega f(x, u(x), \nabla u(x)) \, dx,$$

$$f(x, s, \xi) \in C^1(\bar{\Omega}, \mathbb{R}^N, \mathbb{R}^{N \times n}),$$

and $\Omega \subset \mathbb{R}^n$ is an open bounded connected set.

Proposition 6.8.1 *Assume*

$$U = \{u \in W^{1,2}(\Omega; \mathbb{R}^N); u = u_0 \text{ on } \Gamma_0\},$$

where $\Gamma_0 \subset \partial\Omega$ is closed and $\partial\Omega = \Gamma = \Gamma_0 \cup \Gamma_1$ being Γ_1 open in Γ and $\Gamma_0 \cap \Gamma_1 = \emptyset$. Thus if $\partial\Omega \in C^1$, $f \in C^2(\bar{\Omega}, \mathbb{R}^N, \mathbb{R}^{N \times n})$ and $u \in C^2(\bar{\Omega}; \mathbb{R}^N)$, and also

$$\delta F(u, \varphi) = 0, \forall \varphi \in C^1(\bar{\Omega}; \mathbb{R}^N), \text{ such that } \varphi = 0 \text{ on } \Gamma_0,$$

then u is a extremal of F which satisfies the following natural boundary conditions,

$$n_\alpha f_{x_{i\alpha}}(x,u(x)\nabla u(x)) = 0, \ a.e. \ on \ \Gamma_1, \forall i \in \{1,...,N\}.$$

Proof 6.8   Observe that $\delta F(u,\varphi) = 0, \forall \varphi \in C_c^\infty(\Omega;\mathbb{R}^N)$, thus u is a extremal of F and through integration by parts and the fundamental lemma of calculus of variations, we obtain

$$L_f(u) = 0, \ in \ \Omega,$$

where

$$L_f(u) = f_s(x,u(x),\nabla u(x)) - div(f_\xi(x,u(x),\nabla u(x))).$$

Defining

$$\mathcal{V} = \{\varphi \in C^1(\Omega;\mathbb{R}^N) \mid \varphi = 0 \ on \ \Gamma_0\},$$

for an arbitrary $\varphi \in \mathcal{V}$, we obtain

$$
\begin{aligned}
\delta F(u,\varphi) &= \int_\Omega L_f(u) \cdot \varphi \, dx \\
&\quad + \int_{\Gamma_1} n_\alpha f_{x_{i\alpha}}(x,u(x),\nabla u(x))\varphi^i(x) \, d\Gamma \\
&= \int_{\Gamma_1} n_\alpha f_{x_{i\alpha}}(x,u(x),\nabla u(x))\varphi^i(x) \, d\Gamma \\
&= 0, \forall \varphi \in \mathcal{V}. \quad (6.23)
\end{aligned}
$$

Suppose, to obtain contradiction, that

$$n_\alpha f_{x_{i\alpha}}(x_0,u(x_0),\nabla u(x_0)) = \beta > 0,$$

for some $x_0 \in \Gamma_1$ and some $i \in \{1,...,N\}$. Defining

$$G(x) = n_\alpha f_{x_{i\alpha}}(x,u(x),\nabla u(x)),$$

by the continuity of G, there exists $r > 0$ such that

$$G(x) > \beta/2, \ in \ B_r(x_0),$$

and in particular

$$G(x) > \beta/2, \ in \ B_r(x_0) \cap \Gamma_1.$$

Choose $0 < r_1 < r$ such that $B_{r_1}(x_0) \cap \Gamma_0 = \emptyset$. This is possible since Γ_0 is closed and $x_0 \in \Gamma_1$.

Choose $\varphi^i \in C_c^\infty(B_{r_1}(x_0))$ such that $\varphi^i \geq 0$ in $B_{r_1}(x_0)$ and $\varphi^i > 0$ in $B_{r_1/2}(x_0)$. Therefore

$$\int_{\Gamma_1} G(x)\varphi^i(x) \, dx > \frac{\beta}{2} \int_{\Gamma_1} \varphi^i \, dx > 0,$$

and this contradicts (6.23). Thus

$$G(x) \leq 0, \forall x \in \Gamma_1,$$

and by analogy

$$G(x) \geq 0, \forall x \in \Gamma_1,$$

so that

$$G(x) = 0, \forall x \in \Gamma_1.$$

The proof is complete.

Chapter 7

Convex Analysis and Duality Theory

7.1 Convex sets and functions

For this section the most relevant reference is Ekeland and Temam [34].

Definition 7.1.1 (Convex Functional) *Let U be a vector space and let $S \subset U$ be a convex set. A functional $F : S \to \bar{\mathbb{R}} = \mathbb{R} \cup \{+\infty, -\infty\}$ is said to be convex, if*

$$F(\lambda u + (1 - \lambda)v) \leq \lambda F(u) + (1 - \lambda)F(v), \forall u, v \in S, \lambda \in [0, 1]. \qquad (7.1)$$

7.2 Weak lower semi-continuity

We start with the definition of Epigraph.

Definition 7.2.1 (Epigraph) *Let U be a Banach space and let $F : U \to \bar{\mathbb{R}}$ be a functional.*
    We define the Epigraph of F, denoted by $Epi(F)$, by

$$Epi(F) = \{(u, a) \in U \times \mathbb{R} \mid a \geq F(u)\}.$$

Definition 7.2.2 *Let U be a Banach space. Consider the weak topology $\sigma(U, U^*)$ for U and let $F : U \to \mathbb{R} \cup \{+\infty\}$ be a functional. Let $u \in U$. We say that F is wekly lower semi-continuous at $u \in U$ if for each $\lambda < F(u)$, there exists a weak neighborhood $V_\lambda(u) \in \sigma(U, U^*)$ such that*

$$F(v) > \lambda, \ \forall v \in V_\lambda(u).$$

If F is weakly lower semi-continuous (w.l.s.c.) on U, we write simply F is w.l.s.c.

Theorem 7.2.3 *Let U be a Banach space and let $F : U \to \mathbb{R} \cup \{+\infty\}$ be a functional. Under such hypotheses, the following properties are equivalent.*

1. *F is w.l.s.c.*

2. *$Epi(F)$ is closed for $U \times \mathbb{R}$ with the product topology between $\sigma(U, U^*)$ and the usual topology for $\mathbb{R}$.*

3. *$H_\gamma^F = \{u \in U \mid F(u) \leq \gamma\}$ is closed for $\sigma(U, U^*)$, $\forall \gamma \in \mathbb{R}$.*

4. *The set $G_\gamma^F = \{u \in U \mid F(u) > \gamma\}$ is open for $\sigma(U, U^*)$, $\forall \gamma \in \mathbb{R}$.*

5.
$$\liminf_{v \to u} F(v) \geq F(u), \forall u \in U,$$

where
$$\liminf_{v \to u} F(v) = \sup_{V(u) \in \sigma(U, U^*)} \inf_{v \in V(u)} F(v).$$

Proof 7.1 Assume F is w.l.s.c.. We are going to show that $Epi(F)^c$ is open for $\sigma(U, U^*) \times \mathbb{R}$. Choose $(u, r) \in Epi(F)^c$. Thus $(u, r) \notin Epi(F)$, so that $r < F(u)$. Select λ such that $r < \lambda < F(u)$. Since F is w.l.s.c. at u, there exists a a weak neighborhood $V_\lambda(u)$ such that

$$F(v) > \lambda, \forall v \in V_\lambda(u).$$

Thus,
$$V_\lambda(u) \times (-\infty, \lambda) \subset Epi(F)^c$$

so that (u, r) is an interior point of $Epi(F)^c$ and hence, since such a point is arbitrary in $Epi(F)^c$, we may infer that $Epi(F)^c$ is open so that $Epi(F)$ is closed for the topology in question. Assume now (2). Observe that

$$H_\gamma^F \times \{\gamma\} = Epi(F) \cap (U \times \{\gamma\}).$$

From the hypotheses $Epi(F)$ is closed, that is, $H_\gamma^F \times \{\gamma\}$ is closed and thus H_γ^F is closed.

Assume (3). To obtain (4), it suffices to consider the complement of H_γ^F. Suppose (4) is valid. Let $u \in U$ and let $\gamma \in \mathbb{R}$ be such that

$$\gamma < F(u).$$

Since G_γ^F is open for $\sigma(U, U^*)$ there exists a weak neighborhood $V(u)$ such that

$$V(u) \subset G_\gamma^F,$$

so that
$$F(v) > \gamma, \forall v \in V(u),$$

and hence

$$\inf_{v \in V(u)} F(v) \geq \gamma.$$

In particular, we have

$$\liminf_{v \to u} F(v) \geq \gamma.$$

Letting $\gamma \to F(u)$, we obtain

$$\liminf_{v \to u} F(v) \geq F(u).$$

Finally assume

$$\liminf_{v \to u} F(v) \geq F(u).$$

Let $\lambda < F(u)$ and let $0 < \varepsilon < F(u) - \lambda$.

Observe that

$$\liminf_{v \to u} F(v) = \sup_{V(u) \in \sigma(U, U^*)} \inf_{v \in V(u)} F(v).$$

Thus, there exists a weak neighborhood $V(u)$ such that $F(v) \geq F(u) - \varepsilon > \lambda, \forall v \in V(u)$.

The proof is complete.

Remark 7.2.4 *A similar result is valid for the strong topology (in norm) of a Banach space U so that a $F : U \to \mathbb{R} \cup \{+\infty\}$ is strongly lower semi-continuous (s.c.i.) at $u \in U$, if*

$$\liminf_{v \to u} F(v) \geq F(u). \tag{7.2}$$

Corollary 7.2.5 *All convex s.c.i. functional $F : U \to \overline{\mathbb{R}}$ is also w.l.s.c.*

Proof 7.2 The result follows from the fact for F s.c.i, its epigraph being convex and strongly closed, it is also weakly closed.

Definition 7.2.6 (Affine-continuous functionals) *Let U be a Banach space. A functional $F : U \to \mathbb{R}$ is said to be affine-continuous, if there exist $u^* \in U^*$ and $\alpha \in \mathbb{R}$ such that*

$$F(u) = \langle u, u^* \rangle_U + \alpha, \forall u \in U. \tag{7.3}$$

Definition 7.2.7 ($\Gamma(U)$) *Let U be a Banach space. We say that $F : U \to \overline{\mathbb{R}}$ is a functional in $\Gamma(U)$ and write $F \in \Gamma(U)$ if F may be represented point-wise as the supremum of a family of affine-continuous functionals. If $F \in \Gamma(U)$, and $F(u) \in \mathbb{R}$ for some $u \in U$ we write $F \in \Gamma_0(U)$.*

Definition 7.2.8 (Convex envelop) *Let U be a Banach space. Let $F : U \to \overline{\mathbb{R}}$, be a functional. we define its convex envelop, denoted by $CF : U \to \overline{\mathbb{R}}$, as*

$$CF(u) = \sup_{(u^*, \alpha) \in A^*} \{\langle u, u^* \rangle + \alpha\}, \tag{7.4}$$

where

$$A^* = \{(u^*, \alpha) \in U^* \times \mathbb{R} \mid \langle v, u^* \rangle_U + \alpha \le F(v), \forall v \in U\} \qquad (7.5)$$

7.3 Polar functionals and related topics on convex analysis

Definition 7.3.1 (Polar functional) *Let U be a Banach space and let $F : U \to \bar{\mathbb{R}}$, be a functional. We define the polar functional functional related to F, denoted by $F^* : U^* \to \bar{\mathbb{R}}$, by*

$$F^*(u^*) = \sup_{u \in U}\{\langle u, u^* \rangle_U - F(u)\}, \forall u^* \in U^*. \qquad (7.6)$$

Definition 7.3.2 (Bipolar functional) *Let U be a Banach space and let $F : U \to \bar{\mathbb{R}}$ be a functional. We define the bi-polar functional related to F, denoted by $F^{**} : U \to \bar{\mathbb{R}}$, as*

$$F^{**}(u) = \sup_{u^* \in U^*}\{\langle u, u^* \rangle_U - F^*(u^*)\}, \forall u \in U. \qquad (7.7)$$

Proposition 7.3.3 *Let U be a Banach space and let $F : U \to \bar{\mathbb{R}}$ be a functional. Under such hypotheses $F^{**}(u) = CF(u)$, $\forall u \in U$ and in particular, if $F \in \Gamma(U)$, then $F^{**}(u) = F(u), \forall u \in U$.*

Proof 7.3 By the definition, the convex envelop of F is the supremum of affine-continuous functionals bounded by F at the point in question. In fact we need only to consider those which are maximal, that is, only those in the form

$$u \mapsto \langle u, u^* \rangle_U - F^*(u^*). \qquad (7.8)$$

Thus,

$$CF(u) = \sup_{u^* \in U^*}\{\langle u, u^* \rangle_U - F^*(u^*)\} = F^{**}(u). \qquad (7.9)$$

Corollary 7.3.4 *Let U be a Banach space and let $F : U \to \bar{\mathbb{R}}$ be a functional. Under such hypotheses, $F^* = F^{***}$.*

Proof 7.4 Since $F^{**} \le F$, we obtain

$$F^* \le F^{***}. \qquad (7.10)$$

On the other hand,

$$F^{**}(u) \ge \langle u, u^* \rangle_U - F^*(u^*), \qquad (7.11)$$

so that

$$F^{***}(u^*) = \sup_{u \in U}\{\langle u, u^* \rangle_U - F^{**}(u)\} \leq F^*(u^*). \tag{7.12}$$

From (7.10) and (7.12), we have $F^*(u^*) = F^{***}(u^*), \; \forall u^* \in U^*$.

At this point, we recall the definition of Gâteaux differentiability.

Definition 7.3.5 (Gâteaux differentiability) *Let U be a Banach space. A functional $F : U \to \bar{\mathbb{R}}$ is said to be Gâteaux differentiable at $u \in U$, if there exists $u^* \in U^*$ such that*

$$\lim_{\lambda \to 0} \frac{F(u + \lambda h) - F(u)}{\lambda} = \langle h, u^* \rangle_U, \;\; \forall h \in U. \tag{7.13}$$

The vector u^* is said to be the Gâteaux derivative of $F : U \to \mathbb{R}$ at u and may denoted by

$$u^* = \frac{\partial F(u)}{\partial u} \; or \; u^* = \delta F(u) \tag{7.14}$$

Definition 7.3.6 (Sub-gradients) *Let U be a Banach space and let $F : U \to \bar{\mathbb{R}}$ be a functional. We define the set of sub-gradients of F at u, denoted by $\partial F(u)$, by*

$$\partial F(u) \;\; = \;\; \{u^* \in U^*, \; such \; that$$
$$\langle v - u, u^* \rangle_U + F(u) \leq F(v), \forall v \in U\}. \tag{7.15}$$

Lemma 7.3.7 (Continuity of convex functions) *Let U be a Banach space and let $F : U \to \mathbb{R}$ be a convex functional. Let $u \in U$ and suppose there exists $a > 0$ and a neighborhood V of u such that*

$$F(v) < a < +\infty, \; \forall v \in V.$$

From the hypotheses, F is continuous at u.

Proof 7.5 Redefining the problem with $G(v) = F(v + u) - F(u)$ we need only consider the case in which $u = \mathbf{0}$ and $F(u) = 0$. Let $\mathscr{V}$ be a neighborhood of $\mathbf{0}$ such that $F(v) \leq a < +\infty, \forall v \in \mathscr{V}$. Define $\mathscr{W} = \mathscr{V} \cap (-\mathscr{V})$. Choose $\varepsilon \in (0,1)$. Let $v \in \varepsilon \mathscr{W}$, thus

$$\frac{v}{\varepsilon} \in \mathscr{V} \tag{7.16}$$

and since F is convex, we have that

$$F(v) = F\left((1-\varepsilon)\mathbf{0} + \varepsilon\frac{v}{\varepsilon}\right) \leq (1-\varepsilon)F(\mathbf{0}) + \varepsilon F(v/\varepsilon) \leq \varepsilon a. \tag{7.17}$$

Also

$$\frac{-v}{\varepsilon} \in \mathcal{V}. \tag{7.18}$$

Hence,

$$F(\theta) = F\left(\frac{v}{1+\varepsilon} + \varepsilon\frac{(-v/\varepsilon)}{1+\varepsilon}\right) \le \frac{F(v)}{1+\varepsilon} + \frac{\varepsilon}{1+\varepsilon}F(-v/\varepsilon),$$

so that

$$F(v) \ge (1+\varepsilon)F(\theta) - \varepsilon F(-v/\varepsilon) \ge -\varepsilon a. \tag{7.19}$$

Therefore

$$|F(v)| \le \varepsilon a, \forall v \in \varepsilon \mathcal{W}, \tag{7.20}$$

that is, F is continuous at $u = \mathbf{0}$.

Proposition 7.3.8 *Let U be a Banach space and let $F : U \to \bar{\mathbb{R}}$ be a convex functional, which is finite and continuous at $u \in U$. Under such hypotheses, $\partial F(u) \neq \emptyset$.*

Proof 7.6   Since F is convex, $Epi(F)$ is convex. Since F is continuous at u, we have that $Epi(F)^0$ is non-empty. Observe that $(u, F(u))$ is on the boundary of $Epi(F)$. Therefore, denoting $A = Epi(F)$, from the Hahn-Banach theorem there exists a closed hyperplane H which separates $(u, F(u))$ and A^0, where H

$$H = \{(v,a) \in U \times \mathbb{R} \mid \langle v,u^*\rangle_U + \alpha a = \beta\}, \tag{7.21}$$

for some fixed $\alpha, \beta \in \mathbb{R}$ and $u^* \in U^*$, so that

$$\langle v,u^*\rangle_U + \alpha a \ge \beta, \forall (v,a) \in Epi(F), \tag{7.22}$$

and

$$\langle u,u^*\rangle_U + \alpha F(u) = \beta, \tag{7.23}$$

where $(\alpha, \beta, u^*) \neq (0,0,\mathbf{0})$. Suppose, to obtain contradiction, that $\alpha = 0$.
    Thus,

$$\langle v-u,u^*\rangle_U \ge 0, \forall v \in U, \tag{7.24}$$

and therefore we obtain, $u^* = \mathbf{0}$ and $\beta = 0$, a contradiction. Hence, we may assume $\alpha > 0$ (considering (7.22)) and thus $\forall v \in U$ we have

$$\frac{\beta}{\alpha} - \langle v,u^*/\alpha\rangle_U \le F(v), \tag{7.25}$$

and

$$\frac{\beta}{\alpha} - \langle u,u^*/\alpha\rangle_U = F(u), \tag{7.26}$$

that is,

$$\langle v - u, -u^*/\alpha \rangle_U + F(u) \leq F(v), \forall v \in U, \tag{7.27}$$

so that

$$-u^*/\alpha \in \partial F(u). \tag{7.28}$$

The proof is complete.

Definition 7.3.9 (Carathéodory function) *Let* $S \subset \mathbb{R}^n$ *be an open set. We say that* $g : S \times \mathbb{R}^l \to \mathbb{R}$ *is a Carathéodory function if*

$$\forall \xi \in \mathbb{R}^l, \, x \mapsto g(x, \xi) \text{ is a measurable function,}$$

e

$$\text{a.e. in } S, \, \xi \mapsto g(x, \xi) \text{ is a continuous function.}$$

The proof of next results may be found in Ekeland and Temam [34].

Proposition 7.3.10 *Let E and F be two Banach spaces, let S be a Boreal subset of* $\mathbb{R}^n$, *and* $g : S \times E \to F$ *be a Carathéodory function. For each measurable function* $u : S \to E$, *let* $G_1(u)$ *be the measurable function* $x \mapsto g(x, u(x)) \in F$.

Under such hypotheses, if G_1 *maps* $L^p(S, E)$ *on* $L^r(S, F)$ *for* $1 \leq p, r < \infty$, *then* G_1 *is strongly continuous.*

For the functional $G : U \to \mathbb{R}$, defined by $G(u) = \int_S g(x, u(x)) dS$, where $U = U^* = [L^2(S)]^l$, we have also the following result.

Proposition 7.3.11 *Considering the statement in the last proposition we may express* $G^* : U^* \to \bar{\mathbb{R}}$ *by*

$$G^*(u^*) = \int_S g^*(x, u^*(x)) dx, \tag{7.29}$$

onde $g^*(x, y) = \sup_{\eta \in \mathbb{R}^l} (y \cdot \eta - g(x, \eta))$, *for almost all* $x \in S$.

7.4 The Legendre transform and the Legendre functional

For non-convex functionals, in some cases, the global extremal through which the polar functional obtained corresponds to a local extremal point at which the analytical expression is possible.

This fact motivates the definition of the Legendre Transform, which is obtained through a local extremal point.

Definition 7.4.1 (Legendre transform and associated functional) *Consider the function of C^2 class, $g : \mathbb{R}^n \to \mathbb{R}$. Its Legendre transform, denoted by $g_L^* : R_L^n \to \mathbb{R}$, is expressed by*

$$g_L^*(y^*) = \sum_{i=1}^{n} x_{0i} \cdot y_i^* - g(x_0), \tag{7.30}$$

where x_0 is a solution of the system

$$y_i^* = \frac{\partial g(x_0)}{\partial x_i}, \tag{7.31}$$

and $R_L^n = \{y^* \in \mathbb{R}^n \text{ such an equation (7.31) has a unique solution}\}$.
  Moreover, considering the functional $G : Y \to \mathbb{R}$ defined by $G(v) = \int_S g(v)dS$, we also define the associated Legendre functional, denoted by $G_L^* : Y_L^* \to \mathbb{R}$ as

$$G_L^*(v^*) = \int_S g_L^*(v^*) \, dx, \tag{7.32}$$

where $Y_L^* = \{v^* \in Y^* \mid v^*(x) \in R_L^n, \text{ a.e. in } S\}$.

About the Legendre transform we still have the following results:

Proposition 7.4.2 *Considering the last definitions, suppose that for each $y^* \in R_L^n$ at least in a neighborhood (of y^*) it is possible to define a differentiable function by the expression*

$$x_0(y^*) = [\frac{\partial g}{\partial x}]^{-1}(y^*). \tag{7.33}$$

Then, $\forall\ i \in \{1,...,n\}$ we may write:

$$y_i^* = \frac{\partial g(x_0)}{\partial x_i} \Leftrightarrow x_{0i} = \frac{\partial g_L^*(y^*)}{\partial y_i^*} \tag{7.34}$$

Proof 7.7   Suppose firstly that:

$$y_i^* = \frac{\partial g(x_0)}{\partial x_i}, \forall i \in \{1,...,n\}, \tag{7.35}$$

thus:

$$g_L^*(y^*) = y_i^* x_{0i} - g(x_0) \tag{7.36}$$

and taking derivatives for this expression we have:

$$\frac{\partial g_L^*(y^*)}{\partial y_i^*} = y_j^* \frac{\partial x_{0j}}{\partial y_i^*} + x_{0i} - \frac{\partial g(x_0)}{\partial x_j} \frac{\partial x_{0j}}{\partial y_i^*}, \tag{7.37}$$

or

$$\frac{\partial g_L^*(y^*)}{\partial y_i^*} = (y_j^* - \frac{\partial g(x_0)}{\partial x_j})\frac{\partial x_{0j}}{\partial y_i^*} + x_{0i} \tag{7.38}$$

which from (7.35) implies that:

$$\frac{\partial g_L^*(y^*)}{\partial y_i^*} = x_{0i}, \ \forall i \in \{1,...,n\}. \tag{7.39}$$

This completes the first half of the proof. Conversely, suppose now that:

$$x_{0i} = \frac{\partial g_L^*(y^*)}{\partial y_i^*}, \ \forall i \in \{1,...,n\}. \tag{7.40}$$

As $y^* \in R_L^n$ there exists $\bar{x}_0 \in \mathbb{R}^n$ such that:

$$y_i^* = \frac{\partial g(\bar{x}_0)}{\partial x_i} \ \forall i \in \{1,...,n\}, \tag{7.41}$$

and,

$$g_L^*(y^*) = y_i^* \bar{x}_{0i} - g(\bar{x}_0) \tag{7.42}$$

and therefore taking derivatives for this expression we can obtain:

$$\frac{\partial g_L^*(y^*)}{\partial y_i^*} = y_j^* \frac{\partial \bar{x}_{0j}}{\partial y_i^*} + \bar{x}_{0i} - \frac{\partial g(\bar{x}_0)}{\partial x_j}\frac{\partial \bar{x}_{0j}}{\partial y_i^*}, \tag{7.43}$$

$\forall i \in \{1,...,n\}$, so that:

$$\frac{\partial g_L^*(y^*)}{\partial y_i^*} = (y_j^* - \frac{\partial g(\bar{x}_0)}{\partial x_j})\frac{\partial \bar{x}_{0j}}{\partial y_i^*} + \bar{x}_{0i} \tag{7.44}$$

$\forall i \in \{1,...,n\}$, which from (7.40) and (7.41), implies that:

$$\bar{x}_{0i} = \frac{\partial g_L^*(y^*)}{\partial y_i^*} = x_{0i}, \ \forall \ i \in \{1,...,n\}, \tag{7.45}$$

from this and (7.41) we have:

$$y_i^* = \frac{\partial g(\bar{x}_0)}{\partial x_i} = \frac{\partial g(x_0)}{\partial x_i} \ \forall \ i \in \{1,...,n\}. \tag{7.46}$$

Theorem 7.4.3 *Consider the functional* $J : U \to \bar{\mathbb{R}}$ *defined as* $J(u) = (G \circ \Lambda)(u) - \langle u, f \rangle_U$ *where* $\Lambda(= \{\Lambda_i\}) : U \to Y$ $(i \in \{1,...,n\})$ *is a continuous linear operator and,* $G : Y \to \mathbb{R}$ *is a functional that can be expressed as* $G(v) = \int_S g(v)dS, \ \forall v \in Y$ *(here* $g : \mathbb{R}^n \to \mathbb{R}$ *is a differentiable function that admits Legendre Transform denoted by* $g_L^* : R_L^n \to \mathbb{R}$. *That is, the hypothesis mentioned at Proposition 7.4.2 are satisfied).*

Under these assumptions we have:

$$\delta J(u_0) = \theta \Leftrightarrow \delta(-G_L^*(v_0^*) + \langle u_0, \Lambda^* v_0^* - f \rangle_U) = \theta, \qquad (7.47)$$

where $v_0^* = \frac{\partial G(\Lambda(u_0))}{\partial v}$ is supposed to be such that $v_0^*(x) \in R_L^n$, a.e. in S and in this case:

$$J(u_0) = -G_L^*(v_0^*). \qquad (7.48)$$

Proof 7.8   Suppose first that $\delta J(u_0) = \theta$, that is:

$$\Lambda^* \frac{\partial G(\Lambda u_0)}{\partial v} - f = \theta \qquad (7.49)$$

which, as $v_0^* = \frac{\partial G(\Lambda u_0)}{\partial v}$ implies that:

$$\Lambda^* v_0^* - f = \theta, \qquad (7.50)$$

and

$$v_{0i}^* = \frac{\partial g(\Lambda u_0)}{\partial x_i}. \qquad (7.51)$$

Thus from the last proposition we can write:

$$\Lambda_i(u_0) = \frac{\partial g_L^*(v_0^*)}{\partial y_i^*}, \text{ for } i \in \{1,..,n\} \qquad (7.52)$$

which means:

$$\Lambda u_0 = \frac{\partial G_L^*(v_0^*)}{\partial v^*}. \qquad (7.53)$$

Therefore from (7.50) and (7.53) we have:

$$\delta(-G_L^*(v_0^*) + \langle u_0, \Lambda^* v_0^* - f \rangle_U) = \theta. \qquad (7.54)$$

This completes the first part of the proof.
    Conversely, suppose now that:

$$\delta(-G_L^*(v_0^*) + \langle u_0, \Lambda^* v_0^* - f \rangle_U) = \theta, \qquad (7.55)$$

that is:

$$\Lambda^* v_0^* - f = \theta \qquad (7.56)$$

and

$$\Lambda u_0 = \frac{\partial G_L^*(v_0^*)}{\partial v^*}. \qquad (7.57)$$

Clearly, from (7.57), the last proposition and (7.56) we can write:

$$v_0^* = \frac{\partial G(\Lambda(u_0))}{\partial v} \tag{7.58}$$

and

$$\Lambda^* \frac{\partial G(\Lambda u_0)}{\partial v} - f = \theta, \tag{7.59}$$

which implies:

$$\delta J(u_0) = \theta. \tag{7.60}$$

Finally, we have:

$$J(u_0) = G(\Lambda u_0) - \langle u_0, f \rangle_U \tag{7.61}$$

From this, (7.56) and (7.58) we have

$$J(u_0) = G(\Lambda u_0) - \langle u_0, \Lambda^* v_0^* \rangle_U = G(\Lambda u_0) - \langle \Lambda u_0, v_0^* \rangle_Y$$
$$= -G_L^*(v_0^*). \tag{7.62}$$

7.5 Duality in convex optimization

Let U be a Banach space. Given $F : U \to \bar{\mathbb{R}}$ ($F \in \Gamma_0(U)$) we define the problem $\mathscr{P}$ as

$$\mathscr{P} : \text{ minimize } F(u) \text{ on } U. \tag{7.63}$$

We say that $u_0 \in U$ is a solution of problem $\mathscr{P}$ if $F(u_0) = \inf_{u \in U} F(u)$. Consider a function $\phi(u, p)$ ($\phi : U \times Y \to \bar{\mathbb{R}}$) such that

$$\phi(u, 0) = F(u), \tag{7.64}$$

we define the problem $\mathscr{P}^*$, as

$$\mathscr{P}^* : \text{ maximize } -\phi^*(0, p^*) \text{ on } Y^*. \tag{7.65}$$

Observe that

$$\phi^*(0, p^*) = \sup_{(u,p) \in U \times Y} \{\langle 0, u \rangle_U + \langle p, p^* \rangle_Y - \phi(u, p)\} \geq -\phi(u, 0), \tag{7.66}$$

or

$$\inf_{u \in U} \{\phi(u, 0)\} \geq \sup_{p^* \in Y^*} \{-\phi^*(0, p^*)\}. \tag{7.67}$$

Proposition 7.5.1 *Consider* $\phi \in \Gamma_0(U \times Y)$. *If we define*

$$h(p) = \inf_{u \in U}\{\phi(u,p)\}, \tag{7.68}$$

then h is convex.

Proof 7.9   We have to show that given $p, q \in Y$ and $\lambda \in (0,1)$, we have

$$h(\lambda p + (1 - \lambda)q) \le \lambda h(p) + (1 - \lambda)h(q). \tag{7.69}$$

If $h(p) = +\infty$ or $h(q) = +\infty$ we are done. Thus let us assume $h(p) < +\infty$ and $h(q) < +\infty$. For each $a > h(p)$ there exists $u \in U$ such that

$$h(p) \le \phi(u,p) \le a, \tag{7.70}$$

and, if $b > h(q)$, there exists $v \in U$ such that

$$h(q) \le \phi(v,q) \le b. \tag{7.71}$$

Thus

$$h(\lambda p + (1 - \lambda)q) \le \inf_{w \in U}\{\phi(w, \lambda p + (1 - \lambda)q)\}$$
$$\le \phi(\lambda u + (1 - \lambda)v, \lambda p + (1 - \lambda)q) \le \lambda\phi(u,p) + (1 - \lambda)\phi(v,q)$$
$$\le \lambda a + (1 - \lambda)b. \tag{7.72}$$

Letting $a \to h(p)$ and $b \to h(q)$ we obtain

$$h(\lambda p + (1 - \lambda)q) \le \lambda h(p) + (1 - \lambda)h(q). \; \square \tag{7.73}$$

Proposition 7.5.2 *For h as above, we have* $h^*(p^*) = \phi^*(0, p^*)$, $\forall p^* \in Y^*$, *so that*

$$h^{**}(0) = \sup_{p^* \in Y^*}\{-\phi^*(0,p^*)\}. \tag{7.74}$$

Proof. Observe that

$$h^*(p^*) = \sup_{p \in Y}\{\langle p, p^*\rangle_Y - h(p)\} = \sup_{p \in Y}\{\langle p, p^*\rangle_Y - \inf_{u \in U}\{\phi(u,p)\}\}, \tag{7.75}$$

so that

$$h^*(p^*) = \sup_{(u,p) \in U \times Y}\{\langle p, p^*\rangle_Y - \phi(u,p)\} = \phi^*(0,p^*). \tag{7.76}$$

Proposition 7.5.3 *The set of solutions of the problem* $\mathscr{P}^*$ *(the dual problem) is identical to* $\partial h^{**}(0)$.

Proof 7.10   Consider $p_0^* \in Y^*$ a solution of Problem $\mathscr{P}^*$, that is,

$$-\phi^*(0, p_0^*) \geq -\phi^*(0, p^*), \forall p^* \in Y^*, \tag{7.77}$$

which is equivalent to

$$-h^*(p_0^*) \geq -h^*(p^*), \forall p^* \in Y^*, \tag{7.78}$$

which is equivalent to

$$-h(p_0^*) = \sup_{p^* \in Y^*} \{\langle 0, p^* \rangle_Y - h^*(p^*)\} \Leftrightarrow -h^*(p_0^*) = h^{**}(0)$$

$$\Leftrightarrow p_0^* \in \partial h^{**}(0). \quad (7.79)$$

Theorem 7.5.4 *Consider $\phi : U \times Y \to \bar{\mathbb{R}}$ convex. Assume $\inf_{u \in U} \{\phi(u, 0)\} \in \mathbb{R}$ and there exists $u_0 \in U$ such that $p \mapsto \phi(u_0, p)$ is finite and continuous at $0 \in Y$, then*

$$\inf_{u \in U} \{\phi(u, 0)\} = \sup_{p^* \in Y^*} \{-\phi^*(0, p^*)\}, \tag{7.80}$$

and the dual problem has at least one solution.

Proof 7.11   By hypothesis $h(0) \in \mathbb{R}$ and as was shown above, h is convex. As the function $p \mapsto \phi(u_0, p)$ is convex and continuous at $0 \in Y$, there exists a neighborhood $\mathscr{V}$ of zero in Y such that

$$\phi(u_0, p) \leq M < +\infty, \forall p \in \mathscr{V}, \tag{7.81}$$

for some $M \in \mathbb{R}$. Thus, we may write

$$h(p) = \inf_{u \in U} \{\phi(u, p)\} \leq \phi(u_0, p) \leq M, \forall p \in \mathscr{V}. \tag{7.82}$$

Hence, from Lemma 7.3.7, h is continuous at 0. Thus by Proposition 7.3.8, h is sub-differentiable at 0, which means $h(0) = h^{**}(0)$. Therefore by Proposition 7.5.3, the dual problem has solutions and

$$h(0) = \inf_{u \in U} \{\phi(u, 0)\} = \sup_{p^* \in Y^*} \{-\phi^*(0, p^*)\} = h^{**}(0). \tag{7.83}$$

Now we apply the last results to $\phi(u, p) = G(\Lambda u + p) + F(u)$, where $\Lambda : U \to Y$ is a continuous linear operator whose adjoint operator is denoted by $\Lambda^* : Y^* \to U^*$. We may enunciate the following theorem.

Theorem 7.5.5 *Suppose U is a reflexive Banach space and define $J : U \to \mathbb{R}$ by*

$$J(u) = G(\Lambda u) + F(u) = \phi(u, 0), \tag{7.84}$$

where $\lim J(u) = +\infty$ *as* $\|u\|_U \to \infty$ *and* $F \in \Gamma_0(U)$, $G \in \Gamma_0(Y)$. *Also suppose there exists* $\hat{u} \in U$ *such that* $J(\hat{u}) < +\infty$ *with the function* $p \mapsto G(p)$ *continuous at* $\Lambda \hat{u}$. *Under such hypothesis, there exist* $u_0 \in U$ *and* $p_0^* \in Y^*$ *such that*

$$J(u_0) = \min_{u \in U}\{J(u)\} = \max_{p^* \in Y^*}\{-G^*(p^*) - F^*(-\Lambda^* p^*)\}$$

$$= -G^*(p_0^*) - F^*(-\Lambda^* p_0^*). \quad (7.85)$$

Proof 7.12   The existence of solutions for the primal problem follows from the direct method of calculus of variations. That is, considering a minimizing sequence, from above (coercivity hypothesis), such a sequence is bounded and has a weakly convergent subsequence to some $u_0 \in U$. Finally, from the lower semi-continuity of primal formulation, we may conclude that u_0 is a minimizer. The other conclusions follow from Theorem 7.5.4, observing that

$$\phi^*(0, p^*) = \sup_{u \in U, p \in Y}\{\langle p, p^* \rangle_Y - G(\Lambda u + p) - F(u)\}$$

$$= \sup_{u \in U, q \in Y}\{\langle q, p^* \rangle - G(q) - \langle \Lambda u, p^* \rangle - F(u)\}, \quad (7.86)$$

so that

$$\phi^*(0, p^*) = G^*(p^*) + \sup_{u \in U}\{-\langle u, \Lambda^* p^* \rangle_U - F(u)\}$$

$$= G^*(p^*) + F^*(-\Lambda^* p^*). \quad (7.87)$$

Thus,

$$\inf_{u \in U}\{\phi(u, 0)\} = \sup_{p^* \in Y^*}\{-\phi^*(0, p^*)\} \quad (7.88)$$

and solutions u_0 and p_0^* for the primal and dual problems, respectively, imply that

$$J(u_0) = \min_{u \in U}\{J(u)\} = \max_{p^* \in Y^*}\{-G^*(p^*) - F^*(-\Lambda^* p^*)\}$$

$$= -G^*(p_0^*) - F^*(-\Lambda^* p_0^*). \quad (7.89)$$

7.6   The min-max theorem

Our main objective in this section is to state and prove the min-max theorem.

Definition 7.1   Let U, Y be Banach spaces, $A \subset U$ and $B \subset Y$ and let $L : A \times B \to \mathbb{R}$ be a functional. We say that $(u_0, v_0) \in A \times B$ is a saddle point for L if

$$L(u_0, v) \le L(u_0, v_0) \le L(u, v_0), \ \forall u \in A, \ v \in B.$$

Proposition 7.1

Let U, Y be Banach spaces, $A \subset U$ and $B \subset Y$. A functional $L : U \times Y \to \mathbb{R}$ has a saddle point if and only if

$$\max_{v \in B} \inf_{u \in A} L(u, v) = \min_{u \in A} \sup_{v \in B} L(u, v).$$

Proof 7.13 Suppose $(u_0, v_0) \in A \times B$ is a saddle point of L. Thus,

$$L(u_0, v) \leq L(u_0, v_0) \leq L(u, v_0), \forall u \in A, \ v \in B. \tag{7.90}$$

Define

$$F(u) = \sup_{v \in B} L(u, v).$$

Observe that

$$\inf_{u \in A} F(u) \leq F(u_0),$$

so that

$$\inf_{u \in A} \sup_{v \in B} L(u, v) \leq \sup_{v \in B} L(u_0, v). \tag{7.91}$$

Define

$$G(v) = \inf_{u \in A} L(u, v).$$

Thus

$$\sup_{v \in B} G(v) \geq G(v_0),$$

so that

$$\sup_{v \in B} \inf_{u \in A} L(u, v) \geq \inf_{u \in A} L(u, v_0). \tag{7.92}$$

From (7.90), (7.91) and (7.92) we obtain

$$\begin{aligned}
\inf_{u \in A} \sup_{v \in B} L(u, v) &\leq \sup_{v \in B} L(u_0, v) \\
&\leq L(u_0, v_0) \\
&\leq \inf_{u \in A} L(u, v_0) \\
&\leq \sup_{v \in B} \inf_{u \in A} L(u, v). \tag{7.93}
\end{aligned}$$

Hence

$$\begin{aligned}
\inf_{u \in A} \sup_{v \in B} L(u, v) &\leq L(u_0, v_0) \\
&\leq \sup_{v \in B} \inf_{u \in A} L(u, v). \tag{7.94}
\end{aligned}$$

On the other hand

$$\inf_{u \in A} L(u,v) \leq L(u,v), \forall u \in A, \ v \in B,$$

so that

$$\sup_{v \in B} \inf_{u \in A} L(u,v) \leq \sup_{v \ln B} L(u,v), \forall u \in A,$$

and hence

$$\sup_{v \in B} \inf_{u \in A} L(u,v) \leq \inf_{u \in A} \sup_{v \in B} L(u,v). \tag{7.95}$$

From (7.90), (7.94), (7.95) we obtain

$$\begin{aligned} \inf_{u \in A} \sup_{v \in B} L(u,v) &= \sup_{v \in B} L(u_0,v) \\ &= L(u_0,v_0) \\ &= \inf_{u \in A} L(u,v_0) \\ &= \sup_{v \in B} \inf_{u \in A} L(u,v). \end{aligned} \tag{7.96}$$

Conversely suppose

$$\max_{v \in B} \inf_{u \in A} L(u,v) = \min_{u \in A} \sup_{v \in B} L(u,v).$$

As above defined,

$$F(u) = \sup_{v \in B} L(u,v),$$

and

$$G(v) = \inf_{u \in A} L(u,v).$$

From the hypotheses, there exists $(u_0,v_0) \in A \times B$ such that

$$\sup_{v \in B} G(v) = G(v_0) = F(u_0) = \inf_{u \in A} F(u).$$

so that

$$F(u_0) = \sup_{v \in B} L(u_0,v) = \inf_{u \in U} L(u,v_0) = G(v_0).$$

In particular

$$L(u_0,v_0) \leq \sup_{v \in B} L(u_0,v) = \inf_{u \in U} L(u,v_0) \leq L(u_0,v_0).$$

Therefore

$$\sup_{v \in B} L(u_0,v) = L(u_0,v_0) = \inf_{u \in U} L(u,v_0).$$

The proof is complete.

Proposition 7.2

Let U,Y be Banach spaces, $A \subset U$, $B \subset Y$ and let $L : A \times B \to \mathbb{R}$ be a functional. Assume there exist $u_0 \in A$, $v_0 \in B$ and $\alpha \in \mathbb{R}$ such that

$$L(u_0, v) \leq \alpha, \ \forall v \in B,$$

and

$$L(u, v_0) \geq \alpha, \ \forall u \in A.$$

Under such hypotheses (u_0, v_0) *is a saddle point of L, that is,*

$$L(u_0, v) \leq L(u_0, v_0) \leq L(u, v_0), \ \forall u \in A, \ v \in B.$$

Proof 7.14   Observe, from the hypotheses we have

$$L(u_0, v_0) \leq \alpha,$$

and

$$L(u_0, v_0) \geq \alpha,$$

so that

$$L(u_0, v) \leq \alpha = L(u_0, v_0) \leq L(u, v_0), \ \forall u \in A, \ v \in B.$$

This completes the proof.

In the next lines we state and prove the min $-$ max theorem.

Theorem 7.1
Let U, Y *be reflexive Banach spaces,* $A \subset U$, $B \subset Y$ *and let* $L : A \times B \to \mathbb{R}$ *be a functional.*
Suppose that

1. $A \subset U$ *is convex, closed and non-empty.*

2. $B \subset Y$ *is convex, closed and non-empty.*

3. *For each* $u \in A$, $F_u(v) = L(u, v)$ *is concave and upper semi-continuous.*

4. *For each* $v \in B$, $G_v(u) = L(u, v)$ *is convex and lower semi-continuous.*

5. *The set A and B are bounded.*

Under such hypotheses L has at least one saddle point $(u_0, v_0) \in A \times B$ *such that*

$$
\begin{aligned}
L(u_0, v_0) &= \min_{u \in A} \max_{v \in B} L(u, v) \\
&= \max_{v \in B} \min_{u \in A} L(u, v).
\end{aligned}
\tag{7.97}
$$

Proof 7.15   Fix $v \in B$. Observe that $G_v(u) = L(u, v)$ is convex and lower semi-continuous, therefore it is weakly lower semi-continuous on the weak compact set A. At first we assume the additional hypothesis that $G_v(u)$ is strictly convex, $\forall v \in B$. Hence $F_v(u)$ attains a unique minimum on A. We denote the optimal $u \in A$ by $u(v)$

Define
$$G(v) = \min_{u \in A} G_v(u) = \min_{u \in U} L(u,v).$$

Thus,
$$G(v) = L(u(v),v).$$

The function $G(v)$ is expressed as the minimum of a family of concave weakly upper semi-continuous functions, and hence it is also concave and upper semi-continuous.

Moreover, $G(v)$ is bounded above on the weakly compact set B, so that there exists $v_0 \in B$ such that

$$G(v_0) = \max_{v \in B} G(v) = \max_{v \in B} \min_{u \in A} L(u,v).$$

Observe that
$$G(v_0) = \min_{u \in A} L(u,v_0) \leq L(u,v_0), \ \forall u \in U.$$

Observe that, from the concerned concavity, for $u \in A$, $v \in B$ and $\lambda \in (0,1)$ we have

$$L(u,(1-\lambda)v_0 + \lambda v) \geq (1-\lambda)L(u,v_0) + \lambda L(u,v).$$

In particular denote $u((1-\lambda)v_0 + \lambda v) = u_\lambda$, where u_λ is such that

$$
\begin{aligned}
G((1-\lambda)v_0 + \lambda v) &= \min_{u \in A} L(u,(1-\lambda)v_0 + \lambda v) \\
&= L(u_\lambda,(1-\lambda)v_0 + \lambda v).
\end{aligned}
\tag{7.98}
$$

Therefore,

$$
\begin{aligned}
G(v_0) &= \max_{v \in B} G(v) \\
&\geq G((1-\lambda)v_0 + \lambda v) \\
&= L(u_\lambda,(1-\lambda)v_0 + \lambda v) \\
&\geq (1-\lambda)L(u_\lambda,v_0) + \lambda L(u_\lambda,v) \\
&\geq (1-\lambda)\min_{u \in A} L(u,v_0) + \lambda L(u_\lambda,v) \\
&= (1-\lambda)G(v_0) + \lambda L(u_\lambda,v).
\end{aligned}
\tag{7.99}
$$

From this, we obtain
$$G(v_0) \geq L(u_\lambda,v). \tag{7.100}$$

Let $\{\lambda_n\} \subset (0,1)$ be such that $\lambda_n \to 0$.
Let $\{u_n\} \subset A$ be such that

$$
\begin{aligned}
G((1-\lambda_n)v_0 + \lambda_n v) &= \min_{u \in A} L(u,(1-\lambda_n)v_0 + \lambda_n v) \\
&= L(u_n,(1-\lambda_n)v_0 + \lambda_n v).
\end{aligned}
\tag{7.101}
$$

Since A is weakly compact, there exists a subsequence $\{u_{n_k}\} \subset \{u_n\} \subset A$ and $u_0 \in A$ such that

$$u_{n_k} \rightharpoonup u_0, \text{ weakly in } U, \text{ as } k \to \infty.$$

Observe that

$$
\begin{aligned}
(1 - \lambda_{n_k})L(u_{n_k}, v_0) + \lambda_{n_k}L(u_{n_k}, v) &\leq L(u_{n_k}, (1 - \lambda_{n_k})v_0 + \lambda_{n_k}v) \\
&= \min_{u \in A} L(u, (1 - \lambda_{n_k})v_0 + \lambda_{n_k}v) \\
&\leq L(u, (1 - \lambda_{n_k})v_0 + \lambda_{n_k}v), \quad (7.102)
\end{aligned}
$$

$\forall u \in A$, $k \in \mathbb{N}$.

Recalling that $\lambda_{n_k} \to 0$, from this and (7.102) we obtain

$$
\begin{aligned}
L(u_0, v_0) &\leq \liminf_{k \to \infty} L(u_{n_k}, v_0) \\
&= \liminf_{k \to \infty}((1 - \lambda_{n_k})L(u_{n_k}, v_0) + \lambda_{n_k}L(u, v)) \\
&\leq \limsup_{k \to \infty} L(u, (1 - \lambda_{n_k})v_0 + \lambda_{n_k}v) \\
&\leq L(u, v_0), \; \forall u \in U. \quad (7.103)
\end{aligned}
$$

Hence, $L(u_0, v_0) = \min_{u \in A} L(u, v_0)$.

Observe that from (7.100) we have

$$G(v_0) \geq L(u_{n_k}, v),$$

so that

$$G(v_0) \geq \liminf_{k \to \infty} L(u_{n_k}, v) \geq L(u_0, v), \forall v \in B.$$

Denoting $\alpha = G(v_0)$ we have

$$\alpha = G(v_0) \geq L(u_0, v), \forall v \in B,$$

and

$$\alpha = G(v_0) = \min_{u \in U} L(u, v_0) \leq L(u, v_0), \; \forall u \in A.$$

From these last two results and Proposition 7.2 we have that (u_0, v_0) is a saddle point for L. Now assume that

$$G_v(u) = L(u, v)$$

is convex but not strictly convex $\forall v \in B$.

For each $n \in \mathbb{N}$ Define L_n by

$$L_n(u, v) = L(u, v) + \|u\|_U / n.$$

In such a case

$$(F_v)_n(u) = L_n(u,v)$$

is strictly convex for all $n \in \mathbb{N}$.

From above we main obtain $(u_n, v_n) \in A \times B$ such that

$$
\begin{aligned}
L(u_n,v) + \|u_n\|_U/n &\leq L(u_n,v_n) + \|u_n\|_U/n \\
&\leq L(u,v_n) + \|u\|/n.
\end{aligned}
\tag{7.104}
$$

Since $A \times B$ is weakly compact and $\{(u_n,v_n)\} \subset A \times B$, up to subsequence not relabeled, there exists $(u_0,v_0) \in A \times B$ such that

$$u_n \rightharpoonup u_0, \text{ weakly in } U,$$

$$v_n \rightharpoonup v_0, \text{ weakly in } Y,$$

so that,

$$
\begin{aligned}
L(u_0,v) &\leq \liminf_{n\to\infty}(L(u_n,v) + \|u_n\|_U/n) \\
&\leq \limsup_{n\to\infty} L(u,v_n) + \|u\|_U/n \\
&\leq L(u,v_0).
\end{aligned}
\tag{7.105}
$$

Hence,

$$L(u_0,v) \leq L(u,v_0), \ \forall u \in A, \ v \in B,$$

so that

$$L(u_0,v) \leq L(u_0,v_0) \leq L(u,v_0), \ \forall u \in A, \ v \in B.$$

This completes the proof.

In the next result we deal with more general situations.

Theorem 7.2
Let U,Y be reflexive Banach spaces, $A \subset U$, $B \subset Y$ and let $L : A \times B \to \mathbb{R}$ be a functional.

Suppose that

1. *$A \subset U$ is convex, closed and non-empty.*

2. *$B \subset Y$ is convex, closed and non-empty.*

3. *For each $u \in A$, $F_u(v) = L(u,v)$ is concave and upper semi-continuous.*

4. *For each $v \in B$, $G_v(u) = L(u,v)$ is convex and lower semi-continuous.*

5. *Either the set A is bounded or there exists $\tilde{v} \in B$ such that*

$$L(u,\tilde{v}) \to +\infty, \text{ as } \|u\| \to +\infty, \ u \in A.$$

6. *Either the set B is bounded or there exists $\tilde{u} \in A$ such that*

$$L(\tilde{u}, v) \to -\infty, \text{ as } \|v\| \to +\infty, \; v \in B.$$

Under such hypotheses L has at least one saddle point $(u_0, v_0) \in A \times B$.

Proof 7.16   We prove the result just for the special case such that there exists $\tilde{v} \in B$ such that

$$L(u, \tilde{v}) \to +\infty, \text{ as } \|u\| \to +\infty, \; u \in A,$$

and B is bounded. The proofs of remaining cases are similar.

For each $n \in \mathbb{N}$ denote

$$A_n = \{u \in A \; : \; \|u\|_U \leq n\}.$$

Fix $n \in \mathbb{N}$. The sets A_n and B are closed, convex and bounded, so that from the last Theorem 7.1 there exists a saddle point $(u_n, v_n) \in A_n \times B$ for

$$L : A_n \times B \to \mathbb{R}.$$

Hence,

$$L(u_n, v) \leq L(u_n, v_n) \leq L(u, v_n), \forall u \in A_n, \; v \in B.$$

For a fixed $\tilde{u} \in A_1$ we have

$$
\begin{aligned}
L(u_n, \tilde{v}) &\leq & L(u_n, v_n) \\
&\leq & L(\tilde{u}, v_n) \\
&\leq & \sup_{v \in B} L(\tilde{u}, v) \equiv b \in \mathbb{R}.
\end{aligned}
\tag{7.106}
$$

On the other hand, from the hypotheses,

$$G_{\tilde{v}}(u) = L(u, \tilde{v})$$

is convex, lower semi-continuous and coercive, so that it is bounded below. Thus there exists $a \in \mathbb{R}$ such that

$$-\infty < a \leq F_{\tilde{v}}(u) = L(u, \tilde{v}), \; \forall u \in A.$$

Hence

$$a \leq L(u_n, \tilde{v}) \leq L(u_n, v_n) \leq b, \forall n \in \mathbb{N}.$$

Therefore $\{L(u_n, v_n)\}$ is bounded.

Moreover, from the coercivity hypotheses and

$$a \leq L(u_n, \tilde{v}) \leq b, \forall n \in \mathbb{N},$$

we may infer that $\{u_n\}$ is bounded.

Summarizing, $\{u_n\}, \{v_n\}$, and $\{L(u_n, v_n)\}$ are bounded sequences, and thus there exists a subsequence $\{n_k\}$, $u_0 \in A$, $v_0 \in B$ and $\alpha \in \mathbb{R}$ such that

$$u_{n_k} \rightharpoonup u_0, \text{ weakly in } U,$$

$$v_{n_k} \rightharpoonup v_0, \text{ weakly in } Y,$$

$$L(u_{n_k}, v_{n_k}) \to \alpha \in \mathbb{R},$$

as $k \to \infty$. Fix $(u, v) \in A \times B$. Observe that if $n_k > n_0 = \|u\|_U$, then

$$L(u_{n_k}, v) \le L(u_{n_k}, v_{n_k}) \le L(u, v_{n_k}),$$

so that letting $k \to \infty$, we obtain

$$
\begin{aligned}
L(u_0, v) & \le \liminf_{k \to \infty} L(u_{n_k}, v) \\
& \le \lim_{k \to \infty} L(u_{n_k}, v_{n_k}) = \alpha \\
& \le \limsup_{k \to \infty} L(u, v_{n_k}) \\
& \le L(u, v_0), \quad (7.107)
\end{aligned}
$$

that is,

$$L(u_0, v) \le \alpha \le L(u, v_0), \ \forall u \in A, \ v \in B.$$

From this and Proposition 7.2 we may conclude that (u_0, v_0) is a saddle point for $L : A \times B \to \mathbb{R}$.

The proof is complete.

7.7 Relaxation for the scalar case

In this section, $\Omega \subset \mathbb{R}^N$ denotes a bounded open set with a locally Lipschitz boundary. That is, for each point $x \in \partial\Omega$ there exists a neighborhood $\mathscr{U}_x$ whose the intersection with $\partial\Omega$ is the graph of a Lipschitz continuous function.

We start with the following definition.

Definition 7.7.1 *A function $u : \Omega \to \mathbb{R}$ is said to be affine if ∇u is constant on Ω. Furthermore, we say that $u : \Omega \to \mathbb{R}$ is piecewise affine if it is continuous and there exists a partition of Ω into a set of zero measure and finite number of open sets on which u is affine.*

The proof of next result is found in [34].

Theorem 7.7.2 *Let $r \in \mathbb{N}$ and let u_k, $1 \le k \le r$ be piecewise affine functions from Ω into $\mathbb{R}$ and $\{\alpha_k\}$ such that $\alpha_k > 0, \forall k \in \{1, ..., r\}$ and $\sum_{k=1}^r \alpha_k = 1$. Given $\varepsilon > 0$, there exists a locally Lipschitz function $u : \Omega \to \mathbb{R}$ and r disjoint open sets Ω_k, $1 \le k \le r$, such that*

$$|m(\Omega_k) - \alpha_k m(\Omega)| < \alpha_k \varepsilon, \ \forall k \in \{1, ..., r\}, \quad (7.108)$$

$$\nabla u(x) = \nabla u_k(x), \ a.e. \ on \ \Omega_k, \tag{7.109}$$

$$|\nabla u(x)| \leq \max_{1 \leq k \leq r}\{|\nabla u_k(x)|\}, \ a.e. \ on \ \Omega, \tag{7.110}$$

$$\left| u(x) - \sum_{k=1}^{r} \alpha_k u_k \right| < \varepsilon, \ \forall x \in \Omega, \tag{7.111}$$

$$u(x) = \sum_{k=1}^{r} \alpha_k u_k(x), \forall x \in \partial\Omega. \tag{7.112}$$

The next result is also found in [34].

Proposition 7.7.3 *Let $r \in \mathbb{N}$ and let u_k, $1 \leq k \leq r$ be piecewise affine functions from Ω into $\mathbb{R}$. Consider a Carathéodory function $f : \Omega \times \mathbb{R}^N \rightarrow \mathbb{R}$ and a positive function $c \in L^1(\Omega)$ which satisfy*

$$c(x) \geq \sup\{|f(x,\xi)| \ | \ |\xi| \leq \max_{1 \leq k \leq r}\{\|\nabla u_k\|_\infty\}\}. \tag{7.113}$$

Given $\varepsilon > 0$, there exists a locally Lipschitz function $u : \Omega \rightarrow \mathbb{R}$ such that

$$\left| \int_\Omega f(x, \nabla u)dx - \sum_{k=1}^{r} \alpha_k \int_\Omega f(x, \nabla u_k)dx \right| < \varepsilon, \tag{7.114}$$

$$|\nabla u(x)| \leq \max_{1 \leq k \leq r}\{|\nabla u_k(x)|\}, \ a.e. \ in \ \Omega, \tag{7.115}$$

$$\left| u(x) - \sum_{k=1}^{r} \alpha_k u_k(x) \right| < \varepsilon, \forall x \in \Omega \tag{7.116}$$

$$u(x) = \sum_{k=1}^{r} \alpha_k u_k(x), \forall x \in \partial\Omega. \tag{7.117}$$

Proof 7.17 It is sufficient to establish the result for functions u_k affine over Ω, since Ω can be divided into pieces on which u_k are affine, and such pieces can be put together through (7.117). Let $\varepsilon > 0$ be given. We know that simple functions are dense in $L^1(\Omega)$, concerning the L^1 norm. Thus there exists a partition of Ω into a finite number of open sets $\mathscr{O}_i$, $1 \leq i \leq N_1$ and a negligible set, and there exists $\bar{f}_k$ constant functions over each $\mathscr{O}_i$ such that

$$\int_\Omega |f(x, \nabla u_k(x)) - \bar{f}_k(x)|dx < \varepsilon, \ 1 \leq k \leq r. \tag{7.118}$$

Now choose $\delta > 0$ such that

$$\delta \leq \frac{\varepsilon}{N_1(1 + \max_{1 \leq k \leq r}\{\|\bar{f}_k\|_\infty\})} \tag{7.119}$$

and if B is a measurable set

$$m(B) < \delta \Rightarrow \int_B c(x)dx \leq \varepsilon/N_1. \tag{7.120}$$

Now we apply Theorem 7.7.2, to each of the open sets $\mathcal{O}_i$, therefore there exists a locally Lipschitz function $u : \mathcal{O}_i \to \mathbb{R}$ and there exist r open disjoints spaces Ω_k^i, $1 \leq k \leq r$, such that

$$|m(\Omega_k^i) - \alpha_k m(\mathcal{O}_i)| \leq \alpha_k \delta, \text{ for } 1 \leq k \leq r, \tag{7.121}$$

$$\nabla u = \nabla u_k, \text{ a.e. in } \Omega_k^i, \tag{7.122}$$

$$|\nabla u(x)| \leq \max_{1 \leq k \leq r}\{|\nabla u_k(x)|\}, \text{ a.e. } \mathcal{O}_i, \tag{7.123}$$

$$\left| u(x) - \sum_{k=1}^{r} \alpha_k u_k(x) \right| \leq \delta, \forall x \in \mathcal{O}_i \tag{7.124}$$

$$u(x) = \sum_{k=1}^{r} \alpha_k u_k(x), \forall x \in \partial \mathcal{O}_i. \tag{7.125}$$

We can define $u = \sum_{k=1}^{r} \alpha_k u_k$ on $\Omega - \cup_{i=1}^{N_1} \mathcal{O}_i$. Therefore u is continuous and locally Lipschitz. Now observe that

$$\int_{\mathcal{O}_i} f(x, \nabla u(x))dx - \sum_{k=1}^{r} \int_{\Omega_k^i} f(x, \nabla u_k(x))dx$$
$$= \int_{\mathcal{O}_i - \cup_{k=1}^{r}\Omega_k^i} f(x, \nabla u(x))dx. \tag{7.126}$$

From $|f(x, \nabla u(x))| \leq c(x)$, $m(\mathcal{O}_i - \cup_{k=1}^{r}\Omega_k^i) \leq \delta$ and (7.120) we obtain

$$\left| \int_{\mathcal{O}_i} f(x, \nabla u(x))dx - \sum_{k=1}^{r} \int_{\Omega_k^i} f(x, \nabla u_k(x)dx \right|$$
$$= \left| \int_{\mathcal{O}_i - \cup_{k=1}^{r}\Omega_k^i} f(x, \nabla u(x))dx \right| \leq \varepsilon/N_1. \tag{7.127}$$

Considering that $\bar{f}_k$ is constant in $\mathcal{O}_i$, from (7.119), (7.120) and (7.121) we obtain

$$\sum_{k=1}^{r} |\int_{\Omega_k^i} \bar{f}_k(x)dx - \alpha_k \int_{\mathcal{O}_i} \bar{f}_k(x)dx| < \varepsilon/N_1. \tag{7.128}$$

We recall that $\Omega_k = \cup_{i=1}^{N_1} \Omega_k^i$ so that

$$\left| \int_\Omega f(x, \nabla u(x))dx - \sum_{k=1}^r \alpha_k \int_\Omega f(x, \nabla u_k(x))dx \right|$$

$$\leq \left| \int_\Omega f(x, \nabla u(x))dx - \sum_{k=1}^r \int_{\Omega_k} f(x, \nabla u_k(x))dx \right|$$

$$+ \sum_{k=1}^r \int_{\Omega_k} |f(x, \nabla u_k(x)) - \bar{f}_k(x)|dx$$

$$+ \sum_{k=1}^r \left| \int_{\Omega_k} \bar{f}_k(x)dx - \alpha_k \int_\Omega \bar{f}_k(x)dx \right|$$

$$+ \sum_{k=1}^r \alpha_k \int_\Omega |\bar{f}_k(x) - f(x, \nabla u_k(x))|dx. \tag{7.129}$$

From (7.127), (7.118), (7.128) and (7.118) again, we obtain

$$\left| \int_\Omega f(x, \nabla u(x))dx - \sum_{k=1}^r \alpha_k \int_\Omega f(x, \nabla u_k)dx \right| < 4\varepsilon. \tag{7.130}$$

The next result we do not prove it. It is a well known result from the finite element theory.

Proposition 7.7.4 *If* $u \in W_0^{1,p}(\Omega)$ *there exists a sequence* $\{u_n\}$ *of piecewise affine functions over* Ω, *null on* $\partial\Omega$, *such that*

$$u_n \to u, \text{ in } L^p(\Omega) \tag{7.131}$$

and

$$\nabla u_n \to \nabla u, \text{ in } L^p(\Omega; \mathbb{R}^N). \tag{7.132}$$

Proposition 7.7.5 *For p such that* $1 < p < \infty$, *suppose that* $f : \Omega \times \mathbb{R}^N \to \mathbb{R}$ *is a Carathéodory function, for which there exist* $a_1, a_2 \in L^1(\Omega)$ *and constants* $c_1 \geq c_2 > 0$ *such that*

$$a_2(x) + c_2|\xi|^p \leq f(x, \xi) \leq a_1(x) + c_1|\xi|^p, \forall x \in \Omega, \; \xi \in \mathbb{R}^N. \tag{7.133}$$

Then, given $u \in W^{1,p}(\Omega)$ *piecewise affine,* $\varepsilon > 0$ *and a neighborhood* $\mathscr{V}$ *of zero in the topology* $\sigma(L^p(\Omega, \mathbb{R}^N), L^q(\Omega, \mathbb{R}^N))$ *there exists a function* $v \in W^{1,p}(\Omega)$ *such that*

$$\nabla v - \nabla u \in \mathscr{V}, \tag{7.134}$$

$$u = v \text{ on } \partial\Omega,$$

$$\|v - u\|_\infty < \varepsilon, \tag{7.135}$$

and

$$\left| \int_\Omega f(x, \nabla v(x))dx - \int_\Omega f^{**}(x, \nabla u(x))dx \right| < \varepsilon. \tag{7.136}$$

Proof 7.18 Suppose given $\varepsilon > 0$, $u \in W^{1,p}(\Omega)$ piecewise affine continuous, and a neighborhood $\mathcal{V}$ of zero, which may be expressed as

$$\mathcal{V} = \{w \in L^p(\Omega, \mathbb{R}^N) \mid \left| \int_\Omega h_m \cdot w \, dx \right| < \eta,$$

$$\forall m \in \{1, ..., M\}\}, \quad (7.137)$$

where $M \in \mathbb{N}$, $h_m \in L^q(\Omega, \mathbb{R}^N)$, $\eta \in \mathbb{R}^+$. By hypothesis, there exists a partition of Ω into a negligible set Ω_0 and open subspaces Δ_i, $1 \le i \le r$, over which $\nabla u(x)$ is constant. From standard results of convex analysis in $\mathbb{R}^N$, for each $i \in \{1, ..., r\}$ we can obtain $\{\alpha_k \ge 0\}_{1 \le k \le N+1}$, and ξ_k such that $\sum_{k=1}^{N+1} \alpha_k = 1$ and

$$\sum_{k=1}^{N+1} \alpha_k \xi_k = \nabla u, \forall x \in \Delta_i, \quad (7.138)$$

and

$$\sum_{k=1}^{N+1} \alpha_k f(x, \xi_k) = f^{**}(x, \nabla u(x)). \quad (7.139)$$

Define $\beta_i = \max_{k \in \{1, ..., N+1\}} \{|\xi_k| \text{ on } \Delta_i\}$, and $\rho_1 = \max_{i \in \{1, ..., r\}} \{\beta_i\}$, and $\rho = \max\{\rho_1, \|\nabla u\|_\infty\}$. Now, observe that we can obtain functions $\hat{h}_m \in C_0^\infty(\Omega; \mathbb{R}^N)$ such that

$$\max_{m \in \{1, ..., M\}} \|\hat{h}_m - h_m\|_{L^q(\Omega, \mathbb{R}^N)} < \frac{\eta}{4\rho m(\Omega)}. \quad (7.140)$$

Define $C = \max_{m \in \{1, ..., M\}} \|div(\hat{h}_m)\|_{L^q(\Omega)}$ and we can also define

$$\varepsilon_1 = \min\{\varepsilon/4, 1/(m(\Omega)^{1/p}), \eta/(2Cm(\Omega)^{1/p}), 1/m(\Omega)\} \quad (7.141)$$

We recall that ρ does not depend on ε. Furthermore, for each $i \in \{1, ..., r\}$ there exists a compact subset $K_i \subset \Delta_i$ such that

$$\int_{\Delta_i - K_i} [a_1(x) + c_1(x) \max_{|\xi| \le \rho} \{|\xi|^p\}] dx < \frac{\varepsilon_1}{r}. \quad (7.142)$$

Also, observe that the sets K_i may be obtained such that the restrictions of f and f^{**} to $K_i \times \rho B$ are continuous, so that from this and from the compactness of ρB, for all $x \in K_i$, we can find an open ball ω_x with center in x and contained in Ω, such that

$$|f^{**}(y, \nabla u(x)) - f^{**}(x, \nabla u(x))| < \frac{\varepsilon_1}{m(\Omega)}, \forall y \in \omega_x \cap K_i, \quad (7.143)$$

and

$$|f(y, \xi) - f(x, \xi)| < \frac{\varepsilon_1}{m(\Omega)}, \forall y \in \omega_x \cap K_i, \forall \xi \in \rho B. \quad (7.144)$$

Therefore, from this and (7.139) we may write

$$\left| f^{**}(y, \nabla u(x)) - \sum_{k=1}^{N+1} \alpha_k f(y, \xi_k) \right| < \frac{2\varepsilon_1}{m(\Omega)}, \forall y \in \omega_x \cap K_i. \tag{7.145}$$

We can cover the compact set K_i with a finite number of those open balls ω_x, denoted by ω_j, $1 \leq j \leq l$. Consider the open sets $\omega'_j = \omega_j - \cup_{i=1}^{j-1} \bar{\omega}_i$, we have that $\cup_{j=1}^{l} \bar{\omega}'_j = \cup_{j=1}^{l} \bar{\omega}_j$. Defining functions u_k, for $1 \leq k \leq N+1$ such that $\nabla u_k = \xi_k$ and $u = \sum_{k=1}^{N+1} \alpha_k u_k$ we may apply Proposition 7.7.3 to each of the open sets ω'_j, so that we obtain functions $v_i \in W^{1,p}(\Omega)$ such that

$$\left| \int_{\omega'_j} f(x, \nabla v_i(x)) dx - \sum_{k=1}^{N+1} \alpha_k \int_{\omega'_j} f(x, \xi_k) dx \right| < \frac{\varepsilon_1}{rl}, \tag{7.146}$$

$$|\nabla v_i| < \rho, \forall x \in \omega'_j, \tag{7.147}$$

$$|v_i(x) - u(x)| < \varepsilon_1, \forall x \in \omega'_j, \tag{7.148}$$

and

$$v_i(x) = u(x), \forall x \in \partial \omega'_j. \tag{7.149}$$

Finally we set

$$v_i = u \text{ on } \Delta_i - \cup_{j=1}^{l} \omega_j. \tag{7.150}$$

We may define a continuous mapping $v : \Omega \to \mathbb{R}$ by

$$v(x) = v_i(x), \text{ if } x \in \Delta_i, \tag{7.151}$$

$$v(x) = u(x), \text{ if } x \in \Omega_0. \tag{7.152}$$

We have that $v(x) = u(x), \forall x \in \partial \Omega$ and $\|\nabla v\|_\infty < \rho$. Also, from (7.142)

$$\int_{\Delta_i - K_i} |f^{**}(x, \nabla u(x)| dx < \frac{\varepsilon_1}{r} \tag{7.153}$$

and

$$\int_{\Delta_i - K_i} |f(x, \nabla v(x)| dx < \frac{\varepsilon_1}{r}. \tag{7.154}$$

On the other hand, from (7.145) and (7.146)

$$\left| \int_{K_i \cap \omega'_j} f(x, \nabla v(x)) dx - \int_{K_i \cap \omega'_j} f^{**}(x, \nabla u(x)) dx \right|$$

$$\leq \frac{\varepsilon_1}{rl} + \frac{\varepsilon_1 m(\omega'_j \cap K_i)}{m(\Omega)} \tag{7.155}$$

so that

$$\left| \int_{K_i} f(x, \nabla v(x)) dx - \int_{K_i} f^{**}(x, \nabla u(x)) dx \right|$$

$$\leq \frac{\varepsilon_1}{r} + \frac{\varepsilon_1 m(K_i)}{m(\Omega)}. \quad (7.156)$$

Now summing up in i and considering (7.153) and (7.154) we obtain (7.136), that is

$$\left| \int_{\Omega} f(x, \nabla v(x)) dx - \int_{\Omega} f^{**}(x, \nabla u(x)) dx \right| < 4\varepsilon_1 \leq \varepsilon. \quad (7.157)$$

Also, observe that from above, we have

$$\|v - u\|_{\infty} < \varepsilon_1, \quad (7.158)$$

and thus

$$\left| \int_{\Omega} \hat{h}_m \cdot (\nabla v(x) - \nabla u(x)) dx \right| = \left| - \int_{\Omega} div(\hat{h}_m)(v(x) - u(x)) dx \right|$$

$$\leq \|div(\hat{h}_m)\|_{L^q(\Omega)} \|v - u\|_{L^p(S)}$$

$$\leq C\varepsilon_1 m(\Omega)^{1/p}$$

$$< \frac{\eta}{2}. \quad (7.159)$$

Also we have that

$$\left| \int_{\Omega} (\hat{h}_m - h_m) \cdot (\nabla v - \nabla u) dx \right|$$

$$\leq \|\hat{h}_m - h_m\|_{L^q(\Omega, \mathbb{R}^N)} \|\nabla v - \nabla u\|_{L^p(\Omega, \mathbb{R}^N)} \leq \frac{\eta}{2}. \quad (7.160)$$

Thus

$$\left| \int_{\Omega} h_m \cdot (\nabla v - \nabla u) dx \right| < \eta, \forall m \in \{1, ..., M\}. \quad (7.161)$$

Theorem 7.7.6 *Assuming the hypothesis of last theorem, given a function* $u \in W_0^{1,p}(\Omega)$*, given* $\varepsilon > 0$ *and a neighborhood of zero* $\mathcal{V}$ *in* $\sigma(L^p(\Omega, \mathbb{R}^N), L^q(\Omega, \mathbb{R}^N))$*, we have that there exists a function* $v \in W_0^{1,p}(\Omega)$ *such that*

$$\nabla v - \nabla u \in \mathcal{V}, \quad (7.162)$$

and

$$\left| \int_{\Omega} f(x, \nabla v(x)) dx - \int_{\Omega} f^{**}(x, \nabla u(x)) dx \right| < \varepsilon. \quad (7.163)$$

Proof 7.19   We can approximate u by a function w which is piecewise affine and null on the boundary. Thus, there exists $\delta > 0$ such that we can obtain $w \in W_0^{1,p}(\Omega)$ piecewise affine such that

$$\|u - w\|_{1,p} < \delta \tag{7.164}$$

so that

$$\nabla w - \nabla u \in \frac{1}{2}\mathcal{V}, \tag{7.165}$$

and

$$\left| \int_\Omega f^{**}(x, \nabla w(x)) dx - \int_\Omega f^{**}(x, \nabla u(x)) dx \right| < \frac{\varepsilon}{2}. \tag{7.166}$$

From Proposition 7.7.5 we may obtain $v \in W_0^{1,p}(\Omega)$ such that

$$\nabla v - \nabla w \in \frac{1}{2}\mathcal{V}, \tag{7.167}$$

and

$$\left| \int_\Omega f^{**}(x, \nabla w(x)) dx - \int_\Omega f(x, \nabla v(x)) dx \right| < \frac{\varepsilon}{2}. \tag{7.168}$$

From (7.166) and (7.168)

$$\left| \int_\Omega f^{**}(x, \nabla u(x)) dx - \int_\Omega f(x, \nabla v(x)) dx \right| < \varepsilon. \tag{7.169}$$

Finally, from (7.165), (7.167) and from the fact the weak neighborhoods are convex, we have

$$\nabla v - \nabla u \in \mathcal{V}. \tag{7.170}$$

To finish this chapter, we present two theorems which summarize the last results.

Theorem 7.7.7 *Let f be a Carathéodory function from $\Omega \times \mathbb{R}^N$ into $\mathbb{R}$ which satisfies*

$$a_2(x) + c_2|\xi|^p \leq f(x,\xi) \leq a_1(x) + c_1|\xi|^p \tag{7.171}$$

where $a_1, a_2 \in L^1(\Omega)$, $1 < p < +\infty$, $b \geq 0$ and $c_1 \geq c_2 > 0$. Under such assumptions, defining $\hat{U} = W_0^{1,p}(\Omega)$, we have

$$\inf_{u \in \hat{U}} \left\{ \int_\Omega f(x, \nabla u) dx \right\} = \min_{u \in \hat{U}} \left\{ \int_\Omega f^{**}(x, \nabla u) dx \right\} \tag{7.172}$$

The solutions of relaxed problem are weak cluster points in $W_0^{1,p}(\Omega)$ of the minimizing sequences of primal problem.

Proof 7.20 The existence of solutions for the convex relaxed formulation is a consequence of the reflexivity of U and coercivity hypothesis, which allows an application of the direct method of calculus of variations. That is, considering a minimizing sequence, from above (coercivity hypothesis), such a sequence is bounded and has a weakly convergent subsequence to some $\hat{u} \in W^{1,p}(\Omega)$. Finally, from the lower semi-continuity of relaxed formulation, we may conclude that $\hat{u}$ is a minimizer. The relation (7.172) follows from last theorem.

Theorem 7.7.8 *Let f be a Carathéodory function from $\Omega \times \mathbb{R}^N$ into $\mathbb{R}$ which satisfies*

$$a_2(x) + c_2|\xi|^p \le f(x, \xi) \le a_1(x) + c_1|\xi|^p \tag{7.173}$$

where $a_1, a_2 \in L^1(\Omega)$, $1 < p < +\infty$, $b \ge 0$ and $c_1 \ge c_2 > 0$. Let $u_0 \in W^{1,p}(\Omega)$. Under such assumptions, defining $\hat{U} = \{u \mid u - u_0 \in W_0^{1,p}(\Omega)\}$, we have

$$\inf_{u \in \hat{U}} \left\{ \int_\Omega f(x, \nabla u) dx \right\} = \min_{u \in \hat{U}} \left\{ \int_\Omega f^{**}(x, \nabla u) dx \right\} \tag{7.174}$$

The solutions of relaxed problem are weak cluster points in $W^{1,p}(\Omega)$ of the minimizing sequences of primal problem.

Proof 7.21 Just apply the last theorem to the integrand $g(x, \xi) = f(x, \xi + \nabla u_0)$. For details see [34].

7.8 Duality suitable for the vectorial case

7.8.1 The Ekeland variational principle

In this section, we present and prove the Ekeland variational principle. This proof may be found in Giusti [46], pages 160–161.

Theorem 7.8.1 (Ekeland variational principle) *Let (U, d) be a complete metric space and let $F : U \to \overline{\mathbb{R}}$ be a lower semi continuous bounded below functional taking a finite value at some point.*
Let $\varepsilon > 0$. Assume for some $u \in U$ we have

$$F(u) \le \inf_{u \in U} \{F(u)\} + \varepsilon.$$

Under such hypotheses, there exists $v \in U$ such that

1. *$d(u, v) \le 1$,*

2. *$F(v) \le F(u)$,*

3. *$F(v) \le F(w) + \varepsilon d(v, w), \forall w \in U$.*

Proof 7.22   Define the sequence $\{u_n\} \subset U$ by:

$$u_1 = u,$$

and having $u_1, ..., u_n$, select u_{n+1} as specified in the next lines. First, define

$$S_n = \{w \in U \mid F(w) \le F(u_n) - \varepsilon d(u_n, w)\}.$$

Observe that $u_n \in S_n$ so that S_n in non-empty.

On the other hand, from the definition of infimum, we may select $u_{n+1} \in S_n$ such that

$$F(u_{n+1}) \le \frac{1}{2} \left\{ F(u_n) + \inf_{w \in S_n} \{F(w)\} \right\}. \tag{7.175}$$

Since $u_{n+1} \in S_n$ we have

$$\varepsilon d(u_{n+1}, u_n) \le F(u_n) - F(u_{n+1}). \tag{7.176}$$

and hence

$$\varepsilon d(u_{n+m}, u_n) \le \sum_{i=1}^{m} d(u_{n+i}, u_{n+i-1}) \le F(u_n) - F(u_{n+m}). \tag{7.177}$$

From (7.176) $\{F(u_n)\}$ is decreasing sequence bounded below by $\inf_{u \in U} F(u)$ so that there exists $\alpha \in \mathbb{R}$ such that

$$F(u_n) \to \alpha \text{ as } n \to \infty.$$

From this and (7.177), $\{u_n\}$ is a Cauchy sequence , converging to some $v \in U$. Since F is lower semi-continuous we get,

$$\alpha = \liminf_{m \to \infty} F(u_{n+m}) \ge F(v),$$

so that letting $m \to \infty$ in (7.177) we obtain

$$\varepsilon d(u_n, v) \le F(u_n) - F(v), \tag{7.178}$$

and, in particular for $n = 1$ we get

$$0 \le \varepsilon d(u, v) \le F(u) - F(v) \le F(u) - \inf_{u \in U} F(u) \le \varepsilon.$$

Thus, we have proven 1 and 2.

Suppose, to obtain contradiction, that 3 does not hold.

Hence, there exists $w \in U$ such that

$$F(w) < F(v) - \varepsilon d(w, v).$$

In particular, we have

$$w \neq v. \tag{7.179}$$

Thus, from this and (7.178) we have

$$F(w) < F(u_n) - \varepsilon(u_n, v) - \varepsilon d(w, v) \leq F(u_n) - \varepsilon d(u_n, w), \forall n \in \mathbb{N}.$$

Now observe that $w \in S_n, \forall n \in \mathbb{N}$ so that

$$\inf_{w \in S_n} \{F(w)\} \leq F(w), \forall n \in \mathbb{N}.$$

From this and (7.175) we obtain,

$$2F(u_{n+1}) - F(u_n) \leq F(w) < F(v) - \varepsilon d(v, w),$$

so that

$$2 \liminf_{n \to \infty} \{F(u_{n+1})\} \leq F(v) - \varepsilon d(v, w) + \liminf_{n \to \infty} \{F(u_n)\}.$$

Hence,

$$F(v) \leq \liminf_{n \to \infty} \{F(u_{n+1})\} \leq F(v) - \varepsilon d(v, w),$$

so that

$$0 \leq -\varepsilon d(v, w),$$

which contradicts (7.179).

Thus 3 holds.

Remark 7.8.2 *We may introduce in U a new metric given by $d_1 = \varepsilon^{1/2} d$. We highlight that the topology remains the same and also F remains lower semi-continuous. Under the hypotheses of the last theorem, if there exists $u \in U$ such that $F(u) < \inf_{u \in U} F(u) + \varepsilon$, then there exists $v \in U$ such that*

1. $d(u, v) \leq \varepsilon^{1/2}$,

2. $F(v) \leq F(u)$,

3. $F(v) \leq F(w) + \varepsilon^{1/2} d(u, w), \forall w \in U$.

Remark 7.8.3 *Observe that, if U is a Banach space,*

$$F(v) - F(v + tw) \leq \varepsilon^{1/2} t \|w\|_U, \forall t \in [0, 1], \ w \in U, \tag{7.180}$$

so that, if F is Gâteaux differentiable, we obtain

$$-\langle \delta F(v), w \rangle_U \leq \varepsilon^{1/2} \|w\|_U. \tag{7.181}$$

Similarly

$$F(v) - F(v + t(-w)) \leq \varepsilon^{1/2} t \|w\|_U \leq, \forall t \in [0, 1], \ w \in U, \tag{7.182}$$

so that, if F is Gâteaux differentiable, we obtain

$$\langle \delta F(v), w \rangle_U \leq \varepsilon^{1/2} \|w\|_U. \tag{7.183}$$

Thus

$$\|\delta F(v)\|_{U^*} \leq \varepsilon^{1/2}. \tag{7.184}$$

We have thus obtained, from the last theorem and remarks, the following result.

Theorem 7.8.4 *Let U be a Banach space. Let $F : U \to \mathbb{R}$ be a lower semi-continuous Gâteaux differentiable functional. Given $\varepsilon > 0$ suppose that $u \in U$ is such that*

$$F(u) \leq \inf_{u \in U} \{F(u)\} + \varepsilon. \tag{7.185}$$

Then there exists $v \in U$ such that

$$F(v) \leq F(u), \tag{7.186}$$

$$\|u - v\|_U \leq \sqrt{\varepsilon}, \tag{7.187}$$

and

$$\|\delta F(v)\|_{U^*} \leq \sqrt{\varepsilon}. \tag{7.188}$$

The next theorem easily follows from above results.

Theorem 7.8.5 *Let $J : U \to \mathbb{R}$, be defined by*

$$J(u) = G(\nabla u) - \langle f, u \rangle_{L^2(S; \mathbb{R}^N)}, \tag{7.189}$$

where

$$U = W_0^{1,2}(S; \mathbb{R}^N), \tag{7.190}$$

We suppose G is a l.s.c and Gâteaux-differentiable so that J is bounded below. Then, given $\varepsilon > 0$, there exists $u_\varepsilon \in U$ such that

$$J(u_\varepsilon) < \inf_{u \in U} \{J(u)\} + \varepsilon, \tag{7.191}$$

and

$$\|\delta J(u_\varepsilon)\|_{U^*} < \sqrt{\varepsilon}. \ \square \tag{7.192}$$

We finish this Chapter with an important result for vectorial problems in the Calculus of Variations, namely:

Theorem 7.8.6 *Let U be a reflexive Banach space. Consider $(G \circ \Lambda) : U \to \mathbb{R}$ and $(F \circ \Lambda_1) : U \to \mathbb{R}$ l.s.c. functionals such that $J : U \to \mathbb{R}$ defined as*

$$J(u) = (G \circ \Lambda)(u) - (F \circ \Lambda_1)(u) - \langle u, f \rangle_U$$

is below bounded. (Here $\Lambda : U \to Y$ and $\Lambda_1 : U \to Y_1$ are continuous linear operators whose adjoint operators are denoted by $\Lambda^* : Y^* \to U^*$ and $\Lambda_1^* : Y^* \to U^*$, respectively). Also we suppose the existence of $L : Y_1 \to Y$ continuous and linear operator such that L^* is onto and

$$\Lambda(u) = L(\Lambda_1(u)), \forall u \in U.$$

Under such assumptions, we have

$$\inf_{u \in U} \{J(u)\} \geq \sup_{v^* \in A^*} \{ \inf_{z^* \in Y_1^*} \{F^*(L^*z^*) - G^*(v^* + z^*)\}\},$$

where

$$A^* = \{v^* \in Y^* \mid \Lambda^* v^* = f\}.$$

In addition we assume $(F \circ \Lambda_1) : U \to \mathbb{R}$ is convex and Gâteaux differentiable, and suppose there exists a solution (v_0^*, z_0^*) of the dual formulation, so that,

$$L \left(\frac{\partial F^*(L^* z_0^*)}{\partial v^*} \right) \in \partial G^*(v_0^* + z_0^*),$$

$$\Lambda^* v_0^* - f = 0.$$

Suppose $u_0 \in U$ is such that

$$\frac{\partial F^*(L^* z_0^*)}{\partial v^*} = \Lambda_1 u_0,$$

so that

$$\Lambda u_0 \in \partial G^*(v_0^* + z_0^*).$$

Also we assume that there exists a sequence $\{u_n\} \subset U$ such that $u_n \rightharpoonup u_0$ weakly in U and

$$G(\Lambda u_n) \to G^{**}(\Lambda u_0) \text{ as } n \to \infty.$$

Under these additional assumptions we have

$$\inf_{u \in U} \{J(u)\} = \max_{v^* \in A^*} \{ \inf_{z^* \in Y_1^*} \{F^*(L^*z^*) - G^*(v^* + z^*)\}\}$$

$$= F^*(L^* z_0^*) - G^*(v_0^* + z_0^*).$$

Proof 7.23   Observe that

$$G^*(v^* + z^*) \geq \langle \Lambda u, v^* \rangle_Y + \langle \Lambda u, z^* \rangle_Y - G(\Lambda u), \forall u \in U,$$

that is,

$$-F^*(L^* z^*) + G^*(v^* + z^*) \geq \langle u, f \rangle_U - F^*(L^* z^*) + \langle \Lambda_1 u, L^* z^* \rangle_{Y_1}$$
$$- G(\Lambda u), \; \forall u \in U, v^* \in A^*$$

so that

$$\sup_{z^* \in Y_1^*} \{ -F^*(L^* z^*) + G^*(v^* + z^*) \}$$
$$\geq \sup_{z^* \in Y_1^*} \{ \langle u, f \rangle_U - F^*(L^* z^*) + \langle \Lambda_1 u, L^* z^* \rangle_{Y_1} - G(\Lambda u) \},$$

$\forall v^* \in A^*$, $u \in U$ and therefore

$$G(\Lambda u) - F(\Lambda_1 u) - \langle u, f \rangle_U \geq \inf_{z^* \in Y_1^*} \{ F^*(L^* z^*) - G^*(v^* + z^*) \},$$

$$\forall v^* \in A^*, \; u \in U$$

which means

$$\inf_{u \in U} \{ J(u) \} \geq \sup_{v^* \in A^*} \{ \inf_{z^* \in Y_1^*} \{ F^*(L^* z^*) - G^*(v^* + z^*) \} \},$$

where

$$A^* = \{ v^* \in Y^* \mid \Lambda^* v^* = f \}.$$

Now suppose

$$L \left(\frac{\partial F^*(L^* z_0^*)}{\partial v^*} \right) \in \partial G^*(v_0^* + z_0^*),$$

and $u_0 \in U$ is such that

$$\frac{\partial F^*(L^* z_0^*)}{\partial v^*} = \Lambda_1 u_0.$$

Observe that

$$\Lambda u_0 = L(\Lambda_1 u_0) \in \partial G(v_0^* + z_0^*)$$

implies that

$$G^*(v_0^* + z_0^*) = \langle \Lambda u_0, v_0^* \rangle_Y + \langle \Lambda u_0, z_0^* \rangle_Y - G^{**}(\Lambda u_0).$$

From the hypothesis

$$u_n \rightharpoonup u_0 \text{ weakly in } U$$

and

$$G(\Lambda u_n) \to G^{**}(\Lambda u_0) \text{ as } n \to \infty.$$

Thus, given $\varepsilon > 0$, there exists $n_0 \in \mathbb{N}$ such that if $n \geq n_0$ then

$$G^*(v_0^* + z_0^*) - \langle \Lambda u_n, v_0^* \rangle_Y - \langle \Lambda u_n, z_0^* \rangle_Y + G(\Lambda u_n) < \varepsilon/2.$$

On the other hand, since $F(\Lambda_1 u)$ is convex and l.s.c. we have

$$\limsup_{n \to \infty} \{-F(\Lambda_1 u_n)\} \leq -F(\Lambda_1 u_0).$$

Hence, there exists $n_1 \in \mathbb{N}$ such that if $n \geq n_1$ then

$$\langle \Lambda u_n, z_0^* \rangle_Y - F(\Lambda_1 u_n) \leq \langle \Lambda u_0, z_0^* \rangle_Y - F(\Lambda_1 u_0) + \frac{\varepsilon}{2} = F^*(L^* z_0^*) + \frac{\varepsilon}{2},$$

so that for all $n \geq \max\{n_0, n_1\}$ we obtain

$$G^*(v_0^* + z_0^*) - F^*(L^* z_0^*) - \langle u_n, f \rangle_U - F(\Lambda_1 u_n) + G(\Lambda u_n) < \varepsilon.$$

Since ε is arbitrary, the proof is complete.

7.9 Some examples of duality theory in convex and non-convex analysis

We start with a well known result of Toland, published in 1979.

Theorem 7.9.1 (Toland, 1979) *Let U be a Banach space and let $F, G : U \to \mathbb{R}$ be functionals such that*

$$\inf_{u \in U} \{G(u) - F(u)\} = \alpha \in \mathbb{R}.$$

Under such hypotheses

$$F^*(u^*) - G^*(u^*) \geq \alpha, \ \forall u^* \in U^*.$$

Moreover, suppose that $u_0 \in U$ is such that

$$G(u_0) - F(u_0) = \min_{u \in U} \{G(u) - F(u)\} = \alpha.$$

Assume also $u_0^* \in \partial F(u_0)$.
Under such hypotheses,

$$F^*(u_0^*) - G^*(u_0^*) = \alpha,$$

so that

$$
\begin{aligned}
G(u_0) - F(u_0) &= \min_{u \in U} \{G(u) - F(u)\} \\
&= \min_{u^* \in U^*} \{F^*(u^*) - G^*(u^*)\} \\
&= F^*(u_0^*) - G^*(u_0^*).
\end{aligned}
\tag{7.193}
$$

Proof 7.24   Under such hypotheses,

$$\inf_{u \in U} \{G(u) - F(u)\} = \alpha \in \mathbb{R}.$$

Thus

$$G(u) - F(u) \geq \alpha, \ \forall u \in U.$$

Therefore, for $u^* \in U^*$, we have

$$-\langle u, u^* \rangle_U + G(u) + \langle u, u^* \rangle_U - F(u) \geq \alpha, \ \forall u \in U.$$

Thus,

$$-\langle u, u^* \rangle_U + G(u) + \sup_{u \in U}\{\langle u, u^* \rangle_U - F(u)\} \geq \alpha, \ \forall u \in U,$$

that is,

$$-\langle u, u^* \rangle_U + G(u) + F^*(u^*) \geq \alpha, \ \forall u \in U,$$

so that

$$\inf_{u \in U}\{-\langle u, u^* \rangle_U + G(u)\} + F^*(u^*) \geq \alpha,$$

that is,

$$-G^*(u^*) + F^*(u^*) \geq \alpha, \ \forall u^* \in U^*. \tag{7.194}$$

Also from the hypotheses,

$$G(u_0) - F(u_0) \leq G(u) - F(u), \ \forall u \in U. \tag{7.195}$$

On the other hand, from $u_0^* \in \partial F(u_0)$, we obtain

$$\langle u_0, u_0^* \rangle_U - F(u_0) \geq \langle u, u_0^* \rangle_U - F(u), \ \forall u \in U$$

so that

$$-F(u) \leq \langle u_0 - u, u_0^* \rangle_U - F(u_0), \ \forall u \in U.$$

From this and (7.195), we get,

$$G(u_0) - F(u_0) \leq G(u) + \langle u_0 - u, u_0^* \rangle_U - F(u_0), \ \forall u \in U. \tag{7.196}$$

so that,

$$\langle u_0, u_0^* \rangle_U - G(u_0) \geq \langle u, u_0^* \rangle_U - G(u), \ \forall u \in U,$$

that is

$$\begin{aligned} G^*(u_0^*) &= \sup_{u \in U}\{\langle u, u_0^* \rangle_U - G(u)\} \\ &= \langle u_0, u_0^* \rangle_U - G(u_0). \end{aligned} \tag{7.197}$$

Summarizing, we have got

$$G^*(u_0^*) = \langle u_0, u_0^* \rangle_U - G(u_0),$$

and

$$F^*(u_0^*) = \langle u_0, u_0^* \rangle_U - F(u_0).$$

Hence,

$$F^*(u_0) - G^*(u_0^*) = G(u_0) - F(u_0) = \alpha.$$

From this and (7.194), we have

$$
\begin{aligned}
G(u_0) - F(u_0) &= \min_{u \in U} \{ G(u) - F(u) \} \\
&= \min_{u^* \in U^*} \{ F^*(u^*) - G^*(u^*) \} \\
&= F^*(u_0^*) - G^*(u_0^*). \quad (7.198)
\end{aligned}
$$

The proof is complete.

Exercise 7.9.2 *Let $\Omega \subset \mathbb{R}^2$ be a set of $\hat{C}^1$ class. Let $V = C^1(\overline{\Omega})$ and let $J : D \subset V \to \mathbb{R}$ where*

$$J(u) = \frac{\gamma}{2} \int_\Omega \nabla u \cdot \nabla u \, dx - \int_\Omega f u \, dx, \; \forall u \in U$$

and where

$$D = \{ u \in V \; : \; u = 0 \text{ on } \partial\Omega \}.$$

1. *Prove that J is convex.*

2. *Prove that $u_0 \in D$ such that*

   $$\gamma \nabla^2 u_0 + f = 0, \; in \; \Omega,$$

   minimizes J on D.

3. *Prove that*

   $$\inf_{u \in U} J(u) \geq \sup_{v^* \in Y^*} \{ -G^*(v^*) - F^*(-\Lambda^* v^*) \},$$

   where

   $$G(\nabla u) = \frac{1}{2} \int_\Omega \nabla u \cdot \nabla u \, dx,$$

   and

   $$G^*(v^*) = \sup_{v \in Y} \{ \langle v, v^* \rangle_Y - G(v) \},$$

   where $Y = Y^* = L^2(\Omega)$.

4. *Defining* $\Lambda : U \to Y$ *by*

$$\Lambda u = \nabla u,$$

and

$$F : D \to \mathbb{R}$$

as

$$F(u) = \int_\Omega fu\,dx,$$

so that

$$
\begin{aligned}
F^*(-\Lambda^*v^*) &= \sup_{u \in D}\{-\langle \nabla u, v^*\rangle_Y - F(u)\} \\
&= \sup_{u \in D}\left\{\langle u, div\,v^*\rangle_Y + \int_\Omega fu\,dx\right\} \\
&= \sup_{u \in D}\left\{\int_\Omega (div\,v^* + f)\,u\,dx\right\} \\
&= \begin{cases} 0, & if\,div(v^*) + f = 0,\,in\,\Omega \\ +\infty, & otherwise, \end{cases}
\end{aligned}
\tag{7.199}
$$

prove que $v_0^* = \gamma\nabla u_0$ *is such that*

$$
\begin{aligned}
J(u_0) &= \min_{u \in D} J(u) \\
&= \min_{u \in D}\{G(\Lambda u) + F(u)\} \\
&= \max_{v^* \in Y^*}\{-G^*(v^*) - F^*(-\Lambda^*v^*)\} \\
&= -G^*(v_0^*) - F^*(-\Lambda^*v_0^*).
\end{aligned}
\tag{7.200}
$$

Solution: Let $u \in D$ *and*

$$v \in V_a = \{v \in V : v = 0\,on\,\partial\Omega\}.$$

Thus,

$$
\begin{aligned}
&\delta J(u;v) \\
&= \lim_{\varepsilon \to 0}\frac{J(u + \varepsilon v) - J(u)}{\varepsilon} \\
&= \lim_{\varepsilon \to 0}\frac{(\gamma/2)\int_\Omega(\nabla u + \varepsilon\nabla v)\cdot(\nabla u + \varepsilon\nabla v)\,dx - (\gamma/2)\int_\Omega \nabla u\cdot\nabla u\,dx - \int_\Omega(u + \varepsilon v - u)f\,dx}{\varepsilon} \\
&= \lim_{\varepsilon \to 0}\left(\gamma\int_\Omega \nabla u\cdot\nabla v\,dx - \int_\Omega fv\,dx + \varepsilon(\gamma/2)\int_\Omega \nabla v\cdot\nabla v\,dx\right) \\
&= \gamma\int_\Omega \nabla u\cdot\nabla v\,dx - \int_\Omega fv\,dx.
\end{aligned}
\tag{7.201}
$$

Hence,

$$
\begin{aligned}
J(u+v)-J(u) &= (\gamma/2)\int_\Omega (\nabla u+\nabla v)\cdot(\nabla u+\nabla v)\,dx - (\gamma/2)\int_\Omega \nabla u\cdot\nabla u\,dx \\
&\quad - \int_\Omega (u+v-u)f\,dx \\
&= \gamma\int_\Omega \nabla u\cdot\nabla v\,dx - \int_\Omega fv\,dx + (\gamma/2)\int_\Omega \nabla v\cdot\nabla v\,dx \\
&\geq \gamma\int_\Omega \nabla u\cdot\nabla v\,dx - \int_\Omega fv\,dx \\
&= \delta J(u;v) \quad\quad (7.202)
\end{aligned}
$$

$\forall u \in D,\ v \in V_a$.

From this we may infer that J is convex.
From the hypotheses $u_0 \in D$ is such that

$$\gamma\nabla^2 u_0 + f = 0,\ in\ \Omega.$$

Let $v \in V_a$.
Therefore, we have

$$
\begin{aligned}
\delta J(u_0;v) &= \gamma\int_\Omega \nabla u_0\cdot\nabla v\,dx - \int_\Omega fv\,dx \\
&= \gamma\int_\Omega \nabla u_0\cdot\nabla v\,dx + \gamma\int_\Omega \nabla^2 u_0 v\,dx \\
&= \gamma\int_\Omega \nabla u_0\cdot\nabla v\,dx - \gamma\int_\Omega \nabla u_0\cdot\nabla v\,dx + \int_{\partial\Omega} \nabla u_0\cdot\mathbf{n}\,v\,ds \\
&= 0 \quad\quad (7.203)
\end{aligned}
$$

where $\mathbf{n}$ denotes the unit outward field to $\partial\Omega$.
Summarizing, we got $\delta J(u_0;v) = 0,\ \forall v \in V_a$.
Since J is convex,from this we may conclude that u_0 minimizes J on D.
Observe now that,

$$
\begin{aligned}
J(u) &= G(\nabla u)+F(u) \\
&= -\langle \nabla u,v^*\rangle_Y + G(\nabla u) + \langle \nabla u\cdot v^*\rangle_Y + F(u) \\
&\geq \inf_{v\in Y}\{-\langle v,v^*\rangle_Y + G(v)\} \\
&\quad + \inf_{u\in U}\{\langle \nabla u,v^*\rangle_Y + F(u)\} \\
&= -G^*(v^*) - F^*(-\Lambda^* v^*),\ \forall u \in U,\ v^* \in Y^*. \quad (7.204)
\end{aligned}
$$

Summarizing,

$$\inf_{u\in D} J(u) \geq \sup_{v^*\in Y^*}\{-G^*(v^*)-F^*(-\Lambda^* v^*)\}. \quad\quad (7.205)$$

Also from the hypotheses we have $v_0^* = \gamma\nabla u_0$.
Thus,

$$v_0^* = \frac{\partial G(\nabla u_0)}{\partial v},$$

so that

$$
\begin{aligned}
G^*(v_0^*) &= \sup_{v \in Y}\{\langle v, v_0^* \rangle_Y - G(v)\} \\
&= \langle \nabla u_0, v_0^* \rangle_Y - G(\nabla u_0) \\
&= -\langle u_0, div\, v_0^* \rangle_{L^2} - G(\nabla u_0).
\end{aligned}
\tag{7.206}
$$

On the other hand, from de $v_0^* = \gamma \nabla u_0$, we have

$$
div\, v_0^* = \gamma div(\nabla u_0) = \gamma \nabla^2 u_0 = -f
$$

From this and (7.206), we obtain,

$$
G^*(v_0^*) = -\langle u_0, f \rangle_{L^2} - G(\nabla u_0),
$$

and from

$$
div\, v_0^* + f = 0
$$

we get

$$
F^*(-\Lambda^* v_0^*) = 0.
$$

hence

$$
G(\nabla u_0) - \langle u_0, f \rangle_{L^2} = -G^*(v_0^*) - F^*(-\Lambda^* v_0^*),
$$

so that from this and (7.205) we have

$$
\begin{aligned}
J(u_0) &= \min_{u \in D} J(u) \\
&= \min_{u \in D}\{G(\Lambda u) + F(u)\} \\
&= \max_{v^* \in Y^*}\{-G^*(v^*) - F^*(-\Lambda^* v^*)\} \\
&= -G^*(v_0^*) - F^*(-\Lambda^* v_0^*).
\end{aligned}
\tag{7.207}
$$

A solution is complete.

Chapter 8

Constrained Variational Optimization

8.1 Basic concepts

For this chapter, the most relevant reference is the excellent book of Luenberger, [55], where more details may be found. We start with the definition of cone:

Definition 8.1.1 (Cone) *Given U a Banach space, we say that $C \subset U$ is a cone with vertex at origin, if given $u \in C$, we have that $\lambda u \in C$, $\forall \lambda \geq 0$. By analogy we define a cone with vertex at $p \in U$ as $P = p + C$, where C is any cone with vertex at origin.*

Definition 8.1.2 *Let P be a convex cone in U. For $u, v \in U$ we write $u \geq v$ (with respect to P) if $u - v \in P$. In particular $u \geq \theta$ if and only if $u \in P$. Also*

$$P^+ = \{u^* \in U^* \mid \langle u, u^* \rangle_U \geq 0, \forall u \in P\}. \tag{8.1}$$

If $u^* \in P^+$ we write $u^* \geq \theta^*$.

Proposition 8.1.3 *Let U be a Banach space and P be a convex closed cone in U. If $u \in U$ satisfies $\langle u, u^* \rangle_U \geq 0$, $\forall u^* \geq \theta^*$, then $u \geq \theta$.*

Proof 8.1 We prove the contrapositive. Assume $u \notin P$. Then by the separating hyperplane theorem there is an $u^* \in U^*$ such that $\langle u, u^* \rangle_U < \langle p, u^* \rangle_U, \forall p \in P$. Since P is cone we must have $\langle p, u^* \rangle_U \geq 0$, otherwise we would have $\langle u, u^* \rangle > \langle \alpha p, u^* \rangle_U$ for some $\alpha > 0$. Thus $u^* \in P^+$. Finally, since $\inf_{p \in P}\{\langle p, u^* \rangle_U\} = 0$, we obtain $\langle u, u^* \rangle_U < 0$ which completes the proof.

Definition 8.1.4 (Convex Mapping) *Let U, Z be vector spaces. Let $P \subset Z$ be a convex cone. A mapping $G : U \to Z$ is said to be convex if the domain of G is convex and*

$$G(\alpha u_1 + (1 - \alpha)u_2) \leq \alpha G(u_1) + (1 - \alpha)G(u_2),$$

$$\forall u_1, u_2 \in U, \alpha \in [0, 1]. \quad (8.2)$$

Consider the problem $\mathscr{P}$, defined as

Problem $\mathscr{P}$: Minimize $F : U \to \mathbb{R}$ subject to $u \in \Omega$, and $G(u) \leq \theta$

Define

$$\omega(z) = \inf\{F(u) \mid u \in \Omega \text{ and } G(u) \leq z\}. \quad (8.3)$$

For such a functional we have the following result.

Proposition 8.1.5 *If F is a real convex functional and G is convex, then ω is convex.*

Proof 8.2 Observe that

$$\omega(\alpha z_1 + (1 - \alpha)z_2) = \inf\{F(u) \mid u \in \Omega$$
$$\text{and } G(u) \leq \alpha z_1 + (1 - \alpha)z_2\}$$

$$(8.4)$$

$$\leq \inf\{F(u) \mid u = \alpha u_1 + (1 - \alpha)u_2 \ u_1, u_2 \in \Omega$$
$$\text{and } G(u_1) \leq z_1, \ G(u_2) \leq z_2\}$$

$$(8.5)$$

$$\leq \alpha \inf\{F(u_1) \mid u_1 \in \Omega, \ G(u_1) \leq z_1\}$$
$$+ (1 - \alpha) \inf\{F(u_2) \mid u_2 \in \Omega, \ G(u_2) \leq z_2\}$$

$$(8.6)$$

$$\leq \alpha \omega(z_1) + (1 - \alpha)\omega(z_2). \quad (8.7)$$

Now we establish the Lagrange multiplier theorem for convex global optimization.

Theorem 8.1.6 *Let U be a vector space, Z a Banach space, Ω a convex subset of U, P a positive convex closed cone of Z. Assume that P contains an interior point. Let F be a real convex functional on Ω and G a convex mapping from Ω into Z. Assume the existence of $u_1 \in \Omega$ such that $G(u_1) < \theta$. Defining*

$$\mu_0 = \inf_{u \in \Omega}\{F(u) \mid G(u) \leq \theta\}, \quad (8.8)$$

then there exists $z_0^* \geq \theta$, $z_0^* \in Z^*$ such that

$$\mu_0 = \inf_{u \in \Omega}\{F(u) + \langle G(u), z_0^* \rangle_Z\}. \quad (8.9)$$

Furthermore, if the infimum in (8.8) is attained by $u_0 \in U$ such that $G(u_0) \leq \theta$, it is also attained in (8.9) by the same u_0 and also $\langle G(u_0), z_0^* \rangle_Z = 0$. We refer to z_0^* as the Lagrangian Multiplier.

Proof 8.3   Consider the space $W = \mathbb{R} \times Z$ and the sets A, B where

$$A = \{(r,z) \in \mathbb{R} \times Z \mid r \geq F(u), \ z \geq G(u) \ \text{ for some } \ u \in \Omega\}, \tag{8.10}$$

and

$$B = \{(r,z) \in \mathbb{R} \times Z \mid r \leq \mu_0, \ z \leq \theta\}, \tag{8.11}$$

where $\mu_0 = \inf_{u \in \Omega}\{F(u) \mid G(u) \leq \theta\}$. Since F and G are convex, A and B are convex sets. It is clear that A contains no interior point of B, and since $N = -P$ contains an interior point , the set B contains an interior point. Thus, from the separating hyperplane theorem, there is a non-zero element $w_0^* = (r_0, z_0^*) \in W^*$ such that

$$r_0 r_1 + \langle z_1, z_0^* \rangle z \geq r_0 r_2 + \langle z_2, z_0^* \rangle z, \forall (r_1, z_1) \in A, \ (r_2, z_2) \in B. \tag{8.12}$$

From the nature of B it is clear that $w_0^* \geq \theta$. That is, $r_0 \geq 0$ and $z_0^* \geq \theta$. We will show that $r_0 > 0$. The point $(\mu_0, \theta) \in B$, hence

$$r_0 r + \langle z, z_0^* \rangle z \geq r_0 \mu_0, \forall (r,z) \in A. \tag{8.13}$$

If $r_0 = 0$ then $\langle G(u_1), z_0^* \rangle z \geq 0$ and $z_0^* \neq \theta$. Since $G(u_1) < \theta$ and $z_0^* \geq \theta$ we have a contradiction. Therefore $r_0 > 0$ and, without loss of generality we may assume $r_0 = 1$. Since the point (μ_0, θ) is arbitrarily close to A and B, we have

$$\mu_0 = \inf_{(r,z) \in A} \{r + \langle z, z_0^* \rangle z\} \leq \inf_{u \in \Omega} \{F(u) + \langle G(u), z_0^* \rangle z\}$$

$$\leq \inf\{F(u) \mid u \in \Omega, \ G(u) \leq \theta\} = \mu_0. \tag{8.14}$$

Also, if there exists u_0 such that $G(u_0) \leq \theta$, $\mu_0 = F(u_0)$, then

$$\mu_0 \leq F(u_0) + \langle G(u_0), z_0^* \rangle z \leq F(u_0) = \mu_0. \tag{8.15}$$

Hence

$$\langle G(u_0), z_0^* \rangle z = 0. \tag{8.16}$$

Corollary 8.1.7 *Let the hypothesis of the last theorem hold. Suppose*

$$F(u_0) = \inf_{u \in \Omega} \{F(u) \mid G(u) \leq \theta\}. \tag{8.17}$$

Then there exists $z_0^* \geq \theta$ such that the Lagrangian $L : U \times Z^* \to \mathbb{R}$ defined by

$$L(u, z^*) = F(u) + \langle G(u), z^* \rangle z \tag{8.18}$$

has a saddle point at (u_0, z_0^*). That is

$$L(u_0, z^*) \leq L(u_0, z_0^*) \leq L(u, z_0^*), \forall u \in \Omega, z^* \geq \theta. \tag{8.19}$$

Proof 8.4 For z_0^* obtained in the last theorem, we have

$$L(u_0, z_0^*) \leq L(u, z_0^*), \forall u \in \Omega. \tag{8.20}$$

As $\langle G(u_0), z_0^* \rangle_Z = 0$, we have

$$L(u_0, z^*) - L(u_0, z_0^*) = \langle G(u_0), z^* \rangle_Z - \langle G(u_0), z_0^* \rangle_Z$$
$$= \langle G(u_0), z^* \rangle_Z \leq 0. \tag{8.21}$$

We now prove two theorems relevant to develop the subsequent section.

Theorem 8.1.8 *Let $F : \Omega \subset U \to \mathbb{R}$ and $G : \Omega \to Z$. Let $P \subset Z$ be a convex closed cone. Suppose there exists $(u_0, z_0^*) \in U \times Z^*$ where $z_0^* \geq \theta$ and $u_0 \in \Omega$ are such that*

$$F(u_0) + \langle G(u_0), z_0^* \rangle_Z \leq F(u) + \langle G(u), z_0^* \rangle_Z, \forall u \in \Omega. \tag{8.22}$$

Then

$$F(u_0) + \langle G(u_0), z_0^* \rangle_Z$$
$$= \inf\{F(u) \mid u \in \Omega \text{ and } G(u) \leq G(u_0)\}. \tag{8.23}$$

Proof 8.5 Suppose there is a $u_1 \in \Omega$ such that $F(u_1) < F(u_0)$ and $G(u_1) \leq G(u_0)$. Thus

$$\langle G(u_1), z_0^* \rangle_Z \leq \langle G(u_0), z_0^* \rangle_Z \tag{8.24}$$

so that

$$F(u_1) + \langle G(u_1), z_0^* \rangle_Z < F(u_0) + \langle G(u_0), z_0^* \rangle_Z, \tag{8.25}$$

which contradicts the hypothesis of the theorem.

Theorem 8.1.9 *Let F be a convex real functional and $G : \Omega \to Z$ convex and let u_0 and u_1 be solutions to the problems $\mathscr{P}_0$ and $\mathscr{P}_1$ respectively, where*

$$\mathscr{P}_0 : \text{ minimize } F(u) \text{ subject to } u \in \Omega \text{ and } G(u) \leq z_0, \tag{8.26}$$

and

$$\mathscr{P}_1 : \text{ minimize } F(u) \text{ subject to } u \in \Omega \text{ and } G(u) \leq z_1. \tag{8.27}$$

Suppose z_0^* and z_1^* are the Lagrange multipliers related to these problems. Then

$$\langle z_1 - z_0, z_1^* \rangle_Z \leq F(u_0) - F(u_1) \leq \langle z_1 - z_0, z_0^* \rangle_Z. \tag{8.28}$$

Proof 8.6 For u_0, z_0^* we have

$$F(u_0) + \langle G(u_0) - z_0, z_0^* \rangle_Z \leq F(u) + \langle G(u) - z_0, z_0^* \rangle_Z, \forall u \in \Omega, \tag{8.29}$$

and, particularly for $u = u_1$ and considering that $\langle G(u_0) - z_0, z_0^* \rangle_Z = 0$, we obtain

$$F(u_0) - F(u_1) \leq \langle G(u_1) - z_0, z_0^* \rangle_Z \leq \langle z_1 - z_0, z_0^* \rangle_Z. \tag{8.30}$$

A similar argument applied to u_1, z_1^* provides us the other inequality.

8.2 Duality

Consider the basic convex programming problem:

$$\text{Minimize } F(u) \text{ subject to } G(u) \leq \theta, \ u \in \Omega, \tag{8.31}$$

where $F : U \to \mathbb{R}$ is a convex functional, $G : U \to Z$ is convex mapping, and Ω is a convex set. We define $\varphi : Z^* \to \mathbb{R}$ by

$$\varphi(z^*) = \inf_{u \in \Omega} \{F(u) + \langle G(u), z^* \rangle_Z\}. \tag{8.32}$$

Proposition 8.2.1 φ *is concave and*

$$\varphi(z^*) = \inf_{z \in \Gamma} \{\omega(z) + \langle z, z^* \rangle_Z\}, \tag{8.33}$$

where

$$\omega(z) = \inf_{u \in \Omega} \{F(u) \mid G(u) \leq z\}, \tag{8.34}$$

and

$$\Gamma = \{z \in Z \mid G(u) \leq z \text{ for some } u \in \Omega\}.$$

Proof 8.7   Observe that

$$
\begin{aligned}
\varphi(z^*) &= \inf_{u \in \Omega} \{F(u) + \langle G(u), z^* \rangle_Z\} \\
&\leq \inf_{u \in \Omega} \{F(u) + \langle z, z^* \rangle_Z \mid G(u) \leq z\} \\
&= \omega(z) + \langle z, z^* \rangle_Z, \forall z^* \geq \theta, z \in \Gamma.
\end{aligned} \tag{8.35}
$$

On the other hand, for any $u_1 \in \Omega$, defining $z_1 = G(u_1)$, we obtain

$$F(u_1) + \langle G(u_1), z^* \rangle_Z \geq \inf_{u \in \Omega} \{F(u) + \langle z_1, z^* \rangle_Z \mid G(u) \leq z_1\}$$

$$= \omega(z_1) + \langle z_1, z^* \rangle_Z, \tag{8.36}$$

so that

$$\varphi(z^*) \geq \inf_{z \in \Gamma} \{\omega(z) + \langle z, z^* \rangle_Z\}. \tag{8.37}$$

Theorem 8.2.2 (Lagrange duality)   *Consider $F : \Omega \subset U \to \mathbb{R}$ a convex functional, Ω a convex set, and $G : U \to Z$ a convex mapping. Suppose there exists a u_1 such that $G(u_1) < \theta$ and that $\inf_{u \in \Omega} \{F(u) \mid G(u) \leq \theta\} < \infty$, with such order related to a convex closed cone in Z. Under such assumptions, we have*

$$\inf_{u \in \Omega} \{F(u) \mid G(u) \leq \theta\} = \max_{z^* \geq \theta} \{\varphi(z^*)\}. \tag{8.38}$$

If the infimum on the left side in (8.38) is achieved at some $u_0 \in U$ and the max on the right side at $z_0^* \in Z^*$, then

$$\langle G(u_0), z_0^* \rangle_Z = 0 \tag{8.39}$$

and u_0 minimizes $F(u) + \langle G(u), z_0^* \rangle_Z$ on Ω.

Proof 8.8 For $z^* \geq \theta$ we have

$$\inf_{u \in \Omega} \{F(u) + \langle G(u), z^* \rangle_Z\} \leq \inf_{u \in \Omega, G(u) \leq \theta} \{F(u) + \langle G(u), z^* \rangle_Z\}$$

$$\leq \inf_{u \in \Omega, G(u) \leq \theta} F(u) \leq \mu_0. \quad (8.40)$$

or

$$\varphi(z^*) \leq \mu_0. \quad (8.41)$$

The result follows from Theorem 8.1.6.

8.3 The Lagrange multiplier theorem

Remark 8.3.1 *This section was published in similar form by the journal "Computational and Applied Mathematics, SBMAC-Springer", reference [24].*

In this section, we develop a new and simpler proof of the Lagrange multiplier theorem in a Banach space context. In particular, we address the problem of minimizing a functional $F : U \to \mathbb{R}$ subject to $G(u) = \theta$, where θ denotes the zero vector and $G : U \to Z$ is a Fréchet differentiable transformation. Here U, Z are Banach spaces. General results on Banach spaces may be found in [13, 35] for example. For the theorem in question, among others we would cite [55, 51, 22]. Specially the proof given in [55] is made through the generalized inverse function theorem. We emphasize such a proof is extensive and requires the continuous Fréchet differentiability of F and G. Our approach here is different and the results are obtained through other hypotheses.

The main result is summarized by the following theorem.

Theorem 8.3.2 *Let U and Z be Banach spaces. Assume u_0 is a local minimum of $F(u)$ subject to $G(u) = \theta$, where $F : U \to \mathbb{R}$ is a Gâteaux differentiable functional and $G : U \to Z$ is a Fréchet differentiable transformation such that $G'(u_0)$ maps U onto Z. Finally, assume there exist $\alpha > 0$ and $K > 0$ such that if $\|\varphi\|_U < \alpha$ then,*

$$\|G'(u_0 + \varphi) - G'(u_0)\| \leq K \|\varphi\|_U.$$

Under such assumptions, there exists $z_0^* \in Z^*$ such that

$$F'(u_0) + [G'(u_0)]^*(z_0^*) = \theta,$$

that is,

$$\langle \varphi, F'(u_0) \rangle_U + \langle G'(u_0)\varphi, z_0^* \rangle_Z = 0, \forall \varphi \in U.$$

Proof 8.9 First observe that there is no loss of generality in assuming $0 < \alpha < 1$. Also from the generalized mean value inequality and our hypothesis, if $\|\varphi\|_U < \alpha$, then

$$
\begin{aligned}
&\|G(u_0 + \varphi) - G(u_0) - G'(u_0) \cdot \varphi\| \\
&\leq \quad \sup_{h \in [0,1]} \{\|G'(u_0 + h\varphi) - G'(u_0)\|\} \|\varphi\|_U \\
&\leq \quad K \sup_{h \in [0,1]} \{\|h\varphi\|_U\} \|\varphi\|_U \leq K\|\varphi\|_U^2.
\end{aligned}
\tag{8.42}
$$

For each $\varphi \in U$, define $H(\varphi)$ by

$$
G(u_0 + \varphi) = G(u_0) + G'(u_0) \cdot \varphi + H(\varphi),
$$

that is,

$$
H(\varphi) = G(u_0 + \varphi) - G(u_0) - G'(u_0) \cdot \varphi.
$$

Let $L_0 = N(G'(u_0))$ where $N(G'(u_0))$ denotes the null space of $G'(u_0)$. Observe that U/L_0 is a Banach space for which we define $A : U/L_0 \to Z$ by

$$
A(\bar{u}) = G'(u_0) \cdot u,
$$

where $\bar{u} = \{u + v \mid v \in L_0\}$.

Since $G'(u_0)$ is onto, so is A, so that by the inverse mapping theorem A has a continuous inverse A^{-1}.

Let $\varphi \in U$ be such that $G'(u_0) \cdot \varphi = \theta$. For a given t such that $0 < |t| < \frac{\alpha}{1+\|\varphi\|_U}$, let $\psi_0 \in U$ be such that

$$
G'(u_0) \cdot \psi_0 + \frac{H(t\varphi)}{t^2} = \theta,
$$

Observe that, from (8.42),

$$
\|H(t\varphi)\| \leq Kt^2\|\varphi\|_U^2,
$$

and thus from the boundedness of A^{-1}, $\|\psi_0\|$ as a function of t may be chosen uniformly bounded relating t (that is, despite the fact that ψ_0 may vary with t, there exists $K_1 > 0$ such that $\|\psi_0\|_U < K_1, \forall t$ such that $0 < |t| < \frac{\alpha}{1+\|\varphi\|_U}$).

Now choose $0 < r < 1/4$ and define $g_0 = \theta$.

Also define

$$
\varepsilon = \frac{r}{4(\|A^{-1}\| + 1)(K + 1)(K_1 + 1)(\|\varphi\|_U + 1)}.
$$

Since from the hypotheses $G'(u)$ is continuous at u_0, we may choose $0 < \delta < \alpha$ such that if $\|v\|_U < \delta$ then

$$
\|G'(u_0 + v) - G'(u_0)\| < \varepsilon.
$$

Fix $t \in \mathbb{R}$ such that

$$
0 < |t| < \frac{\delta}{2(1 + \|\varphi\|_U + K_1)}.
$$

Observe that $\psi \in U$ is such that $G(u_0 + t\varphi + t^2\psi) = \theta$ if and only if

$$G'(u_0) \cdot \psi + \frac{H(t\varphi + t^2\psi)}{t^2} = \theta.$$

Define

$$L_1 = A^{-1}\left[G'(u_0) \cdot (\psi_0 - g_0) + \frac{H(t\varphi + t^2(\psi_0 - g_0))}{t^2}\right],$$

so that

$$
\begin{aligned}
L_1 &= A^{-1}[A(\overline{\psi_0 - g_0})] + A^{-1}\left(\frac{H(t\varphi + t^2(\psi_0 - g_0))}{t^2}\right)\\
&= \overline{\psi_0 - g_0} + \overline{w_1}\\
&= \overline{\psi_0 + w_1}\\
&= \{\psi_0 + w_1 + v \mid v \in L_0\}.
\end{aligned}
$$

Here $w_1 \in U$ is such that

$$\overline{w_1} = A^{-1}\left(\frac{H(t\varphi + t^2(\psi_0 - g_0))}{t^2}\right),$$

that is,

$$A(\overline{w_1}) = \frac{H(t\varphi + t^2(\psi_0 - g_0))}{t^2},$$

so that

$$G'(u_0) \cdot w_1 = \frac{H(t\varphi + t^2(\psi_0 - g_0))}{t^2}.$$

Select $g_1 \in L_1$ such that

$$\|g_1 - g_0\|_U \leq 2\|L_1 - L_0\|.$$

This is possible since

$$\|L_1 - L_0\| = \inf_{g \in L_1}\{\|g - g_0\|_U\}.$$

So we have that

$$L_1 = A^{-1}\left[-\frac{H(t\varphi)}{t^2} + \frac{H(t\varphi + t^2(\psi_0 - g_0))}{t^2}\right]. \tag{8.43}$$

However

$$
\begin{aligned}
&H(t\varphi + t^2(\psi_0 - g_0)) - H(t\varphi)\\
&= G(u_0 + t\varphi + t^2(\psi_0)) - G(u_0)\\
&\quad - G'(u_0) \cdot (t\varphi + t^2(\psi_0))\\
&\quad - G(u_0 + t\varphi) + G(u_0)\\
&\quad + G'(u_0) \cdot (t\varphi)\\
&= G(u_0 + t\varphi + t^2(\psi_0)) - G(u_0 + t\varphi)\\
&\quad - G'(u_0) \cdot (t^2(\psi_0)), \tag{8.44}
\end{aligned}
$$

so that by the generalized mean value inequality we may write

$$
\begin{aligned}
\|H(t\varphi + t^2(\psi_0 - g_0)) - H(t\varphi)\| \\
\leq \quad \sup_{h \in [0,1]} \|G'(u_0 + t\varphi + ht^2(\psi_0)) - G'(u_0)\| \|t^2 \psi_0\|_U \\
< \quad \varepsilon t^2 \|\psi_0\|_U.
\end{aligned}
\tag{8.45}
$$

From this and (8.43) we get

$$
\begin{aligned}
\|L_1\| &\leq \|A^{-1}\| \|H(t\varphi + t^2(\psi_0 - g_0)) - H(t\varphi)\|/t^2 \\
&< \|A^{-1}\| \|\varepsilon\| \|\psi_0\|_U \\
&< \|A^{-1}\| K_1 \frac{r}{4(\|A^{-1}\| + 1)(K + 1)(K_1 + 1)(\|\varphi\|_U + 1)} \\
&< \frac{r}{4}.
\end{aligned}
\tag{8.46}
$$

Hence

$$
\|g_1\|_U < 2\|L_1\| < r/2.
$$

Now reasoning by induction, for $n \geq 2$ assume that $\|g_{n-1}\|_U < r$ and $\|g_{n-2}\|_U < r$ and define L_n by

$$
L_n - L_{n-1} = A^{-1} \left[G'(u_0) \cdot (\psi_0 - g_{n-1}) + \frac{H(t\varphi + t^2(\psi_0 - g_{n-1}))}{t^2} \right].
$$

Observe that

$$
\begin{aligned}
L_n &= A^{-1} \left[G'(u_0) \cdot (\psi_0 - g_{n-1}) + \frac{H(t\varphi + t^2(\psi_0 - g_{n-1}))}{t^2} \right] + L_{n-1} \\
&= A^{-1} A(\overline{\psi_0 - g_{n-1}}) + A^{-1} \left[\frac{H(t\varphi + t^2(\psi_0 - g_{n-1}))}{t^2} \right] + \overline{g}_{n-1} \\
&= \overline{\psi_0 - g_{n-1}} + A^{-1} \left[\frac{H(t\varphi + t^2(\psi_0 - g_{n-1}))}{t^2} \right] + \overline{g}_{n-1} \\
&= \overline{\psi_0} + A^{-1} \left[\frac{H(t\varphi + t^2(\psi_0 - g_{n-1}))}{t^2} \right] \\
&= \{\psi_0 + w_n + v \mid v \in L_0\}.
\end{aligned}
$$

Here $w_n \in U$ is such that

$$
\overline{w_n} = A^{-1} \left[\frac{H(t\varphi + t^2(\psi_0 - g_{n-1}))}{t^2} \right],
$$

that is,

$$
A(\overline{w_n}) = \left[\frac{H(t\varphi + t^2(\psi_0 - g_{n-1}))}{t^2} \right],
$$

so that

$$
G'(u_0) \cdot w_n = \left[\frac{H(t\varphi + t^2(\psi_0 - g_{n-1}))}{t^2} \right].
$$

Choose $g_n \in L_n$ such that

$$\|g_n - g_{n-1}\|_U \leq 2\|L_n - L_{n-1}\|.$$

This is possible since

$$\|L_n - L_{n-1}\| = \inf_{g \in L_n} \{\|g - g_{n-1}\|_U\}.$$

Observe that we may write

$$L_{n-1} = A^{-1}[A(\bar{g}_{n-1})] = A^{-1}[G'(u_0) \cdot g_{n-1}].$$

Thus

$$L_n = A^{-1}\left[G'(u_0) \cdot (\psi_0 - g_{n-1}) + \frac{H(t\varphi + t^2(\psi_0 - g_{n-1}))}{t^2} + G'(u_0) \cdot g_{n-1}\right].$$

By analogy

$$L_{n-1} = A^{-1}\left[G'(u_0) \cdot (\psi_0 - g_{n-2}) + \frac{H(t\varphi + t^2(\psi_0 - g_{n-2}))}{t^2} + G'(u_0) \cdot g_{n-2}\right].$$

Observe that

$$
\begin{aligned}
&H(t\varphi + t^2(\psi_0 - g_{n-1})) - H(t\varphi + t^2(\psi_0 - g_{n-2})) \\
= \quad &G(u_0 + t\varphi + t^2(\psi_0 - g_{n-1})) - G(u_0) \\
&-G'(u_0) \cdot (t\varphi + t^2(\psi_0 - g_{n-1})) \\
&-G(u_0 + t\varphi + t^2(\psi_0 - g_{n-2})) + G(u_0) \\
&+G'(u_0) \cdot (t\varphi + t^2(\psi_0 - g_{n-2})) \\
= \quad &G(u_0 + t\varphi + t^2(\psi_0 - g_{n-1})) - G(u_0 + t\varphi + t^2(\psi_0 - g_{n-2})) \\
&-G'(u_0) \cdot (t^2(-g_{n-1} + g_{n-2})),
\end{aligned}
\tag{8.47}
$$

so that by the generalized mean value inequality we may write

$$
\begin{aligned}
&\|H(t\varphi + t^2(\psi_0 - g_{n-1})) - H(t\varphi + t^2(\psi_0 - g_{n-2}))\| \\
\leq \quad &\sup_{h \in [0,1]} \|G'(u_0 + t\varphi + t^2\psi_0 - t^2(h(g_{n-1}) + (1-h)g_{n-2})) - G'(u_0)\| \\
&\times \|t^2(-g_{n-1} + g_{n-2})\|_U \\
< \quad &\varepsilon t^2 \|g_{n-1} - g_{n-2}\|_U.
\end{aligned}
$$

Therefore, similarly as above,

$$
\begin{aligned}
\|L_n - L_{n-1}\| \quad \leq \quad &\frac{\|A^{-1}\|}{t^2}\|H(t\varphi + t^2(\psi_0 - g_{n-1})) - H(t\varphi + t^2(\psi_0 - g_{n-2}))\| \\
< \quad &\varepsilon\|A^{-1}\|\|g_{n-1} - g_{n-2}\|_U \\
< \quad &(r/4)\|g_{n-1} - g_{n-2}\|_U \\
< \quad &\frac{1}{4}\|g_{n-1} - g_{n-2}\|_U.
\end{aligned}
\tag{8.48}
$$

Thus,

$$\|g_n - g_{n-1}\|_U \le 2\|L_n - L_{n-1}\| < \frac{1}{2}\|g_{n-1} - g_{n-2}\|_U.$$

Finally

$$
\begin{aligned}
\|g_n\|_U &= \|g_n - g_{n-1} + g_{n-1} - g_{n-2} + g_{n-2} - \cdots + g_1 - g_0\|_U \\
&\le \|g_1\|_U \left(1 + \frac{1}{2} + \cdots + \frac{1}{2^n}\right) < 2\|g_1\|_U < r.
\end{aligned}
\tag{8.49}
$$

Thus $\|g_n\|_U < r$ and

$$\|g_n - g_{n-1}\|_U < \frac{1}{2}\|g_{n-1} - g_{n-2}\|_U, \forall n \in \mathbb{N},$$

so that $\{g_n\}$ is a Cauchy sequence, and since U is a Banach space there exists $g \in U$ such that

$$g_n \to g, \text{ in norm, as } n \to \infty.$$

Hence

$$L_n \to L = \bar{g}, \text{ in norm, as } n \to \infty,$$

so that,

$$L_n - L_{n-1} \to L - L = \theta = A^{-1}\left[G'(u_0) \cdot (\psi_0 - g) + \frac{H(t\varphi + t^2(\psi_0 - g))}{t^2}\right].$$

Since A^{-1} is a bijection, denoting $\tilde{\psi}_0 = (\psi_0 - g)$, we get

$$G'(u_0) \cdot \tilde{\psi}_0 + \frac{H(t\varphi + t^2(\tilde{\psi}_0))}{t^2} = \theta.$$

Clarifying the dependence on t we denote $\tilde{\psi}_0 = \tilde{\psi}_0(t)$ where as above mentioned, $t \in \mathbb{R}$ is such that

$$0 < |t| < \frac{\delta}{2(1 + \|\varphi\|_U + K_1)}.$$

Therefore

$$G(u_0 + t\varphi + t^2\tilde{\psi}_0(t)) = \theta.$$

Observe also that $\|t^2\tilde{\psi}_0(t)\|_U = \|t^2(\psi_0(t) - g)\|_U \le t^2(K_1 + r) \le t^2(K_1 + 1)$ so that $t^2\tilde{\psi}_0(t) \to \theta$ as $t \to 0$. Thus, by defining $t^2\tilde{\psi}_0(t)|_{t=0} = \theta$ (observe that in principle such a function would not be defined at $t = 0$), we obtain

$$\frac{d(t^2\tilde{\psi}_0(t))}{dt}\Big|_{t=0} = \lim_{t \to 0}\left(\frac{t^2\tilde{\psi}_0(t) - \theta}{t}\right) = \theta,$$

considering that

$$\|t\tilde{\psi}_0(t)\|_U \le |t|(K_1 + 1) \to 0, \text{ as } t \to 0.$$

Finally, defining

$$\phi(t) = F(u_0 + t\varphi + t^2 \tilde{\psi}_0(t)),$$

from the hypotheses we have that there exists a suitable $\tilde{t}_2 > 0$ such that

$$\phi(0) = F(u_0) \le F(u_0 + t\varphi + t^2 \tilde{\psi}_0(t)) = \phi(t), \forall |t| < \tilde{t}_2,$$

also from the hypothesis we get

$$\phi'(0) = \delta F(u_0, \varphi) = 0,$$

that is,

$$\langle \varphi, F'(u_0) \rangle_U = 0, \forall \varphi \text{ such that } G'(u_0) \cdot \varphi = \theta.$$

In the next lines as usual $N[G'(u_0)]$ and $R[G'(u_0)]$ denote the null space and the range of $G'(u_0)$, respectively. Thus $F'(u_0)$ is orthogonal to the null space of $G'(u_0)$, which we denote by

$$F'(u_0) \perp N[G'(u_0)].$$

Since $R[G'(u_0)]$ is closed, we get $F'(u_0) \in R([G'(u_0)]^*)$, that is, there exists $z_0^* \in Z^*$ such that

$$F'(u_0) = [G'(u_0)]^*(-z_0^*).$$

The proof is complete.

8.4   Some examples concerning inequality constraints

In this section, we assume the hypotheses of last theorem for F and G below specified. As an application of this same result, consider the problem of locally minimizing $F(u)$ subject to $G_1(u) = \theta$ and $G_2(u) \le \theta$, where $F : U \to \mathbb{R}$, U being a function Banach space, $G_1 : U \to [L^p(\Omega)]^{m_1}$, $G_2 : U \to [L^p(\Omega)]^{m_2}$ where $1 < p < \infty$, and Ω is an appropriate subset of $\mathbb{R}^N$. We refer to the simpler case in which the partial order in $[L^p(\Omega)]^{m_2}$ is defined by $u = \{u_i\} \ge \theta$ if and only if $u_i \in L^p(\Omega)$ and $u_i(x) \ge 0$ a.e. in $\Omega, \forall i \in \{1, ..., m_2\}$.

Observe that defining

$$\tilde{F}(u, v) = F(u),$$

$$G(u, v) = \left(\{(G_1)_i(u)\}_{m_1 \times 1}, \{(G_2)_i(u) + v_i^2\}_{m_2 \times 1} \right)$$

it is clear that (locally) minimizing $\tilde{F}(u, v)$ subject to $G(u, v) = (\theta, \theta)$ is equivalent to the original problem. We clarify the domain of $\tilde{F}$ is denoted by $U \times Y$, where

$$Y = \{v \text{ measurable such that } v_i^2 \in L^p(\Omega), \forall i \in \{1, ..., m_2\}\}.$$

Therefore, if u_0 is a local minimum for the original constrained problem, then for an appropriate and easily defined v_0 we have that (u_0, v_0) is a point of local minimum for the extended constrained one, so that by the last theorem there exists a Lagrange multiplier $z_0^* = (z_1^*, z_2^*) \in [L^q(\Omega)]^{m_1} \times [L^q(\Omega)]^{m_2}$ where $1/p + 1/q = 1$ and

$$\tilde{F}'(u_0, v_0) + [G'(u_0, v_0)]^*(z_0^*) = (\theta, \theta),$$

that is,

$$F'(u_0) + [G'_1(u_0)]^*(z_1^*) + [G'_2(u_0)]^*(z_2^*) = \theta, \tag{8.50}$$

and

$$(z_2^*)_i v_{0i} = \theta, \forall i \in \{1,...,m_2\}.$$

In particular for almost all $x \in \Omega$, if x is such that $v_{0i}(x)^2 > 0$ then $z_{2i}^*(x) = 0$, and if $v_{0i}(x) = 0$ then $(G_2)_i(u_0(x)) = 0$, so that $(z_2^*)_i(G_2)_i(u_0) = 0$, a.e. in $\Omega, \forall i \in \{1,...,m_2\}$.

Furthermore, consider the problem of minimizing $F_1(v) = \tilde{F}(u_0,v) = F(u_0)$ subject $\{G_{2i}(u_0) + v_i^2\} = \theta$. From above such a local minimum is attained at v_0. Thus, from the stationarity of $F_1(v) + \langle z_2^*, \{(G_2)_i(u_0) + v_i^2\}\rangle_{[L^p(\Omega)]^{m_2}}$ at v_0 and the standard necessary conditions for the case of convex (in fact quadratic) constraints we get $(z_2^*)_i \geq 0$ a.e. in $\Omega, \forall i \in \{1,...,m_2\}$, that is, $z_2^* \geq \theta$.

Summarizing, for the order in question the first order necessary optimality conditions are given by (8.50), $z_2^* \geq \theta$ and $(z_2^*)_i(G_2)_i(u_0) = \theta, \forall i \in \{1,...,m_2\}$ (so that $\langle z_2^*, G_2(u_0)\rangle_{[L^p(\Omega)]^{m_2}} = 0$), $G_1(u_0) = \theta$, and $G_2(u_0) \leq \theta$.

Remark 8.4.1 *For the case $U = \mathbb{R}^n$ and $\mathbb{R}^{m_k}$ replacing $[L^p(\Omega)]^{m_k}$, for $k \in \{1,2\}$ the conditions $(z_2^*)_i v_i = \theta$ means that for the constraints not active (for example $v_i \neq 0$) the corresponding coordinate $(z_2^*)_i$ of the Lagrange multiplier is 0. If $v_i = 0$ then $(G_2)_i(u_0) = 0$, so that in any case $(z_2^*)_i(G_2)_i(u_0) = 0$.*

Summarizing, for this last mentioned case we have obtained the standard necessary optimality conditions: $(z_2^*)_i \geq 0$, and $(z_2^*)_i(G_2)_i(u_0) = 0, \forall i \in \{1,...,m_2\}$.

8.5   The Lagrange multiplier theorem for equality and inequality constraints

In this section, we develop more rigorous results concerning the Lagrange multiplier theorem for the case involving equalities and inequalities.

Theorem 8.1
Let U, Z_1, Z_2 be Banach spaces. Consider a cone C in Z_2 as above specified and such that if $z_1 \leq \theta$ and $z_2 < \theta$ then $z_1 + z_2 < \theta$, where $z \leq \theta$ means that $z \in -C$ and $z < \theta$ means that $z \in (-C)^\circ$. The concerned order is supposed to be also that if $z < \theta$, $z^* \geq \theta^*$ and $z^* \neq \theta$ then $\langle z, z^*\rangle_{Z_2} < 0$. Furthermore, assume $u_0 \in U$ is a point of local minimum for $F : U \to \mathbb{R}$ subject to $G_1(u) = \theta$ and $G_2(u) \leq \theta$, where $G_1 : U \to Z_1, G_2 : U \to Z_2$ and F are Fréchet differentiable at $u_0 \in U$. Suppose also $G'_1(u_0)$ is onto and that there exist $\alpha > 0, K > 0$ such that if $\|\varphi\|_U < \alpha$ then

$$\|G'_1(u_0 + \varphi) - G'_1(u_0)\| \leq K\|\varphi\|_U.$$

Finally, suppose there exists $\varphi_0 \in U$ such that

$$G'_1(u_0) \cdot \varphi_0 = \theta$$

and

$$G'_2(u_0) \cdot \varphi_0 < \theta.$$

Under such hypotheses, there exists a Lagrange multiplier $z_0^* = (z_1^*, z_2^*) \in Z_1^* \times Z_2^*$ such that

$$F'(u_0) + [G'_1(u_0)]^*(z_1^*) + [G'_2(u_0)]^*(z_2^*) = \theta,$$

$$z_2^* \geq \theta^*,$$

and

$$\langle G_2(u_0), z_2^* \rangle_{Z_2} = 0.$$

Proof 8.10 Let $\varphi \in U$ be such that

$$G'_1(u_0) \cdot \varphi = \theta$$

and

$$G'_2(u_0) \cdot \varphi = v - \lambda G_2(u_0),$$

for some $v \leq \theta$ and $\lambda \geq 0$.

Select $\alpha \in (0,1)$ and define

$$\varphi_\alpha = \alpha \varphi_0 + (1-\alpha)\varphi.$$

Observe that $G_1(u_0) = \theta$ and $G'_1(u_0) \cdot \varphi_\alpha = \theta$ so that as in the proof of the Lagrange multiplier theorem 9.2.1 we may find $K_1 > 0$, $\varepsilon > 0$ and $\psi_0(t)$ such that

$$G_1(u_0 + t\varphi_\alpha + t^2\psi_0(t)) = \theta, \ \forall |t| < \varepsilon,$$

and

$$\|\psi_0(t)\|_U < K_1, \forall |t| < \varepsilon.$$

Observe that

$$
\begin{aligned}
& G'_2(u_0) \cdot \varphi_\alpha \\
={} & \alpha G'_2(u_0) \cdot \varphi_0 + (1-\alpha)G'_2(u_0) \cdot \varphi \\
={} & \alpha G'_2(u_0) \cdot \varphi_0 + (1-\alpha)(v - \lambda G_2(u_0)) \\
={} & \alpha G'_2(u_0) \cdot \varphi_0 + (1-\alpha)v - (1-\alpha)\lambda G_2(u_0)) \\
={} & v_0 - \lambda_0 G_2(u_0),
\end{aligned}
\tag{8.51}
$$

where,

$$\lambda_0 = (1-\alpha)\lambda,$$

and

$$v_0 = \alpha G'_2(u_0) \cdot \varphi_0 + (1-\alpha)v < \theta.$$

Hence, for $t > 0$

$$G_2(u_0 + t\varphi_\alpha + t^2\psi_0(t)) = G_2(u_0) + G'_2(u_0) \cdot (t\varphi_\alpha + t^2\psi_0(t)) + r(t),$$

where

$$\lim_{t \to 0^+} \frac{\|r(t)\|}{t} = 0.$$

Therefore from (9.12) we obtain

$$G_2(u_0 + t\varphi_\alpha + t^2 \psi_0(t)) = G_2(u_0) + tv_0 - t\lambda_0 G_2(u_0) + r_1(t),$$

where

$$\lim_{t \to 0^+} \frac{\|r_1(t)\|}{t} = 0.$$

Observe that there exists $\varepsilon_1 > 0$ such that if $0 < t < \varepsilon_1 < \varepsilon$, then

$$v_0 + \frac{r_1(t)}{t} < \theta,$$

and

$$G_2(u_0) - t\lambda_0 G_2(u_0) = (1 - t\lambda_0)G_2(u_0) \le \theta.$$

Hence

$$G_2(u_0 + t\varphi_\alpha + t^2 \psi_0(t)) < \theta, \ \text{if } 0 < t < \varepsilon_1.$$

From this there exists $0 < \varepsilon_2 < \varepsilon_1$ such that

$$
\begin{aligned}
F(u_0 + t\varphi_\alpha + t^2 \psi_0(t)) - F(u_0) \\
= \ \langle t\varphi_\alpha + t^2 \psi_0(t), F'(u_0) \rangle_U + r_2(t) \ge 0,
\end{aligned}
\tag{8.52}
$$

where

$$\lim_{t \to 0^+} \frac{|r_2(t)|}{t} = 0.$$

Dividing the last inequality by $t > 0$ we get

$$\langle \varphi_\alpha + t\psi_0(t), F'(u_0) \rangle_U + r_2(t)/t \ge 0, \forall 0 < t < \varepsilon_2.$$

Letting $t \to 0^+$ we obtain

$$\langle \varphi_\alpha, F'(u_0) \rangle_U \ge 0.$$

Letting $\alpha \to 0^+$, we get

$$\langle \varphi, F'(u_0) \rangle_U \ge 0,$$

if

$$G_1'(u_0) \cdot \varphi = \theta,$$

and

$$G_2'(u_0) \cdot \varphi = v - \lambda G_2(u_0),$$

for some $v \le \theta$ and $\lambda \ge 0$. Define

$$
\begin{aligned}
A \ = \ & \{(\langle \varphi, F'(u_0) \rangle_U + r, G_1'(u_0) \cdot \varphi, G_2'(u_0)\varphi - v + \lambda G(u_0)), \\
& \varphi \in U, \, r \ge 0, v \le \theta, \lambda \ge 0\}.
\end{aligned}
\tag{8.53}
$$

Observe that A is a convex set with a non-empty interior.

If

$$G_1'(u_0) \cdot \varphi = \theta,$$

and

$$G_2'(u_0) \cdot \varphi - v + \lambda G_2(u_0) = \theta,$$

with $v \leq \theta$ and $\lambda \geq 0$ then

$$\langle \varphi, F'(u_0) \rangle_U \geq 0,$$

so that

$$\langle \varphi, F'(u_0) \rangle_U + r \geq 0.$$

Moreover, if

$$\langle \varphi, F'(u_0) \rangle + r = 0,$$

with $r \geq 0$,

$$G_1'(u_0) \cdot \varphi = \theta,$$

and

$$G_2'(u_0) \cdot \varphi - v + \lambda G_2(u_0) = \theta,$$

with $v \leq \theta$ and $\lambda \geq 0$, then we have

$$\langle \varphi, F'(u_0) \rangle_U \geq 0,$$

so that

$$\langle \varphi, F'(u_0) \rangle_U = 0,$$

and $r = 0$. Hence $(0, \theta, \theta)$ is on the boundary of A. Therefore, by the Hahn-Banach theorem, geometric form, there exists

$$(\beta, z_1^*, z_2^*) \in \mathbb{R} \times Z_1^* \times Z_2^*$$

such that

$$(\beta, z_1^*, z_2^*) \neq (0, \theta, \theta)$$

and

$$\beta(\langle \varphi, F'(u_0) \rangle_U + r) \; + \; \langle G_1'(u_0) \cdot \varphi, z_1^* \rangle_{Z_1}$$
$$+ \; \langle G_2'(u_0) \cdot \varphi - v + \lambda G_2(u_0), z_2^* \rangle_{Z_2} \geq 0, \qquad (8.54)$$

$\forall \varphi \in U, r \geq 0, v \leq \theta, \lambda \geq 0$. Suppose $\beta = 0$. Fixing all variable except v we get $z_2^* \geq \theta$. Thus, for $\varphi = c\varphi_0$ with arbitrary $c \in \mathbb{R}$, $v = \theta, \lambda = 0$, if $z_2^* \neq \theta$, then $\langle G_2'(u_0) \cdot \varphi_0, z_2^* \rangle_{Z_2} < 0$, so that we get $z_2^* = \theta$. Since $G_1'(u_0)$ is onto, a similar reasoning lead us to $z_1^* = \theta$, which contradicts $(\beta, z_1^*, z_2^*) \neq (0, \theta, \theta)$.

Hence, $\beta \neq 0$, and fixing all variables except r we obtain $\beta > 0$. There is no loss of generality in assuming $\beta = 1$.

Again fixing all variables except v, we obtain $z_2^* \geq \theta$. Fixing all variables except λ, since $G_2(u_0) \leq \theta$ we get

$$\langle G_2(u_0), z_2^* \rangle_{Z_2} = 0.$$

Finally, for $r = 0$, $v = \theta$, $\lambda = 0$, we get

$$\langle \varphi, F'(u_0) \rangle_U + \langle G'_1(u_0)\varphi, z_1^* \rangle_{Z_1} + \langle G'_2(u_0) \cdot \varphi, z_2^* \rangle_{Z_2} \geq 0, \; \forall \varphi \in U,$$

that is, since obviously such an inequality is valid also for $-\varphi$, $\forall \varphi \in U$, we obtain

$$\langle \varphi, F'(u_0) \rangle_U + \langle \varphi, [G'_1(u_0)]^*(z_1^*) \rangle_U + \langle \varphi, [G'_2(u_0)]^*(z_2^*) \rangle_U = 0, \; \forall \varphi \in U,$$

so that

$$F'(u_0) + [G'_1(u_0)]^*(z_1^*) + [G'_2(u_0)]^*(z_2^*) = \theta.$$

The proof is complete.

8.6 Second order necessary conditions

In this section, we establish second order necessary conditions for a class of constrained problems in Banach spaces. We highlight the next result is particularly applicable to optimization in $\mathbb{R}^n$.

Theorem 8.2
Let U, Z_1, Z_2 be Banach spaces. Consider a cone C in Z_2 as above specified and such that if $z_1 \leq \theta$ and $z_2 < \theta$ then $z_1 + z_2 < \theta$, where $z \leq \theta$ means that $z \in -C$ and $z < \theta$ means that $z \in (-C)^\circ$. The concerned order is supposed to be also that if $z < \theta$, $z^* \geq \theta^*$ and $z^* \neq \theta$ then $\langle z, z^* \rangle_{Z_2} < 0$. Furthermore, assume $u_0 \in U$ is a point of local minimum for $F : U \to \mathbb{R}$ subject to $G_1(u) = \theta$ and $G_2(u_0) \leq \theta$, where $G_1 : U \to Z_1$, $G_2 : U \to (Z_2)^k$ and F are twice Fréchet differentiable at $u_0 \in U$. Assume $G_2(u) = \{(G_2)_i(u)\}$ where $(G_2)_i : U \to Z_2$, $\forall i \in \{1, ..., k\}$ and define

$$A = \{i \in \{1, ..., k\} \; : \; (G_2)_i(u_0) = \theta\},$$

and also suppose that $(G_2)_i(u_0) < \theta$, if $i \notin A$. Moreover, suppose $\{G'_1(u_0), \{(G_2)'_i(u_0)\}_{i \in A}\}$ is onto and that there exist $\alpha > 0, K > 0$ such that if $\|\varphi\|_U < \alpha$ then

$$\|\tilde{G}'(u_0 + \varphi) - \tilde{G}'(u_0)\| \leq K\|\varphi\|_U,$$

where

$$\tilde{G}(u) = \{G_1(u), \{(G_2)_i(u)\}_{i \in A}\}.$$

Finally, suppose there exists $\varphi_0 \in U$ such that

$$G'_1(u_0) \cdot \varphi_0 = \theta$$

and

$$G'_2(u_0) \cdot \varphi_0 < \theta.$$

Under such hypotheses, there exists a Lagrange multiplier $z_0^* = (z_1^*, z_2^*) \in Z_1^* \times (Z_2^*)^k$ such that

$$F'(u_0) + [G_1'(u_0)]^*(z_1^*) + [G_2'(u_0)]^*(z_2^*) = \theta,$$

$$z_2^* \geq (\theta^*, ..., \theta^*) \equiv \theta_k^*,$$

and

$$\langle (G_2)_i(u_0), (z_2^*)_i \rangle_Z = 0, \forall i \in \{1, ..., k\},$$

$$(z_2^*)_i = \theta^*, \ if \ i \notin A,$$

Moreover, defining

$$L(u, z_1^*, z_2^*) = F(u) + \langle G_1(u), z_1^* \rangle_{Z_1} + \langle G_2(u), z_2^* \rangle_{Z_2},$$

we have that

$$\delta_{uu}^2 L(u_0, z_1^*, z_2^*; \varphi) \geq 0, \forall \varphi \in \mathcal{V}_0,$$

where

$$\mathcal{V}_0 = \{\varphi \in U \ : \ G_1'(u_0) \cdot \varphi = \theta, \ (G_2)_i'(u_0) \cdot \varphi = \theta, \ \forall i \in A\}.$$

Proof 8.11 Observe that A is defined by

$$A = \{i \in \{1, ..., k\} \ : \ (G_2)_i(u_0) = \theta\}.$$

Observe also that $(G_2)_i(u_0) < \theta$, if $i \notin A$.

Hence the point $u_0 \in U$ is a local minimum for $F(u)$ under the constraints

$$G_1(u) = \theta, \ and \ (G_2)_i(u) \leq \theta, \forall i \in A.$$

From the last Theorem 9.3.1 for such an optimization problem there exists a Lagrange multiplier $(z_1^*, \{(z_2^*)_{i \in A}\})$ such that $(z_2^*)_i \geq \theta^*$, $\forall i \in A$, and

$$F'(u_0) + [G_1'(u_0)]^*(z_1^*) + \sum_{i \in A} [(G_2)_i'(u_0)]^*((z_2^*)_i) = \theta. \qquad (8.55)$$

The choice $(z_2^*)_i = \theta$, if $i \notin A$ leads to the existence of a Lagrange multiplier $(z_1^*, z_2^*) = (z_1^*, \{(z_2^*)_{i \in A}, (z_2^*)_{i \notin A}\})$ such that

$$z_2^* \geq \theta_k^*$$

and

$$\langle (G_2)_i(u_0), (z_2^*)_i \rangle_Z = 0, \forall i \in \{1, ..., k\}.$$

Let $\varphi \in \mathcal{V}_0$, that is, $\varphi \in U$,

$$G_1'(u_0)\varphi = \theta$$

and

$$(G_2)_i'(u_0) \cdot \varphi = \theta, \ \forall i \in A.$$

Recall that $\tilde{G}(u) = \{G_1(u), (G_2)_{i \in A}(u)\}$ and therefore, similarly as in the proof of the Lagrange multiplier theorem 9.2.1, we may obtain $\psi_0(t), K > 0$ and $\varepsilon > 0$ such that

$$\tilde{G}(u_0 + t\varphi + t^2\psi_0(t)) = \theta, \ \forall |t| < \varepsilon,$$

and

$$\|\psi_0(t)\| \leq K, \forall |t| < \varepsilon.$$

Also, if $i \notin A$, we have that $(G_2)_i(u_0) < \theta$, so that

$$(G_2)_i(u_0 + t\varphi + t^2\psi_0(t)) = (G_2)_i(u_0) + G'_i(u_0) \cdot (t\varphi + t^2\psi_0(t)) + r(t),$$

where

$$\lim_{t \to 0} \frac{\|r(t)\|}{t} = 0,$$

that is,

$$(G_2)_i(u_0 + t\varphi + t^2\psi_0(t)) = (G_2)_i(u_0) + t(G_2)'_i(u_0) \cdot \varphi + r_1(t),$$

where,

$$\lim_{t \to 0} \frac{\|r_1(t)\|}{t} = 0,$$

and hence there exists, $0 < \varepsilon_1 < \varepsilon$, such that

$$(G_2)_i(u_0 + t\varphi + t^2\psi_0(t)) < \theta, \ \forall |t| < \varepsilon_1 < \varepsilon.$$

Therefore, since u_0 is a point of local minimum under the constraint $G(u) \leq \theta$, there exists $0 < \varepsilon_2 < \varepsilon_1$, such that

$$F(u_0 + t\varphi + t^2\psi_0(t)) - F(u_0) \geq 0, \ \forall |t| < \varepsilon_2,$$

so that,

$$
\begin{aligned}
& F(u_0 + t\varphi + t^2\psi_0(t)) - F(u_0) \\
=\ & F(u_0 + t\varphi + t^2\psi_0(t)) - F(u_0) \\
& + \langle G_1(u_0 + t\varphi + t^2\psi_0(t)), z_1^* \rangle_{Z_1} + \sum_{i \in A} \{ \langle (G_2)_i(u_0 + t\varphi + t^2\psi_0(t)), (z_2^*)_i \rangle_{Z_2} \} \\
& - \langle G_1(u_0), z_1^* \rangle_{Z_1} - \sum_{i \in A} \{ \langle (G_2)_i(u_0), (z_2^*)_i \rangle_{Z_2} \} \\
=\ & F(u_0 + t\varphi + t^2\psi_0(t)) - F(u_0) \\
& + \langle G_1(u_0 + t\varphi + t^2\psi_0(t)), z_1^* \rangle_{Z_1} - \langle G_1(u_0), z_1^* \rangle_{Z_1} \\
& + \langle G_2(u_0 + t\varphi + t^2\psi_0(t)), z_2^* \rangle_{Z_2} - \langle G_2(u_0), z_2^* \rangle_{Z_2} \\
=\ & L(u_0 + t\varphi + t^2\psi_0(t)), z_1^*, z_2^*) - L(u_0, z_1^*, z_2^*) \\
=\ & \delta_u L(u_0, z_1^*, z_2^*; t\varphi + t^2\psi_0(t)) + \frac{1}{2} \delta_{uu}^2 L(u_0, z_1^*, z_2^*; t\varphi + t^2\psi_0(t)) + r_2(t) \\
=\ & \frac{t^2}{2} \delta_{uu}^2 L(u_0, z_1^*, z_2^*; \varphi + t\psi_0(t)) + r_2(t) \geq 0, \forall |t| < \varepsilon_2.
\end{aligned}
$$

where

$$\lim_{t \to 0} |r_2(t)|/t^2 = 0.$$

To obtain the last inequality we have used

$$\delta_u L(u_0, z_1^*, z_2^*; t\varphi + t^2 \psi_0(t)) = 0$$

Dividing the last inequality by $t^2 > 0$ we obtain

$$\frac{1}{2} \delta_{uu}^2 L(u_0, z_1^*, z_2^*; \varphi + t\psi_0(t)) + r_2(t)/t^2 \geq 0, \forall 0 < |t| < \varepsilon_2,$$

and finally, letting $t \to 0$ we get

$$\frac{1}{2} \delta_{uu}^2 L(u_0, z_1^*, z_2^*; \varphi) \geq 0.$$

The proof is complete.

8.7 On the Banach fixed point theorem

Now we recall a classical definition namely the Banach fixed theorem also known as the contraction mapping theorem.

Definition 8.7.1 *Let C be a subset of a Banach space U and let $T : C \to C$ be an operator. Thus T is said to be a contraction mapping if there exists $0 \leq \alpha < 1$ such that*

$$\|T(u) - T(v)\|_U \leq \alpha \|u - v\|_U, \forall u, v \in C.$$

Remark 8.7.2 *Observe that if $\|T'(u)\|_U \leq \alpha < 1$, on a convex set C then T is a contraction mapping, since by the mean value inequality,*

$$\|T(u) - T(v)\|_U \leq \sup_{u \in C} \{\|T'(u)\|\} \|u - v\|_U, \forall u, v \in C.$$

The next result is the base of our generalized method of lines.

Theorem 8.7.3 (Contraction Mapping Theorem) *Let C be a closed subset of a Banach space U. Assume T is contraction mapping on C, then there exists a unique $u_0 \in C$ such that $u_0 = T(u_0)$. Moreover, for an arbitrary $u_1 \in C$ defining the sequence*

$$u_2 = T(u_1) \text{ and } u_{n+1} = T(u_n), \forall n \in \mathbb{N}$$

we have

$$u_n \to u_0, \text{ in norm, as } n \to +\infty.$$

Proof 8.12 Let $u_1 \in C$. Let $\{u_n\} \subset C$ be defined by

$$u_{n+1} = T(u_n), \forall n \in \mathbb{N}.$$

Hence, reasoning inductively

$$
\begin{aligned}
\|u_{n+1} - u_n\|_U &= \|T(u_n) - T(u_{n-1})\|_U \\
&\le \alpha \|u_n - u_{n-1}\|_U \\
&\le \alpha^2 \|u_{n-1} - u_{n-2}\|_U \\
&\le \ \ldots\ldots \\
&\le \alpha^{n-1} \|u_2 - u_1\|_U, \forall n \in \mathbb{N}.
\end{aligned}
\tag{8.56}
$$

Thus, for $p \in \mathbb{N}$ we have

$$
\begin{aligned}
&\|u_{n+p} - u_n\|_U \\
&= \|u_{n+p} - u_{n+p-1} + u_{n+p-1} - u_{n+p-2} + \ldots - u_{n+1} + u_{n+1} - u_n\|_U \\
&\le \|u_{n+p} - u_{n+p-1}\|_U + \|u_{n+p-1} - u_{n+p-2}\|_U + \ldots + \|u_{n+1} - u_n\|_U \\
&\le (\alpha^{n+p-2} + \alpha^{n+p-3} + \ldots + \alpha^{n-1}) \|u_2 - u_1\|_U \\
&\le \alpha^{n-1}(\alpha^{p-1} + \alpha^{p-2} + \ldots + \alpha^0) \|u_2 - u_1\|_U \\
&\le \alpha^{n-1} \left(\sum_{k=0}^{\infty} \alpha^k \right) \|u_2 - u_1\|_U \\
&\le \frac{\alpha^{n-1}}{1-\alpha} \|u_2 - u_1\|_U
\end{aligned}
\tag{8.57}
$$

Denoting $n + p = m$, we obtain

$$
\|u_m - u_n\|_U \le \frac{\alpha^{n-1}}{1-\alpha} \|u_2 - u_1\|_U, \forall m > n \in \mathbb{N}.
$$

Let $\varepsilon > 0$. Since $0 \le \alpha < 1$ there exists $n_0 \in \mathbb{N}$ such that if $n > n_0$ then

$$
0 \le \frac{\alpha^{n-1}}{1-\alpha} \|u_2 - u_1\|_U < \varepsilon,
$$

so that

$$
\|u_m - u_n\|_U < \varepsilon, \text{ if } m > n > n_0.
$$

From this we may infer that $\{u_n\}$ is a Cauchy sequence, and since U is a Banach space, there exists $u_0 \in U$ such that

$$
u_n \to u_0, \text{ in norm, as } n \to \infty.
$$

Observe that

$$
\begin{aligned}
\|u_0 - T(u_0)\|_U &= \|u_0 - u_n + u_n - T(u_0)\|_U \\
&\le \|u_0 - u_n\|_U + \|u_n - T(u_0)\|_U \\
&\le \|u_0 - u_n\|_U + \alpha \|u_{n-1} - u_0\|_U \\
&\to 0, \text{ as } n \to \infty.
\end{aligned}
\tag{8.58}
$$

Thus $\|u_0 - T(u_0)\|_U = 0$.

Finally, we prove the uniqueness. Suppose $u_0, v_0 \in C$ are such that

$$u_0 = T(u_0) \text{ and } v_0 = T(v_0).$$

Hence,

$$
\begin{aligned}
\|u_0 - v_0\|_U &= \|T(u_0) - T(v_0)\|_U \\
&\leq \alpha \|u_0 - v_0\|_U.
\end{aligned}
\tag{8.59}
$$

From this we get

$$\|u_0 - v_0\|_U \leq 0,$$

that is

$$\|u_0 - v_0\|_U = 0.$$

The proof is complete.

8.8 Sensitivity analysis

8.8.1 Introduction

In this section, we state and prove the implicit function theorem for Banach spaces. A similar result may be found in Ito and Kunisch [51], page 31.

We emphasize the result found in [51] is more general however, the proof present here is almost the same for a simpler situation. The general result found in [51] is originally from Robinson [61].

8.9 The implicit function theorem

Theorem 8.9.1 *Let V, U, W be Banach spaces. Let $F : V \times U \to W$ be a functions such that*

$$F(x_0, u_0) = \mathbf{0},$$

where $(x_0, u_0) \in V \times U$.

Assume there exists $r > 0$ such that F is Fréchet differentiable and $F_x(x, u)$ is continuous in (x, u) in $B_r(x_0, u_0)$.

Suppose also $[F_x(x_0, u_0)]^{-1}$ exists and is bounded so that there exists $\rho > 0$ such that

$$0 < \|[F_x(x_0, u_0)]^{-1}\| \leq \rho.$$

Under such hypotheses, there exist $0 < \varepsilon_1 < r/2$ and $0 < \varepsilon_2 < 1$ such that for each $u \in B_{\varepsilon_1}(u_0)$, there exists $x \in B_{\varepsilon_2}(x_0)$ such that

$$F(x, u) = \mathbf{0},$$

where we denote $x = x(u)$ so that

$$F(x(u), u) = \mathbf{0}.$$

Moreover, there exists $\delta > 0$ such that $0 < \delta\rho < 1$, such that for each $u, v \in B_{\varepsilon_1}(u_0)$ we have

$$\|x(u) - x(v)\| \leq \frac{\rho^2\delta}{1 - \rho\delta}\|F(x(v), u) - F(x(v), v)\|.$$

Finally, if there exists $K > 0$ such that

$$\|F_u(x, u)\| \leq K, \ \forall (x, u) \in B_{\varepsilon_2}(x_0) \times B_{\varepsilon_1}(u_0)$$

so that

$$\|F(x, u) - F(x, v)\| \leq K\|u - v\|, \ \forall (x, u) \in B_{\varepsilon_2}(x_0) \times B_{\varepsilon_1}(u_0),$$

then

$$\|x(u) - x(v)\| \leq K_1\|u - v\|,$$

where

$$K_1 = K\frac{\rho^2\delta}{1 - \delta\rho}.$$

Proof 8.13   Let $0 < \varepsilon < r/2$. Choose $\delta > 0$ such that

$$0 < \rho\delta < 1.$$

Define

$$T(x) = F(x_0, u_0) + F_x(x_0, u_0)(x - x_0) = F_x(x_0, u_0)(x - x_0),$$

and

$$h(x, u) = F(x_0, u_0) + F_x(x_0, u_0)(x - x_0) - F(x, u).$$

Choose $0 < \varepsilon_3 < r/2$ and $0 < \varepsilon_2 < 1$ such that

$$B_{\varepsilon_2}(u_0) \times B_{\varepsilon_3}(x_0) \subset B_r(x_0, u_0)$$

and if $(x, u) \in B_{\varepsilon_2}(x_0) \times B_{\varepsilon_3}(u_0)$ then

$$\|F_x(x, u) - F_x(x_0, u_0)\| < \frac{\delta}{2}.$$

Select $0 < \varepsilon_1 < \varepsilon_3$ such that if $u \in B_{\varepsilon_1}(u_0)$, then

$$\rho\|F(x_0, u) - F(x_0, u_0)\| < (1 - \rho\delta)\varepsilon_2.$$

For each $u \in B_{\varepsilon_1}(u_0)$ define

$$\phi_u(x) = T^{-1}(h(x, u)).$$

Fix $u \in B_{\varepsilon_1}(u_0)$.

Observe that for $x_1, x_2 \in B_{\varepsilon_2}(x_0)$ we have

$$
\begin{aligned}
\|\phi_u(x_1) - \phi_u(x_2)\| &\leq \|[F_x(x_0, u_0)]^{-1}\| \|h(x_1, u) - h(x_2, u)\| \\
&\leq \rho \|h(x_1, u) - h(x_2, u)\| \\
&\leq \rho \sup_{t \in [0,1]} \|h_x(tx_1 + (1-t)x_2, u)\| \|x_1 - x_2\| \\
&\leq \rho 2 \frac{\delta}{2} \|x_1 - x_2\| \\
&= \rho \delta \|x_1 - x_2\|.
\end{aligned}
\tag{8.60}
$$

Observe that since $0 < \rho \delta < 1$ we have that ϕ_u is a candidate to be a contraction mapping.

Observe also that

$$
x_0 = T^{-1}(\mathbf{0}),
$$

so that

$$
T(x_0) = \mathbf{0} = h(x_0, u_0)
$$

and

$$
x_0 = T^{-1}(h(x_0, u_0)).
$$

Thus,

$$
\begin{aligned}
\|\phi_u(x_0) - x_0\| &\leq \rho \|h(x_0, u) - h(x_0, u_0)\| \\
&= \rho \|h(x_0, u)\| \\
&\leq \rho \|F(x_0, u) - F(x_0, u_0)\| \\
&\leq (1 - \rho \delta) \varepsilon_2.
\end{aligned}
\tag{8.61}
$$

On the other hand, for each $x \in B_{\varepsilon_2}(x_0)$, we have that

$$
\begin{aligned}
\|\phi_u(x) - x_0\| &= \|\phi_u(x) - \phi_u(x_0) + \phi_u(x_0) - x_0\| \\
&\leq \|\phi_u(x) - \phi_u(x_0)\| + \|\phi_u(x_0) - x_0\| \\
&\leq \rho \delta \|x - x_0\| + (1 - \rho \delta) \varepsilon_2 \\
&< \rho \delta \varepsilon_2 + (1 - \rho \delta) \varepsilon_2 \\
&= \varepsilon_2.
\end{aligned}
\tag{8.62}
$$

From this we may infer that

$$
\phi_u(x) \in B_{\varepsilon_2}(x_0), \ \forall x \in B_{\varepsilon_2}(x_0)
$$

so that indeed ϕ_u is a contraction mapping.

Therefore, from the Banach fixed point theorem, there exists a unique fixed point $x = x(u)$ for $\phi_u(x)$, so that

$$
\begin{aligned}
x(u) &= \phi_u(x(u)) \\
&= T^{-1}(h(x(u), u)) \\
&= T^{-1}(F_x(x_0, u_0)(x(u) - x_0) - F(x(u, u))) \\
&= [F_x(x_0, u_0)]^{-1}[F_x(x_0, u_0)(x(u) - x_0) - F(x(u), u)] + x_0 \\
&= x(u) - x_0 + x_0 - [F_x(x_0, u_0)]^{-1}(F(x(u), u)) \\
&= x(u) - [F_x(x_0, u_0)]^{-1}(F(x(u), u)).
\end{aligned}
\tag{8.63}
$$

From this, we have

$$
[F_x(x_0, u_0)]^{-1}(F(x(u), u)) = \mathbf{0},
$$

so that

$$
F(x(u), u) = F_x(x_0, u_0)\mathbf{0} = \mathbf{0},
$$

that is,

$$
F(x(u), u) = \mathbf{0}.
$$

Let $x \in B_{\varepsilon_2}(x_0)$ and $u \in B_{\varepsilon_1}(u_0)$.
Thus

$$
\|\phi_u(x) - x\| \le \|\phi_u(x)\| + \|x\| < 2\varepsilon_2.
$$

Moreover,

$$
\|\phi_u(x_1) - \phi_u(x_2)\| < \rho\delta\|x_2 - x_1\|, \forall x_1, x_2 \in B_{\varepsilon_2}(x_0).
$$

From these last two results we obtain

$$
\|\phi_u^2(x) - \phi_u(x)\| < 2\varepsilon_2(\rho\delta),
$$

and reasoning inductively,

$$
\|\phi_u^{n+1}(x) - \phi_u^n(x)\| < 2\varepsilon_2(\rho\delta)^n, \ \forall n \in \mathbb{N}.
$$

Therefore

$$
\begin{aligned}
\|\phi_u^{n+1}(x) - x\| &= \|\phi_u^{n+1}(x) - \phi_u^n(x) + \phi_u^n(x) - \cdots + \phi_u(x) - x\| \\
&\le \|\phi_u^{n+1}(x) - \phi_u^n(x)\| + \|\phi_u^n(x) - \phi_u^{n-1}(x)\| + \cdots + \|\phi_u(x) - x\| \\
&\le \sum_{j=1}^{n+1}(\rho\delta)^j\|\phi_u(x) - x\| \\
&\le \frac{\rho\delta}{1 - \rho\delta}\|\phi_u(x) - x\|, \ \forall n \in \mathbb{N}
\end{aligned}
\tag{8.64}
$$

Letting $n \to \infty$, we obtain

$$
\|x(u) - x\| \le \frac{\rho\delta}{1 - \rho\delta}\|\phi_u(x) - x\|.
$$

In particular for $v \in B_{\varepsilon_1}(u_0)$ and $x = x(v)$, we get

$$
\begin{aligned}
\|x(u) - x(v)\| &\leq \frac{\rho\delta}{1 - \rho\delta} \|\phi_u(x(v)) - x(v)\| \\
&= \frac{\rho\delta}{1 - \rho\delta} \|\phi_u(x(v)) - \phi_v(x(v))\| \\
&\leq \frac{\rho^2\delta}{1 - \rho\delta} \|h(x(v), u) - h(x(v), v)\| \\
&\leq \frac{\rho^2\delta}{1 - \rho\delta} \|F(x(v), u) - F(x(v), v)\| \\
&\leq K\frac{\rho^2\delta}{1 - \rho\delta} \|u - v\| \\
&= K_1 \|u - v\|, \quad \forall u, v \in B_{\varepsilon_1}(u_0).
\end{aligned}
\tag{8.65}
$$

The proof is complete.

Corollary 8.9.2 *Consider the hypotheses and statements of the last theorem. More-over, assume $F_x : V \times U \to W$ is such that $[F_x(x,u)]^{-1}$ exists and it is bounded in $B_r(x_0, u_0)$.*

Suppose also, F is Fréchet differentiable in $B_r(x_0, u_0)$.

Let $\varphi \in U$.

Under such hypotheses,

$$
x'(u, \varphi) = -[F_x(x(u), u)]^{-1} [F_u(x(u), u)](\varphi),
$$

where

$$
x'(u, \varphi) = \lim_{t \to 0} \frac{x(u + t\varphi) - x(u)}{t}.
$$

Proof 8.14   Observe that

$$
F(x(u), u) = \mathbf{0}, \text{ in } B_{\varepsilon_1}(u_0).
$$

Let $u \in B_{\varepsilon_1}(u_0)$.

Let $t_0 > 0$ be such that

$$
u + t\varphi \in B_{\varepsilon_1}(u_0), \forall |t| < t_0.
$$

Observe that

$$
F(x(u + t\varphi), u + t\varphi) - F(x(u), u) = \mathbf{0}, \forall |t| < t_0.
$$

From the Fréchet differentiability of F at $(x(u), u)$, for $0 < |t| < t_0$, we obtain

$$
\begin{aligned}
&F_x(x(u), u) \cdot (x(u + t\varphi) - x(u)) + F_u(x(u), u)(t\varphi) \\
&+ W(u, \varphi, t)(\|x(u + t\varphi) - x(u)\| + |t| \|\varphi\|) = \mathbf{0},
\end{aligned}
\tag{8.66}
$$

where W is such that

$$W(u,\varphi,t) \to \mathbf{0}, \text{ as } t \to 0,$$

since

$$x(u+t\varphi) - x(u) \to \mathbf{0}$$

and

$$t\varphi \to \mathbf{0}, \text{ as } t \to 0.$$

From this we obtain

$$\frac{x(u+t\varphi) - x(u)}{t} = -[F_x(x(u),u)]^{-1}[[F_u(x(u),u)](\varphi) + \mathbf{r}(u,\varphi,t)]$$

$$\to -[F_x(x(u),u)]^{-1}[F_u(x(u),u)](\varphi), \text{ as } t \to 0, \quad (8.67)$$

since

$$\|\mathbf{r}(u,\varphi,t)\| \leq \|W(u,\varphi,t)\| \left| \frac{\|x(u+t\varphi) - x(u)\|}{t} + \|\varphi\| \right|$$

$$\leq \|W(u,\varphi,t)\|(K_1\|\varphi\| + \|\varphi\|)$$

$$\to 0, \text{ as } t \to 0. \quad (8.68)$$

Summarizing,

$$x'(u,\varphi) = \lim_{t \to 0} \frac{x(u+t\varphi) - x(u)}{t}$$

$$= -[F_x(x(u),u)]^{-1}[F_u(x(u),u)](\varphi). \quad (8.69)$$

The proof is complete.

8.9.1   The main results about Gâteaux differentiability

Again let V, U be Banach spaces and let $F : V \times U \to \mathbb{R}$ be a functional. Fix $u \in U$ and consider the problem of minimizing $F(x,u)$ subject to $G(x,u) \leq \theta$ and $H(x,u) = \theta$. Here, the order and remaining details on the primal formulation are the same as those indicated in Section 8.4.

Hence, for the specific case in which

$$G : V \times U \to [L^p(\Omega)]^{m_1}$$

and

$$H : V \times U \to [L^p(\Omega)]^{m_2},$$

(the cases in which the co-domains of G and H are $\mathbb{R}^{m_1}$ and $\mathbb{R}^{m_2}$ respectively are dealt similarly) we redefine the concerned optimization problem, again for a fixed $u \in U$, by minimizing $F(x,u)$ subject to

$$\{G_i(x,u) + v_i^2\} = \theta,$$

and

$$H(x,u) = \theta.$$

At this point we assume $F(x,u)$, $\tilde{G}(x,u,v) = \{G_i(x,u) + v_i^2\} \equiv G(u) + v^2$ (from now on we use this general notation) and $H(x,u)$ satisfy the hypotheses of the Lagrange multiplier theorem 9.2.1.

Hence, for the fixed $u \in U$ we assume there exists an optimal $x \in V$ which locally minimizes $F(x,u)$ under the mentioned constraints.

From Theorem 9.2.1 there exist Lagrange multipliers λ_1, λ_2 such that denoting $[L^p(\Omega)]^{m_1}$ and $[L^p(\Omega)]^{m_2}$ simply by L^p, and defining

$$\tilde{F}(x,u,\lambda_1,\lambda_2,v) = F(x,u) + \langle \lambda_1, G(u) + v^2 \rangle_{L^p} + \langle \lambda_2, H(x,u) \rangle_{L^p},$$

the following necessary conditions hold,

$$\tilde{F}_x(x,u) = F_x(x,u) + \lambda_1 \cdot G_x(x,u) + \lambda_2 \cdot H_x(x,u) = \theta, \tag{8.70}$$

$$G(x,u) + v^2 = \theta, \tag{8.71}$$

$$\lambda_1 \cdot v = \theta, \tag{8.72}$$

$$\lambda_1 \geq \theta, \tag{8.73}$$

$$H(x,u) = \theta. \tag{8.74}$$

Clarifying the dependence on u, we denote the solution $x, \lambda_1, \lambda_2, v$ by $x(u)$, $\lambda_1(u)$, $\lambda_2(u)$, $v(u)$, respectively. In particular, we assume that for a $u_0 \in U$, $x(u_0), \lambda_1(u_0), \lambda_2(u_0), v(u_0)$ satisfy the hypotheses of the implicit function theorem. Thus, for any u in an appropriate neighborhood of u_0, the corresponding $x(u), \lambda_1(u), \lambda_2(u), v(u)$ are uniquely defined.

We emphasize that from now on the main focus of our analysis is to evaluate variations of the optimal $x(u), \lambda_1(u), \lambda_2(u), v(u)$ with variations of u in a neighborhood of u_0.

For such an analysis, the main tool is the implicit function theorem and its main hypothesis is satisfied through the invertibility of the matrix of Fréchet second derivatives.

Hence, denoting, $x_0 = x(u_0), (\lambda_1)_0 = \lambda_1(u_0), (\lambda_2)_0 = \lambda_2(u_0), v_0 = v(u_0)$, and

$$A_1 = F_x(x_0, u_0) + (\lambda_1)_0 \cdot G_x(x_0, u_0) + (\lambda_2)_0 \cdot H_x(x_0, u_0),$$

$$A_2 = G(x_0, u_0) + v_0^2$$

$$A_3 = H(x_0, u_0),$$

$$A_4 = (\lambda_1)_0 \cdot v_0,$$

we reiterate to assume that

$$A_1 = \theta, A_2 = \theta, A_3 = \theta, A_4 = \theta,$$

and M^{-1} to represent a bounded linear operator, where

$$
M = \begin{bmatrix}
(A_1)_x & (A_1)_{\lambda_1} & (A_1)_{\lambda_2} & (A_1)_v \\
(A_2)_x & (A_2)_{\lambda_1} & (A_2)_{\lambda_2} & (A_2)_v \\
(A_3)_x & (A_3)_{\lambda_1} & (A_3)_{\lambda_2} & (A_3)_v \\
(A_4)_x & (A_4)_{\lambda_1} & (A_4)_{\lambda_2} & (A_4)_v
\end{bmatrix}
\tag{8.75}
$$

where the derivatives are evaluated at $(x_0, u_0, (\lambda_1)_0, (\lambda_2)_0, v_0)$, so that,

$$
M = \begin{bmatrix}
A & G_x(x_0, u_0) & H_x(x_0, u_0) & \theta \\
G_x(x_0, u_0) & \theta & \theta & 2v_0 \\
H_x(x_0, u_0) & \theta & \theta & \theta \\
\theta & v_0 & \theta & (\lambda_1)_0
\end{bmatrix}
\tag{8.76}
$$

where

$$
A = F_{xx}(x_0, u_0) + (\lambda_1)_0 \cdot G_{xx}(x_0, u_0) + (\lambda_2)_0 \cdot H_{xx}(x_0, u_0).
$$

Moreover, also from the implicit function theorem,

$$
\|(x(u), \lambda_1(u), \lambda_2(u), v(u)) - (x(u_0), \lambda_1(u_0), \lambda_2(u_0), v(u_0))\| \le K\|u - u_0\|, \tag{8.77}
$$

for some appropriate $K > 0$, $\forall u \in B_r(u_0)$, for some $r > 0$.

We highlight to have denoted $\lambda(u) = (\lambda_1(u), \lambda_2(u))$.

Let $\varphi \in [C^\infty(\Omega)]^k \cap U$, where k depends on the vectorial expression of U.

At this point we will be concerned with the following Gâteaux variation evaluation

$$
\delta_u \tilde{F}(x(u_0), u_0, \lambda(u_0), v(u_0); \varphi).
$$

Observe that

$$
\delta_u \tilde{F}(x(u_0), u_0, \lambda(u_0), v(u_0); \varphi) =
\lim_{\varepsilon \to 0} \left\{ \frac{\tilde{F}(x(u_0 + \varepsilon\varphi), u_0 + \varepsilon\varphi, \lambda(u_0 + \varepsilon\varphi), v(u_0 + \varepsilon\varphi))}{\varepsilon} \right.
$$
$$
\left. - \frac{\tilde{F}(x(u_0), u_0, \lambda(u_0), v(u_0))}{\varepsilon} \right\},
$$

so that

$$
\delta_u \tilde{F}(x(u_0), u_0, \lambda(u_0), v(u_0); \varphi) =
\lim_{\varepsilon \to 0} \left\{ \frac{\tilde{F}(x(u_0 + \varepsilon\varphi), u_0 + \varepsilon\varphi, \lambda(u_0 + \varepsilon\varphi), v(u_0 + \varepsilon\varphi))}{\varepsilon} \right.
$$
$$
- \frac{\tilde{F}(x(u_0), u_0 + \varepsilon\varphi, \lambda(u_0 + \varepsilon\varphi), v(u_0 + \varepsilon\varphi))}{\varepsilon}
$$
$$
+ \frac{\tilde{F}(x(u_0), u_0 + \varepsilon\varphi, \lambda(u_0 + \varepsilon\varphi), v(u_0 + \varepsilon\varphi))}{\varepsilon}
$$
$$
\left. - \frac{\tilde{F}(x(u_0), u_0, \lambda(u_0), v(u_0))}{\varepsilon} \right\}.
$$

However,

$$\left| \frac{\tilde{F}(x(u_0+\varepsilon\varphi),u_0+\varepsilon\varphi,\lambda(u_0+\varepsilon\varphi),v(u_0+\varepsilon\varphi))}{\varepsilon} \right.$$
$$\left. - \frac{\tilde{F}(x(u_0),u_0+\varepsilon\varphi,\lambda(u_0+\varepsilon\varphi),v(u_0+\varepsilon\varphi))}{\varepsilon} \right|$$
$$\leq \ \|\tilde{F}_x(x(u_0+\varepsilon\varphi),u_0+\varepsilon\varphi,\lambda(u_0+\varepsilon\varphi),v(u_0+\varepsilon\varphi))\| \, K \|\varphi\|$$
$$+ K_1 \|x(u_0+\varepsilon\varphi)-x(u_0)\|$$
$$\leq \ K_1 K \|\varphi\| \varepsilon$$
$$\rightarrow \ 0, \text{ as } \varepsilon \rightarrow 0.$$

In these last inequalities we have used

$$\limsup_{\varepsilon\to 0} \left\| \frac{x(u_0+\varepsilon\varphi)-x(u_0)}{\varepsilon} \right\| \leq K \|\varphi\|,$$

and

$$\tilde{F}_x(x(u_0+\varepsilon\varphi),u_0+\varepsilon\varphi,\lambda(u_0+\varepsilon\varphi),v(u_0+\varepsilon\varphi)) = \theta.$$

On the other hand,

$$\left\{ \frac{\tilde{F}(x(u_0),u_0+\varepsilon\varphi,\lambda(u_0+\varepsilon\varphi),v(u_0+\varepsilon\varphi))}{\varepsilon} \right.$$
$$\left. - \frac{\tilde{F}(x(u_0),u_0,\lambda(u_0),v(u_0))}{\varepsilon} \right\}$$
$$= \ \left\{ \frac{\tilde{F}(x(u_0),u_0+\varepsilon\varphi,\lambda(u_0+\varepsilon\varphi),v(u_0+\varepsilon\varphi))}{\varepsilon} \right.$$
$$- \frac{\tilde{F}(x(u_0),u_0+\varepsilon\varphi,\lambda(u_0),v(u_0))}{\varepsilon}$$
$$+ \frac{\tilde{F}(x(u_0),u_0+\varepsilon\varphi,\lambda(u_0),v(u_0))}{\varepsilon}$$
$$\left. - \frac{\tilde{F}(x(u_0),u_0,\lambda(u_0),v(u_0))}{\varepsilon} \right\}$$

Now observe that

$$\frac{\tilde{F}(x(u_0),u_0+\varepsilon\varphi,\lambda(u_0+\varepsilon\varphi),v(u_0+\varepsilon\varphi))}{\varepsilon}$$
$$- \frac{\tilde{F}(x(u_0),u_0+\varepsilon\varphi,\lambda(u_0),v(u_0))}{\varepsilon}$$
$$= \ \frac{\langle \lambda_1(u_0+\varepsilon\varphi),G(x(u_0),u_0+\varepsilon\varphi)+v(u_0+\varepsilon\varphi)^2 \rangle_{L^p}}{\varepsilon}$$
$$- \frac{\langle \lambda_1(u_0),G(x(u_0),u_0+\varepsilon\varphi)+v(u_0)^2 \rangle_{L^p}}{\varepsilon}$$
$$+ \frac{\langle \lambda_2(u_0+\varepsilon\varphi)-\lambda_2(u_0),H(x(u_0),u_0+\varepsilon\varphi) \rangle_{L^p}}{\varepsilon}. \tag{8.78}$$

Also,

$$
\begin{aligned}
&\left| \frac{\langle \lambda_1(u_0+\varepsilon\varphi), G(x(u_0),u_0+\varepsilon\varphi)+v(u_0+\varepsilon\varphi)^2 \rangle_{L^p}}{\varepsilon} \right. \\
&\left. -\frac{\langle \lambda_1(u_0), G(x(u_0),u_0+\varepsilon\varphi)+v(u_0)^2 \rangle_{L^p}}{\varepsilon} \right| \\[2mm]
\leq\ &\left| \frac{\langle \lambda_1(u_0+\varepsilon\varphi), G(x(u_0),u_0+\varepsilon\varphi)+v(u_0+\varepsilon\varphi)^2 \rangle_{L^p}}{\varepsilon} \right. \\
&\left. -\frac{\langle \lambda_1(u_0), G(x(u_0),u_0+\varepsilon\varphi)+v(u_0+\varepsilon\varphi)^2 \rangle_{L^p}}{\varepsilon} \right| \\[2mm]
+\ &\left| \frac{\langle \lambda_1(u_0), G(x(u_0),u_0+\varepsilon\varphi)+v(u_0+\varepsilon\varphi)^2 \rangle_{L^p}}{\varepsilon} \right. \\
&\left. -\frac{\langle \lambda_1(u_0), G(x(u_0),u_0+\varepsilon\varphi)+v(u_0)^2 \rangle_{L^p}}{\varepsilon} \right| \\[2mm]
\leq\ &\varepsilon\frac{K\|\varphi\|}{\varepsilon}\|G(x(u_0),u_0+\varepsilon\varphi)+v(u_0+\varepsilon\varphi)^2\| \\[2mm]
+\ &\|\lambda_1(u_0)(v(u_0+\varepsilon\varphi)+v(u_0))\|\frac{K\|\varphi\|\varepsilon}{\varepsilon} \\[2mm]
\to\ &0 \text{ as } \varepsilon \to 0.
\end{aligned}
$$

To obtain the last inequalities we have used

$$
\limsup_{\varepsilon\to 0}\left\|\frac{\lambda_1(u_0+\varepsilon\varphi)-\lambda_1(u_0)}{\varepsilon}\right\| \leq K\|\varphi\|,
$$

$$
\lambda_1(u_0)v(u_0)=\theta,
$$

$$
\lambda_1(u_0)v(u_0+\varepsilon\varphi)\to\theta, \text{ as } \varepsilon\to 0,
$$

and

$$
\begin{aligned}
&\left\|\frac{\lambda_1(u_0)(v(u_0+\varepsilon\varphi)^2-v(u_0)^2)}{\varepsilon}\right\| \\[2mm]
=\ &\left\|\frac{\lambda_1(u_0)(v(u_0+\varepsilon\varphi)+v(u_0))(v(u_0+\varepsilon\varphi)-v(u_0))}{\varepsilon}\right\| \\[2mm]
\leq\ &\frac{\|\lambda_1(u_0)(v(u_0+\varepsilon\varphi)+v(u_0))\|K\|\varphi\|\varepsilon}{\varepsilon} \\[2mm]
\to\ &0, \text{ as } \varepsilon\to 0. \tag{8.79}
\end{aligned}
$$

Finally,

$$\left| \frac{\langle \lambda_2(u_0 + \varepsilon \varphi) - \lambda_2(u_0), H(x(u_0), u_0 + \varepsilon \varphi) \rangle_{L^p}}{\varepsilon} \right|$$

$$\leq \quad \frac{K\varepsilon \|\varphi\|}{\varepsilon} \|H(x(u_0), u_0 + \varepsilon \varphi)\|$$

$$\to \quad 0, \text{ as } \varepsilon \to 0.$$

To obtain the last inequalities we have used

$$\limsup_{\varepsilon \to 0} \left\| \frac{\lambda_2(u_0 + \varepsilon \varphi) - \lambda_2(u_0)}{\varepsilon} \right\| \leq K \|\varphi\|,$$

and

$$H(x(u_0), u_0 + \varepsilon \varphi) \to \theta, \text{ as } \varepsilon \to 0.$$

From these last results, we get

$$\delta_u \tilde{F}(x(u_0), u_0, \lambda(u_0), v(u_0); \varphi)$$

$$= \quad \lim_{\varepsilon \to 0} \left\{ \frac{\tilde{F}(x(u_0), u_0 + \varepsilon \varphi, \lambda(u_0), v(u_0))}{\varepsilon} \right.$$

$$\left. - \frac{\tilde{F}(x(u_0), u_0, \lambda(u_0), v(u_0))}{\varepsilon} \right\}$$

$$= \quad \langle F_u(x(u_0), u_0), \varphi \rangle_U + \langle \lambda_1(u_0) \cdot G_u(x(u_0), u_0), \varphi \rangle_{L^p}$$

$$+ \langle \lambda_2(u_0) \cdot H_u(x(u_0), u_0), \varphi \rangle_{L^p}.$$

In the last lines we have proven the following corollary of the implicit function theorem,

Corollary 8.1

Suppose $(x_0, u_0, (\lambda_1)_0, (\lambda_2)_0, v_0)$ is a solution of the system (8.70), (8.71),(8.72), (8.74), and assume the corresponding hypotheses of the implicit function theorem are satisfied. Also assume $\tilde{F}(x, u, \lambda_1, \lambda_2, v)$ is such that the Fréchet second derivative $\tilde{F}_{xx}(x, u, \lambda_1, \lambda_2)$ is continuous in a neighborhood of

$$(x_0, u_0, (\lambda_1)_0, (\lambda_2)_0).$$

Under such hypotheses, for a given $\varphi \in [C^\infty(\Omega)]^k$, *denoting*

$$F_1(u) = \tilde{F}(x(u), u, \lambda_1(u), \lambda_2(u), v(u)),$$

we have

$$\delta(F_1(u); \varphi)|_{u=u_0}$$

$$=$$

$$\langle F_u(x(u_0), u_0), \varphi \rangle_U + \langle \lambda_1(u_0) \cdot G_u(x(u_0), u_0), \varphi \rangle_{L^p}$$
$$+ \langle \lambda_2(u_0) \cdot H_u(x(u_0), u_0), \varphi \rangle_{L^p}.$$

Chapter 9

On Lagrange Multiplier Theorems for Non-Smooth Optimization for a Large Class of Variational Models in Banach Spaces

9.1 Introduction

This article develops optimality conditions for a large class of non-smooth variational models. The main results are based on standard tools of functional analysis and calculus of variations. Firstly, we address a model with equality constraints and, in a second step, a more general model with equality and inequality constraints, always in a general Banach space context. We highlight the results which in general are well known, however, some novelties related to the proof procedures are introduced, which are in general softer than those concerning the present literature.

Remark 9.1.1 *This chapter has been accepted for publication in a similar article format by the Universal Wiser Journal Contemporary Mathematics, reference [19].*

DOI: https://doi.org/10.37256/cm.3420221711 This is an open-access article distributed under a CC BY license (Creative Commons Attribution 4.0 International License) https://creativecommons.org/licenses/by/4.0/.

The main references for this article are [79, 12, 13]. Indeed, the results presented here are, in some sense, extensions of the previous ones found in F. Clarke [79].

We also highlight specific details on the function spaces addressed, and concerning functional analysis and Lagrange multiplier basic results may be found in [1, 77, 26, 24, 79, 78, 12, 13].

Related subjects are addressed in [80, 76]. Specifically in [80], the authors propose an augmented Lagrangian method for the solution of constrained optimization problems suitable for a large class of variational models.

At this point, we highlight that the main novelties mentioned in the abstract are specified in the first three paragraphs of Section 12, and are applied in the statements and proofs of Theorems 9.2.1 and 9.3.1.

Finally, fundamental results on the calculus of variations are addressed in [82].

We start with some preliminary results and basic definitions. The first result we present is the Hahn-Banach Theorem in its analytic form. Concerning our context, we have assumed the hypothesis that the space U is a Banach space but indeed such a result is much more general.

Theorem 9.1.2 (The Hahn-Banach theorem) *Let U be a Banach space. Consider a functional $p : U \to \mathbb{R}$ such that*

$$p(\lambda u) = \lambda p(u), \forall u \in U, \lambda > 0, \qquad (9.1)$$

and

$$p(u+v) \leq p(u) + p(v), \forall u, v \in U. \qquad (9.2)$$

Let $V \subset U$ be a proper subspace of U and let $g : V \to \mathbb{R}$ be a linear functional such that

$$g(u) \leq p(u), \forall u \in V. \qquad (9.3)$$

Under such hypotheses, there exists a linear functional $f : U \to \mathbb{R}$ such that

$$g(u) = f(u), \forall u \in V, \qquad (9.4)$$

and

$$f(u) \leq p(u), \forall u \in U. \qquad (9.5)$$

For a proof, please see [26, 12, 13].

Here we introduce the definition of topological dual space.

Definition 9.1.3 (Topological dual spaces) *Let U be a Banach space. We shall define its dual topological space, as the set of all linear continuous functionals defined* *on U. We suppose such a dual space of U, may be represented by another Banach space U^*, through a bilinear form $\langle \cdot, \cdot \rangle_U : U \times U^* \to \mathbb{R}$ (here we are referring to standard representations of dual spaces of Sobolev and Lebesgue spaces). Thus, given $f : U \to \mathbb{R}$ linear and continuous, we assume the existence of a unique $u^* \in U^*$ such that*

$$f(u) = \langle u, u^* \rangle_U, \forall u \in U. \tag{9.6}$$

The norm of f, denoted by $\|f\|_{U^*}$, is defined as

$$\|f\|_{U^*} = \sup_{u \in U} \{ |\langle u, u^* \rangle_U| : \|u\|_U \leq 1 \} \equiv \|u^*\|_{U^*}. \tag{9.7}$$

At this point we present the Hahn-Banach Theorem in its geometric form.

Theorem 9.1.4 (The Hahn-Banach theorem, the geometric form) *Let U be a Banach space and let $A, B \subset U$ be two non-empty, convex sets such that $A \cap B = \emptyset$ and A is open. Under such hypotheses, there exists a closed hyperplane which separates A and B, that is, there exist $\alpha \in \mathbb{R}$ and $u^* \in U^*$ such that $u^* \neq \mathbf{0}$ and*

$$\langle u, u^* \rangle_U \leq \alpha \leq \langle v, u^* \rangle_U, \forall u \in A, v \in B.$$

For a proof, please see [26, 12, 13].

Another important definition, is the one concerning locally Lipschitz functionals.

Definition 9.1.5 *Let U be a Banach space and let $F : U \to \mathbb{R}$ be a functional. We say that F is locally Lipschitz at $u_0 \in U$ if there exist $r > 0$ and $K > 0$ such that*

$$|F(u) - F(v)| \leq K \|u - v\|_U, \forall u, v \in B_r(u_0).$$

In this definition, we have denoted

$$B_r(u_0) = \{ v \in U : \|u_0 - v\|_U < r \}.$$

The next definition is established as those found in the reference [79]. More specifically, such a next one, corresponds to the definition of generalized directional derivative found in Section 10.1 at page 194, in reference [79].

Definition 9.1.6 *Let U be a Banach space and let $F : U \to \mathbb{R}$ be a locally Lipschitz functional at $u \in U$. Let $\varphi \in U$. Under such statements, we define*

$$H_u(\varphi) = \sup_{(\{u_n\}, \{t_n\}) \subset U \times \mathbb{R}^+} \left\{ \limsup_{n \to \infty} \frac{F(u_n + t_n \varphi) - F(u_n)}{t_n} : u_n \to u \text{ in } U, t_n \to 0^+ \right\}.$$

We also define the generalized local sub-gradient set of F at u, denoted by $\partial^0 F(u)$, by

$$\partial^0 F(u) = \{ u^* \in U^* : \langle \varphi, u^* \rangle_U \leq H_u(\varphi), \forall \varphi \in U \}.$$

We also highlight such a last definition of generalized local sub-gradient is similar as the definition of generalized gradient, which may be found in Section 10.13, at page 196, in the book [79].

In the next lines we prove some relevant auxiliary results.

Proposition 9.1.7 *Considering the context of the last two definitions, we have*

1.
$$H_u(\varphi_1 + \varphi_2) \le H_u(\varphi_1) + H_u(\varphi_2), \ \forall \varphi_1, \varphi_2 \in U.$$

2.
$$H_u(\lambda \varphi) = \lambda H_u(\varphi), \ \forall \lambda > 0, \ \varphi \in U.$$

Proof 9.1 Let $\varphi_1, \varphi_2 \in U$.
Observe that

$$
\begin{aligned}
&H_u(\varphi_1 + \varphi_2) \\
&= \sup_{(\{u_n\},\{t_n\}) \subset U \times \mathbb{R}^+} \left\{ \limsup_{n \to \infty} \frac{F(u_n + t_n(\varphi_1 + \varphi_2)) - F(u_n)}{t_n} : u_n \to u \text{ in } U, \ t_n \to 0^+ \right\} \\
&= \sup_{(\{u_n\},\{t_n\}) \subset U \times \mathbb{R}^+} \left\{ \limsup_{n \to \infty} \frac{F(u_n + t_n(\varphi_1 + \varphi_2) - F(u_u + t_n \varphi_2) + F(u_n + t_n \varphi_2)) - F(u_n)}{t_n} \right. \\
&\qquad\qquad\qquad : u_n \to u \text{ in } U, \ t_n \to 0^+ \Big\} \\
&\le \sup_{(\{v_n\},\{t_n\}) \subset U \times \mathbb{R}^+} \left\{ \limsup_{n \to \infty} \frac{F(v_n + t_n \varphi_1) - F(v_n)}{t_n} : v_n \to u \text{ in } U, \ t_n \to 0^+ \right\} \\
&\quad + \sup_{(\{u_n\},\{t_n\}) \subset U \times \mathbb{R}^+} \left\{ \limsup_{n \to \infty} \frac{F(u_n + t_n \varphi_2) - F(u_n)}{t_n} : u_n \to u \text{ in } U, \ t_n \to 0^+ \right\} \\
&= H_u(\varphi_1) + H_u(\varphi_2).
\end{aligned}
\tag{9.8}
$$

Let $\varphi \in U$ and $\lambda > 0$.
Thus,

$$
\begin{aligned}
&H_u(\lambda \varphi) \\
&= \sup_{(\{u_n\},\{t_n\}) \subset U \times \mathbb{R}^+} \left\{ \limsup_{n \to \infty} \frac{F(u_n + t_n(\lambda \varphi)) - F(u_n)}{t_n} : u_n \to u \text{ in } U, \ t_n \to 0^+ \right\} \\
&= \lambda \sup_{(\{u_n\},\{t_n\}) \subset U \times \mathbb{R}^+} \left\{ \limsup_{n \to \infty} \frac{F(u_n + t_n(\lambda \varphi)) - F(u_n)}{\lambda t_n} : u_n \to u \text{ in } U, \ t_n \to 0^+ \right\} \\
&= \lambda \sup_{(\{u_n\},\{\hat{t}_n\}) \subset U \times \mathbb{R}^+} \left\{ \limsup_{n \to \infty} \frac{F(u_n + \hat{t}_n(\varphi)) - F(u_n)}{\hat{t}_n} : u_n \to u \text{ in } U, \ \hat{t}_n \to 0^+ \right\} \\
&= \lambda H_u(\varphi).
\end{aligned}
\tag{9.9}
$$

The proof is complete.

9.2 The Lagrange multiplier theorem for equality constraints and non-smooth optimization

In this section, we state and prove a Lagrange multiplier theorem for non-smooth optimization. This first one is related to equality constraints.

Here we refer to a related result in the Theorem 10.45 at page 220, in the book [79]. We emphasize that in such a result, in this mentioned book, the author assumes the function which defines the constraints to be continuously differentiable in a neighborhood of the point in question.

In our next result, we do not assume such a hypothesis. Indeed, our hypotheses are different and in some sense weaker. More specifically, we assume the continuity of the Frechét derivative $G'(u)$ of a concerning constraint $G(u)$ only at the optimal point u_0 and not necessarily in a neighborhood, as properly indicated in the next lines.

Theorem 9.2.1 *Let U and Z be Banach spaces. Assume u_0 is a local minimum of $F(u)$ subject to $G(u) = \theta$, where $F : U \to \mathbb{R}$ is locally Lipschitz at u_0 and $G : U \to Z$ is a Frechét differentiable transformation such that $G'(u_0)$ maps U onto Z. Finally, assume there exist $\alpha > 0$ and $K > 0$ such that if $\|\varphi\|_U < \alpha$ then,*

$$\|G'(u_0 + \varphi) - G'(u_0)\| \leq K\|\varphi\|_U.$$

Under such assumptions, there exists $z_0^* \in Z^*$ such that

$$\theta \in \partial^0 F(u_0) + (G'(u_0)^*)(z_0^*),$$

that is, there exist $u^* \in \partial^0 F(u_0)$ and $z_0^* \in Z^*$ such that

$$u^* + [G'(u_0)]^*(z_0^*) = \theta,$$

so that,

$$\langle \varphi, u^* \rangle_U + \langle G'(u_0)\varphi, z_0^* \rangle_Z = 0, \forall \varphi \in U.$$

Proof 9.2 Let $\varphi \in U$ be such that

$$G'(u_0)\varphi = \theta.$$

From the proof of Theorem 11.3.2 at page 292, in [12], there exist $\varepsilon_0 > 0$, $K_1 > 0$ and

$$\{\psi_0(t), \, 0 < |t| < \varepsilon_0\} \subset U$$

such that

$$\|\psi_0(t)\|_U \leq K_1, \, \forall 0 < |t| < \varepsilon_0,$$

and

$$G(u_0 + t\varphi + t^2 \psi_0(t)) = \theta, \, \forall 0 < |t| < \varepsilon_0.$$

From this and the hypotheses on u_0, there exists $0 < \varepsilon_1 < \varepsilon_0$ such that

$$F(u_0 + t\varphi + t^2 \psi_0(t)) \geq F(u_0), \ \forall 0 < |t| < \varepsilon_1,$$

so that

$$\frac{F(u_0 + t\varphi + t^2 \psi_0(t)) - F(u_0)}{t} \geq 0, \ \forall 0 < t < \varepsilon_1.$$

Hence,

$$0 \leq \frac{F(u_0 + t\varphi + t^2 \psi_0(t)) - F(u_0)}{t}$$

$$= \frac{F(u_0 + t\varphi + t^2 \psi_0(t)) - F(u_0 + t^2 \psi_0(t)) + F(u_0 + t^2 \psi_0(t)) - F(u_0)}{t}$$

$$\leq \frac{F(u_0 + t\varphi + t^2 \psi_0(t)) - F(u_0 + t^2 \psi_0(t))}{t} + Kt \| \psi_0(t) \|_U, \ \forall 0 < t < \min\{r, \varepsilon_1\}. \ (9.10)$$

From this, we obtain

$$0 \leq \limsup_{t \to 0^+} \frac{F(u_0 + t\varphi + t^2 \psi_0(t)) - F(u_0)}{t}$$

$$= \limsup_{t \to 0^+} \frac{F(u_0 + t\varphi + t^2 \psi_0(t)) - F(u_0 + t^2 \psi_0(t)) + F(u_0 + t^2 \psi_0(t)) - F(u_0)}{t}$$

$$\leq \limsup_{t \to 0^+} \frac{F(u_0 + t\varphi + t^2 \psi_0(t)) - F(u_0 + t^2 \psi_0(t))}{t} + \limsup_{t \to 0^+} Kt \| \psi_0(t) \|_U$$

$$= \limsup_{t \to 0^+} \frac{F(u_0 + t\varphi + t^2 \psi_0(t)) - F(u_0 + t^2 \psi_0(t))}{t}$$

$$\leq H_{u_0}(\varphi). \tag{9.11}$$

Summarizing,

$$H_{u_0}(\varphi) \geq 0, \ \forall \varphi \in N(G'(u_0)).$$

Hence,

$$H_{u_0}(\varphi) \geq 0 = \langle \varphi, \theta \rangle_U, \ \forall \varphi \in N(G'(u_0)).$$

From the Hahn-Banach Theorem, the functional

$$f \equiv 0$$

defined on $N(G'(u_0))$ may be extended to U through a linear functional $f_1 : U \to \mathbb{R}$ such that

$$f_1(\varphi) = 0, \ \forall \varphi \in N[G'(u_0)]$$

and

$$f_1(\varphi) \leq H_{u_0}(\varphi), \ \forall \varphi \in U.$$

Since from the local Lipschitz property H_{u_0} is bounded, so is f_1.

Therefore, there exists $u^* \in U^*$ such that

$$f_1(\varphi) = \langle \varphi, u^* \rangle_U \leq H_{u_0}(\varphi), \forall \varphi \in U,$$

so that

$$u^* \in \partial^0 F(u_0).$$

Finally, observe that

$$\langle \varphi, u^* \rangle_U = 0, \ \forall \varphi \in N(G'(u_0)).$$

Since $G'(u_0)$ is onto (closed range), from a well known result for linear operators, we have that

$$u^* \in R[G'(u_0)^*].$$

Thus, there exists, $z_0^* \in Z^*$ such that

$$u^* = [G'(u_0)^*](-z_0^*),$$

so that

$$u^* + [G'(u_0)^*](z_0^*) = \theta.$$

From this, we obtain

$$\langle \varphi, u^* \rangle_U + \langle \varphi, [G'(u_0)^*](z_0^*) \rangle_U = 0,$$

that is,

$$\langle \varphi, u^* \rangle_U + \langle G'(u_0)\varphi, (z_0^*) \rangle_Z = 0, \ \forall \varphi \in U$$

The proof is complete.

9.3 The Lagrange multiplier theorem for equality and inequality constraints for non-smooth optimization

In this section, we develop a rigorous result concerning the Lagrange multiplier theorem for the case involving equalities and inequalities.

Theorem 9.3.1 *Let U, Z_1, Z_2 be Banach spaces. Consider a cone C in Z_2 (as specified at Theorem 11.1 in [12]) such that if $z_1 \leq \theta$ and $z_2 < \theta$ then $z_1 + z_2 < \theta$, where $z \leq \theta$ means that $z \in -C$ and $z < \theta$ means that $z \in (-C)°$. The concerned order is supposed to be also that if $z < \theta$, $z^* \geq \theta^*$ and $z^* \neq \theta$ then $\langle z, z^* \rangle_{Z_2} < 0$. Furthermore, assume $u_0 \in U$ is a point of local minimum for $F : U \to \mathbb{R}$ subject to $G_1(u) = \theta$ and $G_2(u) \leq \theta$, where $G_1 : U \to Z_1$, $G_2 : U \to Z_2$ are Fréchet differentiable transformations and F locally Lipschitz at $u_0 \in U$. Suppose also $G_1'(u_0)$ is onto and that there exist $\alpha > 0, K > 0$ such that if $\|\varphi\|_U < \alpha$ then*

$$\|G_1'(u_0 + \varphi) - G_1'(u_0)\| \leq K\|\varphi\|_U.$$

Finally, suppose there exists $\varphi_0 \in U$ such that

$$G_1'(u_0) \cdot \varphi_0 = \theta$$

and

$$G_2'(u_0) \cdot \varphi_0 < \theta.$$

Under such hypotheses, there exists a Lagrange multiplier $z_0^* = (z_1^*, z_2^*) \in Z_1^* \times Z_2^*$ such that

$$\theta \in \partial^0 F(u_0) + [G_1'(u_0)^*](z_1^*) + [G_2'(u_0)^*](z_2^*),$$

$$z_2^* \geq \theta^*,$$

and

$$\langle G_2(u_0), z_2^* \rangle_{Z_2} = 0,$$

that is, there exists $u^* \in \partial^0 F(u_0)$ and a Lagrange multiplier $z_0^* = (z_1^*, z_2^*) \in Z_1^* \times Z_2^*$ such that

$$u^* + [G_1'(u_0)]^*(z_1^*) + [G_2'(u_0)]^*(z_2^*) = \theta,$$

so that

$$\langle \varphi, u^* \rangle_U + \langle \varphi, G_1'(u_0)^*(z_1^*) \rangle_U + \langle \varphi, G_2'(u_0)^*(z_2^*) \rangle_U = 0,$$

that is,

$$\langle \varphi, u^* \rangle_U + \langle G_1'(u_0)\varphi, z_1^* \rangle_{Z_1} + \langle G_2'(u_0)\varphi, z_2^* \rangle_{Z_2} = 0, \ \forall \varphi \in U.$$

Proof 9.3 Let $\varphi \in U$ be such that

$$G_1'(u_0) \cdot \varphi = \theta$$

and

$$G_2'(u_0) \cdot \varphi = v - \lambda G_2(u_0),$$

for some $v \leq \theta$ and $\lambda \geq 0$.

For $\alpha \in (0,1)$ define

$$\varphi_\alpha = \alpha \varphi_0 + (1 - \alpha)\varphi.$$

Observe that $G_1(u_0) = \theta$ and $G_1'(u_0) \cdot \varphi_\alpha = \theta$ so that as in the proof of the Lagrange multiplier Theorem 11.3.2 in [12], we may find $K_1 > 0$, $\varepsilon > 0$ and $\psi_0^\alpha(t)$ such that

$$G_1(u_0 + t\varphi_\alpha + t^2 \psi_0^\alpha(t)) = \theta, \ \forall |t| < \varepsilon, \forall \alpha \in (0,1)$$

and

$$\|\psi_0^\alpha(t)\|_U < K_1, \forall |t| < \varepsilon, \ \forall \alpha \in (0,1).$$

Observe that

$$
\begin{aligned}
&G_2'(u_0) \cdot \varphi_\alpha \\
={} & \alpha G_2'(u_0) \cdot \varphi_0 + (1 - \alpha)G_2'(u_0) \cdot \varphi \\
={} & \alpha G_2'(u_0) \cdot \varphi_0 + (1 - \alpha)(v - \lambda G_2(u_0)) \\
={} & \alpha G_2'(u_0) \cdot \varphi_0 + (1 - \alpha)v - (1 - \alpha)\lambda G_2(u_0)) \\
={} & v_0 - \lambda_0 G_2(u_0),
\end{aligned}
\tag{9.12}
$$

where,

$$\lambda_0 = (1 - \alpha)\lambda,$$

and

$$v_0 = \alpha G_2'(u_0) \cdot \varphi_0 + (1 - \alpha)v < \theta.$$

Hence, for $t > 0$

$$G_2(u_0 + t\varphi_\alpha + t^2 \psi_0^\alpha(t)) = G_2(u_0) + G_2'(u_0) \cdot (t\varphi_\alpha + t^2 \psi_0^\alpha(t)) + r(t),$$

where

$$\lim_{t \to 0^+} \frac{\|r(t)\|}{t} = 0.$$

Therefore from (9.12) we obtain

$$G_2(u_0 + t\varphi_\alpha + t^2 \psi_0^\alpha(t)) = G_2(u_0) + tv_0 - t\lambda_0 G_2(u_0) + r_1(t),$$

where

$$\lim_{t \to 0^+} \frac{\|r_1(t)\|}{t} = 0.$$

Observe that there exists $\varepsilon_1 > 0$ such that if $0 < t < \varepsilon_1 < \varepsilon$, then

$$v_0 + \frac{r_1(t)}{t} < \theta,$$

and

$$G_2(u_0) - t\lambda_0 G_2(u_0) = (1 - t\lambda_0)G_2(u_0) \le \theta.$$

Hence

$$G_2(u_0 + t\varphi_\alpha + t^2 \psi_0^\alpha(t)) < \theta, \text{ if } 0 < t < \varepsilon_1.$$

From this there exists $0 < \varepsilon_2 < \varepsilon_1$ such that

$$F(u_0 + t\varphi_\alpha + t^2 \psi_0^\alpha(t)) \ge F(u_0), \forall 0 < t < \varepsilon_2, \ \alpha \in (0, 1).$$

In particular

$$F(u_0 + t\varphi_t + t^2 \psi_0^t(t)) \ge F(u_0), \forall 0 < t < \min\{1, \varepsilon_2\},$$

so that

$$\frac{F(u_0 + t\varphi_t + t^2 \psi_0^t(t)) - F(u_0)}{t} \ge 0, \forall 0 < t < \min\{1, \varepsilon_2\},$$

that is,

$$\frac{F(u_0 + t\varphi + t^2(\psi_0^t(t) + \varphi_0 - \varphi)) - F(u_0)}{t} \ge 0, \forall 0 < t < \min\{1, \varepsilon_2\}.$$

From this we obtain,

$$0 \leq \limsup_{t \to 0^+} \frac{F(u_0 + t\varphi + t^2(\psi_0^t(t) + \varphi_0 - \varphi)) - F(u_0)}{t}$$

$$= \limsup_{t \to 0^+} \left(\frac{F(u_0 + t\varphi + t^2(\psi_0^t(t) + \varphi_0 - \varphi)) - F(u_0 + t^2(\psi_0^t(t) + \varphi_0 - \varphi))}{t} \right.$$

$$\left. + \frac{F(u_0 + t^2(\psi_0^t(t) + \varphi_0 - \varphi)) - F(u_0)}{t} \right)$$

$$\leq \limsup_{t \to 0^+} \frac{F(u_0 + t\varphi + t^2(\psi_0^t(t) + \varphi_0 - \varphi)) - F(u_0 + t^2(\psi_0^t(t) + \varphi_0 - \varphi))}{t}$$

$$+ \limsup_{t \to 0^+} Kt \| \psi_0^t(t) + \varphi_0 - \varphi \|_U$$

$$= \limsup_{t \to 0^+} \frac{F(u_0 + t\varphi + t^2(\psi_0^t(t) + \varphi_0 - \varphi)) - F(u_0 + t^2(\psi_0^t(t) + \varphi_0 - \varphi))}{t}$$

$$\leq H_{u_0}(\varphi). \tag{9.13}$$

Summarizing, we have

$$H_{u_0}(\varphi) \geq 0,$$

if

$$G_1'(u_0) \cdot \varphi = \theta,$$

and

$$G_2'(u_0) \cdot \varphi = v - \lambda G_2(u_0),$$

for some $v \leq \theta$ and $\lambda \geq 0$.

Define

$$A = \{ H_{u_0}(\varphi) + r, G_1'(u_0) \cdot \varphi, G_2'(u_0)\varphi - v + \lambda G_2(u_0)),$$

$$\varphi \in U, r \geq 0, v \leq \theta, \lambda \geq 0 \}. \tag{9.14}$$

From the convexity of H_{u_0} (and the hypotheses on $G_1'(u_0)$ and $G_2'(u_0)$) we have that A is a convex set (with a non-empty interior).

If

$$G_1'(u_0) \cdot \varphi = \theta,$$

and

$$G_2'(u_0) \cdot \varphi - v + \lambda G_2(u_0) = \theta,$$

with $v \leq \theta$ and $\lambda \geq 0$ then

$$H_{u_0}(\varphi) \geq 0,$$

so that

$$H_{u_0}(\varphi) + r \geq 0, \ \forall r \geq 0.$$

From this and

$$H_{u_0}(\theta) = 0,$$

we have that $(0, \theta, \theta)$ is on the boundary of A. Therefore, by the Hahn-Banach theorem, geometric form, there exists

$$(\beta, z_1^*, z_2^*) \in \mathbb{R} \times Z_1^* \times Z_2^*$$

such that

$$(\beta, z_1^*, z_2^*) \neq (0, \theta, \theta)$$

and

$$\beta(H_{u_0}(\varphi) + r) + \langle G_1'(u_0) \cdot \varphi, z_1^* \rangle_{Z_1}$$
$$+ \langle G_2'(u_0) \cdot \varphi - v + \lambda G_2(u_0), z_2^* \rangle_{Z_2} \geq 0, \qquad (9.15)$$

$\forall \varphi \in U, r \geq 0, v \leq \theta, \lambda \geq 0$. Suppose $\beta = 0$. Fixing all variable except v we get $z_2^* \geq \theta$. Thus, for $\varphi = c\varphi_0$ with arbitrary $c \in \mathbb{R}$, $v = \theta, \lambda = 0$, if $z_2^* \neq \theta$, then $\langle G_2'(u_0) \cdot \varphi_0, z_2^* \rangle_{Z_2} < 0$ so that, letting $c \to +\infty$, we get a contradiction through (9.15), so that $z_2^* = \theta$. Since $G_1'(u_0)$ is onto, a similar reasoning lead us to $z_1^* = \theta$, which contradicts $(\beta, z_1^*, z_2^*) \neq (0, \theta, \theta)$.

Hence, $\beta \neq 0$, and fixing all variables except r we obtain $\beta > 0$. There is no loss of generality in assuming $\beta = 1$.

Again fixing all variables except v, we obtain $z_2^* \geq \theta$. Fixing all variables except λ, since $G_2(u_0) \leq \theta$ we obtain

$$\langle G_2(u_0), z_2^* \rangle_{Z_2} = 0.$$

Finally, for $r = 0$, $v = \theta$, $\lambda = 0$, we get

$$H_{u_0}(\varphi) + \langle G_1'(u_0)\varphi, z_1^* \rangle_{Z_1} + \langle G_2'(u_0) \cdot \varphi, z_2^* \rangle_{Z_2} \geq 0 = \langle \varphi, \theta \rangle_U, \forall \varphi \in U.$$

From this,

$$\theta \in \partial^0(F(u_0) + \langle G_1(u_0), z_1^* \rangle_{Z_1} + \langle G_2(u_0), z_2^* \rangle_{Z_2}) = \partial^0 F(u_0) + [G_1'(u_0)^*](z_1^*) + [G_2'(u_0)^*](z_2^*),$$

so that there exists $u^* \in \partial^0 F(u_0)$, such that

$$u^* + [G_1'(u_0)^*](z_1^*) + [G_2'(u_0)^*](z_2^*) = \theta,$$

so that

$$\langle \varphi, u^* \rangle_U + \langle \varphi, G_1'(u_0)^*(z_1^*) \rangle_U + \langle \varphi, G_2'(u_0)^*(z_2^*) \rangle_U = 0,$$

that is,

$$\langle \varphi, u^* \rangle_U + \langle G_1'(u_0)\varphi, z_1^* \rangle_{Z_1} + \langle G_2'(u_0)\varphi, z_2^* \rangle_{Z_2} = 0, \forall \varphi \in U.$$

The proof is complete.

9.4 Conclusion

In this article, we have presented an approach on Lagrange multiplier theorems for non-smooth variational optimization in a general Banach space context. The results are based on standard tools of functional analysis, calculus of variations and optimization.

We emphasize, in the present article, no hypotheses concerning convexity are assumed and the results indeed are valid for such a more general Banach space context.

III

DUALITY PRINCIPLES AND RELATED NUMERICAL EXAMPLES THROUGH THE GENERALIZED METHOD OF LINES

Chapter 10

A Convex Dual Formulation for a Large Class of Non-Convex Models in Variational Optimization

10.1 Introduction

This short communication develops a convex dual variational formulation for a large class of models in variational optimization. The results are established through basic tools of functional analysis, convex analysis, and the duality theory. The main duality principle is developed as an application to a Ginzburg-Landau type system in superconductivity in the absence of a magnetic field.

Such results are based on the works of J.J. Telega and W.R. Bielski [10, 11, 74, 75], and on a D.C. optimization approach developed in Toland [81].

At this point, we start to describe the primal and dual variational formulations.

Let $\Omega \subset \mathbb{R}^3$ be an open, bounded, connected set with a regular (Lipschitzian) boundary denoted by $\partial\Omega$.

For the primal formulation we consider the functional $J : U \to \mathbb{R}$ where

$$J(u) = \frac{\gamma}{2} \int_\Omega \nabla u \cdot \nabla u \, dx$$
$$+ \frac{\alpha}{2} \int_\Omega (u^2 - \beta)^2 \, dx - \langle u, f \rangle_{L^2}. \tag{10.1}$$

Here we assume $\alpha > 0, \beta > 0, \gamma > 0$, $U = W_0^{1,2}(\Omega)$, $f \in L^2(\Omega)$. Moreover we denote

$$Y = Y^* = L^2(\Omega).$$

Define also $G : U \to \mathbb{R}$ by

$$G(u) = \frac{\alpha}{2} \int_\Omega (u^2 - \beta)^2 \, dx + \frac{K}{2} \int_\Omega u^2 \, dx,$$

and $F : U \to \mathbb{R}$ by

$$F(u) = -\frac{\gamma}{2} \int_\Omega \nabla u \cdot \nabla u \, dx + \frac{K}{2} \int_\Omega u^2 \, dx + \langle u, f \rangle_{L^2},$$

It is worth highlighting that in such a case

$$J(u) = -F(u) + G(u), \quad \forall u \in U.$$

From now and on, we assume a finite dimensional version for this model, in a finite elements of finite differences context, where, for not relabeled operators and spaces, we also assume,

$$\gamma \nabla^2 + K > 0$$

in an appropriate matrices sense.

Furthermore, define

$$A^+ = \{u \in U \ : \ \delta^2 J(u) \geq 0\}$$

$$(A^+)^0 = \{u \in U \ : \ \delta^2 J(u) > 0\},$$

$$C^+ = \{u \in U \ : \ uf \geq 0, \text{ in } \Omega\},$$

$$E^+ = A^+ \cap C^+$$

and the following specific polar functionals specified, namely, $G^* : Y^* \to \mathbb{R}$ by

$$G^*(v_1^*) = \sup_{u \in A^+} \{\langle u, v_1^* \rangle_{L^2} - G(u)\} \tag{10.2}$$

and $F^* : Y^* \to \mathbb{R}$ by

$$F^*(v_1^*) = \sup_{u \in U} \{\langle u, v_1^* \rangle_{L^2} - F(u)\}$$

$$= \frac{1}{2} \int_\Omega \frac{(v_1^* - f)^2}{\gamma \nabla^2 + K} \, dx. \tag{10.3}$$

Define also $J^* : Y^* \to \mathbb{R}$ by

$$J^*(v_1^*) = F^*(v_1^*) - G^*(v_1^*)$$

Observe that there exists a Lagrange multiplier $\lambda \in W_0^{1,2}(\Omega)$ such that

$$
\begin{aligned}
G^*(v_1^*) &= \sup_{u \in U} \left\{ \langle u, v_1^* \rangle_{L^2} - G(u) + \frac{\gamma}{2} \int_\Omega \nabla\lambda \cdot \nabla\lambda \, dx \right. \\
&\left. + \frac{6\alpha}{2} \int_\Omega \lambda^2 u^2 \, dx - \alpha\beta \int_\Omega \lambda^2 \, dx \right\}.
\end{aligned}
\tag{10.4}
$$

Define now $G_2 : Y^* \times U \times U \to \mathbb{R}$ by

$$G_2(v_1^*, u, \lambda) = \langle u, v_1^* \rangle_{L^2} - G(u) + \frac{\gamma}{2} \int_\Omega \nabla\lambda \cdot \nabla\lambda \, dx + \frac{6\alpha}{2} \int_\Omega \lambda^2 u^2 \, dx - \alpha\beta \int_\Omega \lambda^2 \, dx.$$

Observe also that

$$G^*(v_1^*) = G_2(v_1^*, \hat{u}, \hat{\lambda}),$$

where $\hat{u} = u(v_1^*)$ and $\hat{\lambda} = \lambda(v_1^*)$ are such that

$$\frac{\partial G_2(v_1^*, \hat{u}, \hat{\lambda})}{\partial u} = \mathbf{0},$$

and

$$\frac{\partial G_2(v_1^*, \hat{u}, \hat{\lambda})}{\partial \lambda} = \mathbf{0}.$$

On the other hand,

$$\frac{\partial^2 G^*(v_1^*)}{\partial (v_1^*)^2} = \frac{\partial^2 G_2(v_1^*, \hat{u}, \hat{\lambda})}{\partial (v_1^*)^2} + \frac{\partial^2 G_2(v_1^*, \hat{u}, \hat{\lambda})}{\partial v_1^* \partial u} \frac{\partial \hat{u}}{\partial v_1^*} + + \frac{\partial^2 G_2(v_1^*, \hat{u}, \hat{\lambda})}{\partial v_1^* \partial \lambda} \frac{\partial \hat{\lambda}}{\partial v_1^*}.$$

Moreover,

$$\frac{\partial^2 G_2(v_1^*, \hat{u}, \hat{\lambda})}{\partial (v_1^*)^2} = \mathbf{0},$$

$$\frac{\partial^2 G_2(v_1^*, \hat{u}, \hat{\lambda})}{\partial v_1^* \partial u} = 1,$$

and

$$\frac{\partial^2 G_2(v_1^*, \hat{u}, \hat{\lambda})}{\partial v_1^* \partial \lambda} = \mathbf{0}.$$

From these last results we get

$$\frac{\partial^2 G^*(v_1^*)}{\partial (v_1^*)^2} = \frac{\partial \hat{u}}{\partial v_1^*}.$$

However from

$$\frac{\partial G_2(v_1^*, \hat{u}, \hat{\lambda})}{\partial u} = \mathbf{0},$$

we have

$$v_1^* - 2\alpha(\hat{u}^2 - \beta)\hat{u} - K\hat{u} + 6\alpha\hat{\lambda}^2\hat{u} = \mathbf{0}$$

Taking the variation in v_1^* in this last equation, we obtain

$$1 - 6\alpha\hat{u}^2\frac{\partial\hat{u}}{\partial v_1^*} + 2\alpha\beta\frac{\partial\hat{u}}{\partial v_1^*}$$

$$-K\frac{\partial\hat{u}}{\partial v_1^*} + 6\alpha\hat{\lambda}^2\frac{\partial\hat{u}}{\partial v_1^*} + 12\alpha\hat{\lambda}\frac{\partial\hat{\lambda}}{\partial v_1^*}\hat{u} = \mathbf{0}. \tag{10.5}$$

On the other hand we must have also

$$\frac{\gamma}{2}\int_\Omega \nabla\hat{\lambda}\cdot\nabla\hat{\lambda}\,dx + \frac{1}{2}\int_\Omega 6\alpha\hat{\lambda}^2\hat{u}^2\,dx - \int_\Omega \alpha\beta\hat{\lambda}^2\,dx = 0,$$

so that taking the variation in v_1^* for this last equation and considering that

$$-\gamma\nabla^2\hat{\lambda} + 6\alpha\hat{u}^2\hat{\lambda} - 2\alpha\beta\hat{\lambda} = \mathbf{0},$$

we get

$$12\alpha\hat{\lambda}^2\hat{u}\frac{\partial\hat{u}}{\partial v_1^*} = \mathbf{0}.$$

Hence if locally $\hat{\lambda}^2\hat{u} \neq 0$, then locally

$$\frac{\partial\hat{u}}{\partial v_1^*} = 0.$$

On the other hand if $\hat{\lambda}^2\hat{u} = \mathbf{0}$, then from (10.5) we have

$$\frac{\partial\hat{u}}{\partial v_1^*} = \frac{1}{6\alpha\hat{u}^2 - 2\alpha\beta - 6\alpha\hat{\lambda}^2 + K}.$$

Recalling that

$$\frac{\partial^2 G^*(v_1^*)}{\partial(v_1^*)^2} = \frac{\partial\hat{u}}{\partial v_1^*},$$

we have got

$$\frac{\partial^2 G^*(v_1^*)}{\partial(v_1^*)^2} = \begin{cases} 0, & \text{if } \hat{\lambda}^2\hat{u} \neq 0, \\ \frac{1}{6\alpha\hat{u}^2 - 2\alpha\beta - 6\alpha\hat{\lambda}^2 + K}, & \text{if } \hat{\lambda}^2\hat{u} = 0. \end{cases} \tag{10.6}$$

Observe also that

$$\frac{\partial^2 J^*(v_1^*)}{\partial(v_1^*)^2} = \frac{\partial^2 F^*(v_1^*)}{\partial(v_1^*)^2} - \frac{\partial^2 G^*(v_1^*)}{\partial(v_1^*)^2} = \frac{1}{\gamma\nabla^2 + K} - \frac{\partial\hat{u}}{\partial v_1^*},$$

so that, for $\hat{\lambda}^2\hat{u} = 0$ we obtain

$$
\begin{aligned}
\frac{1}{\gamma\nabla^2 + K} - \frac{\partial\hat{u}}{\partial v_1^*} &= \frac{1}{\gamma\nabla^2 + K} - \frac{1}{6\alpha\hat{u}^2 - 2\alpha\beta - 6\alpha\hat{\lambda}^2 + K} \\
&= \frac{-\gamma\nabla^2 - K + 6\alpha\hat{u}^2 - 2\alpha\beta - 6\alpha\hat{\lambda}^2 + K}{(\gamma\nabla^2 + K)(6\alpha\hat{u}^2 - 2\alpha\beta - 6\alpha\hat{\lambda}^2 + K)} \\
&= \frac{\delta^2 J(\hat{u}) - 6\alpha\hat{\lambda}^2}{(\gamma\nabla^2 + K)(6\alpha\hat{u}^2 - 2\alpha\beta - 6\alpha\hat{\lambda}^2 + K)} \\
&\geq 0.
\end{aligned}
$$
(10.7)

Summarizing,

$$\frac{\partial^2 J^*(v_1^*)}{\partial(v_1^*)^2} = \begin{cases} \frac{1}{\gamma\nabla^2 + K}, & \text{if } \hat{\lambda}^2\hat{u} \neq 0, \\ \frac{\delta^2 J(\hat{u}) - 6\alpha\hat{\lambda}^2}{(\gamma\nabla^2 + K)(6\alpha\hat{u}^2 - 2\alpha\beta - 6\alpha\hat{\lambda}^2 + K)}, & \text{if } \hat{\lambda}^2\hat{u} = 0. \end{cases}$$
(10.8)

Hence, in any case, we have obtained

$$\frac{\partial^2 J^*(v_1^*)}{\partial(v_1^*)^2} \geq \mathbf{0}, \ \forall v_1^* \in Y^*$$

so that J^* is convex in Y^*.

10.2 The main duality principle, a convex dual variational formulation

Our main result is summarized by the following theorem.

Theorem 10.2.1 *Considering the definitions and statements in the last section, suppose also $\hat{v}^* \in Y^*$ is such that*

$$\delta J^*(\hat{v}^*) = \mathbf{0}.$$

Assume also

$$u_0 = \frac{\partial F^*(\hat{v}_1^*)}{\partial v_1^*} \in E^+ \cap (A^+)^0.$$

Under such hypotheses, we have

$$\delta J(u_0) = \mathbf{0},$$

and

$$J(u_0) = \inf_{u \in E^+} \{J(u)\}$$

$$= \inf_{v_1^* \in Y^*} J^*(v_1^*)$$

$$= J_1^*(\hat{v}_1^*). \tag{10.9}$$

Proof 10.1   From the hypothesis

$$\frac{\partial J^*(\hat{v}_1^*)}{\partial v_1^*} = \mathbf{0}.$$

so that

$$\frac{\partial J^*(\hat{v}_1^*)}{\partial v_1^*} = \frac{\partial F^*(\hat{v}_1^*)}{\partial v_1^*} - \frac{\partial G_1^*(\hat{v}_1^*)}{\partial v_1^*} = \mathbf{0}. \tag{10.10}$$

Since from the previous section we have got that J^* is convex on Y^*, we may infer that

$$J^*(\hat{v}_1^*) = \inf_{v_1^* \in Y^*} J^*(v_1^*).$$

Also, from these last results,

$$u_0 - \frac{\partial G^*(\hat{v}_1^*)}{\partial v_1^*} = \mathbf{0},$$

so that, since the restriction is not active in a neighborhood of u_0, from the Legendre transform properties, we obtain

$$\hat{v}_1^* = \frac{\partial G(u_0)}{\partial u},$$

and

$$\hat{v}_1^* = \frac{\partial F(u_0)}{\partial u},$$

and thus

$$\mathbf{0} = \hat{v}_1^* - \hat{v}_1^* = -\frac{\partial F(u_0)}{\partial u} + \frac{\partial G(u_0)}{\partial u} = \delta J(u_0).$$

Summarizing $\delta J(u_0) = \mathbf{0}$.

Also from the Legendre transform properties we have

$$F^*(\hat{v}_1^*) = \langle u_0, \hat{v}_1^* \rangle_{L^2} - F(u_0),$$

and

$$G^*(\hat{v}_1^*) = \langle u_0, \hat{v}_1^* \rangle_{L^2} - G(u_0),$$

so that

$$J^*(\hat{v}_1^*) = F^*(\hat{v}_1^*) - G^*(\hat{v}_1^*) = -F(u_0) + G(u_0) = J(u_0).$$

Finally, from similar results in [13], we may infer that E^+ is convex so that from this and $\delta J(u_0) = \mathbf{0}$, we get

$$J(u_0) = \min_{u \in E^+} J(u).$$

Joining the pieces, we have got

$$
\begin{aligned}
J(u_0) &= \inf_{u \in E^+} \{J(u)\} \\
&= \inf_{v_1^* \in Y^*} J^*(v_1^*) \\
&= J^*(\hat{v}_1^*).
\end{aligned}
\tag{10.11}
$$

The proof is complete.

Chapter 11

Duality Principles and Numerical Procedures for a Large Class of Non-Convex Models in the Calculus of Variations

11.1   Introduction

In this section, we establish a dual formulation for a large class of models in non-convex optimization.

The main duality principle is applied to double well models similar as those found in the phase transition theory.

Such results are based on the works of J.J. Telega and W.R. Bielski [10, 11, 74, 75], and on a D.C. optimization approach developed in Toland [81].

About the other references, details on the Sobolev spaces involved are found in [1, 26]. Related results on convex analysis and the duality theory are addressed in [12, 13, 14, 22, 23, 25, 62].

Moreover, related results for phase transition and similar models may be found in [29, 33, 36, 37, 38, 44, 45, 47, 56, 57, 58, 59, 63, 64 ,72, 84].

Concerning results in shape optimization may be found in [57, 68] and basic approaches on analysis and functional analysis are developed in [26, 33, 48, 49, 65, 66].

Finally, in this text we adopt the standard Einstein convention of summing up repeated indices, unless otherwise indicated.

In order to clarify the notation, here we introduce the definition of topological dual space.

Definition 11.1.1 (Topological dual spaces) *Let U be a Banach space. We shall define its dual topological space, as the set of all linear continuous functionals defined on U. We suppose such a dual space of U, may be represented by another Banach space U^*, through a bilinear form $\langle \cdot, \cdot \rangle_U : U \times U^* \to \mathbb{R}$ (here we are referring to standard representations of dual spaces of Sobolev and Lebesgue spaces). Thus, given $f : U \to \mathbb{R}$ linear and continuous, we assume the existence of a unique $u^* \in U^*$ such that*

$$f(u) = \langle u, u^* \rangle_U, \forall u \in U. \tag{11.1}$$

The norm of f, denoted by $\|f\|_{U^*}$, is defined as

$$\|f\|_{U^*} = \sup_{u \in U}\{|\langle u, u^* \rangle_U| : \|u\|_U \leq 1\} \equiv \|u^*\|_{U^*}. \tag{11.2}$$

At this point we start to describe the primal and dual variational formulations.

11.2 A general duality principle non-convex optimization

In this section, we present a duality principle applicable to a model in phase transition.

This case corresponds to the vectorial one in the calculus of variations.

Let $\Omega \subset \mathbb{R}^n$ be an open, bounded, connected set with a regular (Lipschitzian) boundary denoted by $\partial\Omega$.

Consider a functional $J : V \to \mathbb{R}$ where

$$J(u) = F(\nabla u_1, \cdots, \nabla u_N) + G(u_1, \cdots, u_N) - \langle u_i, f_i \rangle_{L^2},$$

and where

$$V = \{u = (u_1, \cdots, u_N) \in W^{1,p}(\Omega; \mathbb{R}^N) : u = u_0 \text{ on } \partial\Omega\},$$

$f \in L^2(\Omega; \mathbb{R}^N)$, and $1 < p < +\infty$.

We assume there exists $\alpha \in \mathbb{R}$ such that

$$\alpha = \inf_{u \in V} J(u).$$

Moreover, suppose F and G are Fréchet differentiable but not necessarily convex. A global optimum point may not be attained for J so that the problem of finding a global minimum for J may not be a solution.

Anyway, one question remains, how the minimizing sequences behave close the infimum of J.

We intend to use duality theory to approximately solve such a global optimization problem.

Denoting $V_0 = W_0^{1,p}(\Omega; \mathbb{R}^N)$, $Y_1 = Y_1^* = L^2(\Omega; \mathbb{R}^{N \times n})$, $Y_2 = Y_2^* = L^2(\Omega; \mathbb{R}^{N \times n})$, $Y_3 = Y_3^* = L^2(\Omega; \mathbb{R}^N)$, at this point we define, $F_1 : V \times V_0 \to \mathbb{R}$, $G_1 : V \to \mathbb{R}$, $G_2 : V \to \mathbb{R}$, $G_3 : V_0 \to \mathbb{R}$ and $G_4 : V \to \mathbb{R}$, by

$$
\begin{aligned}
F_1(\nabla u, \nabla \phi) &= F(\nabla u_1 + \nabla \phi_1, \cdots, \nabla u_N + \nabla \phi_N) + \frac{K}{2} \int_\Omega \nabla u_j \cdot \nabla u_j \, dx \\
&\quad + \frac{K_2}{2} \int_\Omega \nabla \phi_j \cdot \nabla \phi_j \, dx
\end{aligned}
\tag{11.3}
$$

and

$$
G_1(u_1, \cdots, u_n) = G(u_1, \cdots, u_N) + \frac{K_1}{2} \int_\Omega u_j \, u_j \, dx - \langle u_i, f_i \rangle_{L^2},
$$

$$
G_2(\nabla u_1, \cdots, \nabla u_N) = \frac{K_1}{2} \int_\Omega \nabla u_j \cdot \nabla u_j \, dx,
$$

$$
G_3(\nabla \phi_1, \cdots, \nabla \phi_N) = \frac{K_2}{2} \int_\Omega \nabla \phi_j \cdot \nabla \phi_j \, dx,
$$

and

$$
G_4(u_1, \cdots, u_N) = \frac{K_1}{2} \int_\Omega u_j \, u_j \, dx.
$$

Define now $J_1 : V \times V_0 \to \mathbb{R}$,

$$
J_1(u, \phi) = F(\nabla u + \nabla \phi) + G(u) - \langle u_i, f_i \rangle_{L^2}.
$$

Observe that

$$
\begin{aligned}
J_1(u, \phi) &= F_1(\nabla u, \nabla \phi) + G_1(u) - G_2(\nabla u) - G_3(\nabla \phi) - G_4(u) \\
&\leq F_1(\nabla u, \nabla \phi) + G_1(u) - \langle \nabla u, z_1^* \rangle_{L^2} - \langle \nabla \phi, z_2^* \rangle_{L^2} - \langle u, z_3^* \rangle_{L^2} \\
&\quad + \sup_{v_1 \in Y_1} \{ \langle v_1, z_1^* \rangle_{L^2} - G_2(v_1) \} \\
&\quad + \sup_{v_2 \in Y_2} \{ \langle v_2, z_2^* \rangle_{L^2} - G_3(v_2) \} \\
&\quad + \sup_{u \in V} \{ \langle u, z_3^* \rangle_{L^2} - G_4(u) \} \\
&= F_1(\nabla u, \nabla \phi) + G_1(u) - \langle \nabla u, z_1^* \rangle_{L^2} - \langle \nabla \phi, z_2^* \rangle_{L^2} - \langle u, z_3^* \rangle_{L^2} \\
&\quad + G_2^*(z_1^*) + G_3^*(z_2^*) + G_4^*(z_3^*) \\
&= J_1^*(u, \phi, z^*),
\end{aligned}
\tag{11.4}
$$

$\forall u \in V$, $\phi \in V_0$, $z^* = (z_1^*, z_2^*, z_3^*) \in Y^* = Y_1^* \times Y_2^* \times Y_3^*$.

Here we assume K, K_1, K_2 are large enough so that F_1 and G_1 are convex. Hence, from the general results in [81], we may infer that

$$
\inf_{(u, \phi) \in V \times V_0} J(u, \phi) = \inf_{(u, \phi, z^*) \in V \times V_0 \times Y^*} J_1^*(u, \phi, z^*).
\tag{11.5}
$$

On the other hand

$$\inf_{u \in V} J(u) \geq \inf_{(u,\phi) \in V \times V_0} J_1(u,\phi).$$

From these last two results we may obtain

$$\inf_{u \in V} J(u) \geq \inf_{(u,\phi,z^*) \in V \times V_0 \times Y^*} J_1^*(u,\phi,z^*).$$

Moreover, from standards results on convex analysis, we may have

$$
\begin{aligned}
\inf_{u \in V} J_1^*(u,\phi,z^*) &= \inf_{u \in V} \{F_1(\nabla u, \nabla \phi) + G_1(u) \\
&\quad - \langle \nabla u, z_1^* \rangle_{L^2} - \langle \nabla \phi, z_2^* \rangle_{L^2} - \langle u, z_3^* \rangle_{L^2} \\
&\quad + G_2^*(z_1^*) + G_3^*(z_2^*) + G_4^*(z_3^*)\} \\
&= \sup_{(v_1^*, v_2^*) \in C^*} \{-F_1^*(v_1^* + z_1^*, \nabla \phi) - G_1^*(v_2^* + z_3^*) - \langle \nabla \phi, z_2^* \rangle_{L^2} \\
&\quad + G_2^*(z_1^*) + G_3^*(z_2^*) + G_4^*(z_3^*)\},
\end{aligned}
\tag{11.6}
$$

where

$$C^* = \{v^* = (v_1^*, v_2^*) \in Y_1^* \times Y_3^* : -\mathrm{div}(v_1^*)_i + (v_2^*)_i = \mathbf{0}, \forall i \in \{1, \cdots, N\}\},$$

$$F_1^*(v_1^* + z_1^*, \nabla \phi) = \sup_{v_1 \in Y_1} \{\langle v_1, z_1^* + v_1^* \rangle_{L^2} - F_1(v_1, \nabla \phi)\},$$

and

$$G_1^*(v_2^* + z_2^*) = \sup_{u \in V} \{\langle u, v_2^* + z_2^* \rangle_{L^2} - G_1(u)\}.$$

Thus, defining

$$J_2^*(\phi, z^*, v^*) = F_1^*(v_1^* + z_1^*, \nabla \phi) - G_1^*(v_2^* + z_3^*) - \langle \nabla \phi, z_2^* \rangle_{L^2} + G_2^*(z_1^*) + G_3^*(z_2^*) + G_4^*(z_3^*),$$

we have got

$$
\begin{aligned}
\inf_{u \in V} J(u) &\geq \inf_{(u,\phi) \in V \times V_0} J_1(u,\phi) \\
&= \inf_{(u,\phi,z^*) \in V \times V_0 \times Y^*} J_1^*(u,\phi,z^*) \\
&= \inf_{z^* \in Y^*} \left\{ \inf_{\phi \in V_0} \left\{ \sup_{v^* \in C^*} J_2^*(\phi, z^*, v^*) \right\} \right\}.
\end{aligned}
\tag{11.7}
$$

Finally, observe that

$$
\begin{aligned}
\inf_{u \in V} &J(u) \\
&\geq \inf_{z^* \in Y^*} \left\{ \inf_{\phi \in V_0} \left\{ \sup_{v^* \in C^*} J_2^*(\phi, z^*, v^*) \right\} \right\} \\
&\geq \sup_{v^* \in C^*} \left\{ \inf_{(z^*,\phi) \in Y^* \times V_0} J_2^*(\phi, z^*, v^*) \right\}.
\end{aligned}
\tag{11.8}
$$

This last variational formulation corresponds to a concave relaxed formulation in v^* concerning the original primal formulation.

11.3 Another duality principle for a simpler related model in phase transition with a respective numerical example

In this section, we present another duality principle for a related model in phase transition.

Let $\Omega = [0,1] \subset \mathbb{R}$ and consider a functional $J : V \to \mathbb{R}$ where

$$J(u) = \frac{1}{2} \int_{\Omega} ((u')^2 - 1)^2 \, dx + \frac{1}{2} \int_{\Omega} u^2 \, dx - \langle u, f \rangle_{L^2},$$

and where

$$V = \{u \in W^{1,4}(\Omega) \; : \; u(0) = 0 \text{ and } u(1) = 1/2\}$$

and $f \in L^2(\Omega)$.

A global optimum point is not attained for J so that the problem of finding a global minimum for J has no solution.

Anyway, one question remains, how the minimizing sequences behave close the infimum of J.

We intend to use duality theory to approximately solve such a global optimization problem.

Denoting $V_0 = W_0^{1,4}(\Omega)$, at this point we define, $F : V \to \mathbb{R}$ and $F_1 : V \times V_0 \to \mathbb{R}$ by

$$F(u) = \frac{1}{2} \int_{\Omega} ((u')^2 - 1)^2 \, dx,$$

and

$$F_1(u, \phi) = \frac{1}{2} \int_{\Omega} ((u' + \phi')^2 - 1)^2 \, dx.$$

Observe that

$$F(u) \geq \inf_{\phi \in V_0} F_1(u, \phi), \; \forall u \in V.$$

In order to restrict the action of ϕ only on the region where the primal functional is nonconvex, we redefine a not relabeled $V_0 = \{\phi \in W_0^{1,4}(\Omega) : (\phi')^2 \leq 1, \text{ in } (\Omega)\}$ and we also define

$$F_2 : V \times V_0 \to \mathbb{R},$$

$$F_3 : V \times V_0 \to \mathbb{R}$$

and

$$G : V \times V_0 \to \mathbb{R}$$

by

$$F_2(u, \phi) = \frac{1}{2} \int_{\Omega} ((u' + \phi')^2 - 1)^2 \, dx + \frac{1}{2} \int_{\Omega} u^2 \, dx - \langle u, f \rangle_{L^2},$$

$$F_3(u,\phi) = F_2(u,\phi) + \frac{K}{2}\int_\Omega (u')^2\,dx$$
$$+ \frac{K_1}{2}\int_\Omega (\phi')^2\,dx \tag{11.9}$$

and

$$G(u,\phi) = \frac{K}{2}\int_\Omega (u')^2\,dx$$
$$+ \frac{K_1}{2}\int_\Omega (\phi')^2\,dx \tag{11.10}$$

Observe that if $K>0, K_1>0$ is large enough, both F_3 and G are convex.

Denoting $Y=Y^*=L^2(\Omega)$ we also define the polar functional $G^*:Y^*\times Y^*\to\mathbb{R}$ by

$$G^*(v^*,v_0^*) = \sup_{(u,\phi)\in V\times V_0}\{\langle u,v^*\rangle_{L^2} + \langle\phi,v_0^*\rangle_{L^2} - G(u,\phi)\}.$$

Observe that

$$\inf_{u\in U} J(u) \geq \inf_{((u,\phi),(v^*,v_0^*))\in V\times V_0\times[Y^*]^2}\{G^*(v^*,v_0^*) - \langle u,v^*\rangle_{L^2} - \langle\phi,v_0^*\rangle_{L^2} + F_3(u,\phi)\}.$$

With such results in mind, we define a relaxed primal dual variational formulation for the primal problem, represented by $J_1^*:V\times V_0\times[Y^*]^2\to\mathbb{R}$, where

$$J_1^*(u,\phi,v^*,v_0^*) = G^*(v^*,v_0^*) - \langle u,v^*\rangle_{L^2} - \langle\phi,v_0^*\rangle_{L^2} + F_3(u,\phi).$$

Having defined such a functional, we may obtain numerical results by solving a sequence of convex auxiliary sub-problems, through the following algorithm. (Here we highlight at first to have neglected the restriction $(\phi')^2\leq 1$ in (Ω) to obtain the concerning critical points.)

1. Set $K=0.1$, $K_1=120$ and $0<\varepsilon\ll 1$.

2. Choose $(u_1,\phi_1)\in V\times V_0$, such that $\|u_1\|_{1,\infty}<1$ and $\|\phi_1\|_{1,\infty}<1$.

3. Set $n=1$.

4. Calculate $(v_n^*,(v_0^*)_n)$ solution of the system of equations:

$$\frac{\partial J_1^*(u_n,\phi_n,v_n^*,(v_0^*)_n)}{\partial v^*} = 0$$

and

$$\frac{\partial J_1^*(u_n,\phi_n,v_n^*,(v_0^*)_n)}{\partial v_0^*} = 0,$$

that is

$$\frac{\partial G^*(v_n^*,(v_0^*)_n)}{\partial v^*} - u_n = 0$$

and

$$\frac{\partial G^*(v_n^*, (v_0^*)_n)}{\partial v_0^*} - \phi_n = 0$$

so that

$$v_n^* = \frac{\partial G(u_n, \phi_n)}{\partial u}$$

and

$$(v_0^*)_n^* = \frac{\partial G(u_n, \phi_n)}{\partial \phi}$$

5. Calculate (u_{n+1}, ϕ_{n+1}) by solving the system of equations:

$$\frac{\partial J_1^*(u_{n+1}, \phi_{n+1}, v_n^*, (v_0^*)_n)}{\partial u} = \mathbf{0}$$

and

$$\frac{\partial J_1^*(u_{n+1}, \phi_{n+1}, v_n^*, (v_0^*)_n)}{\partial \phi} = \mathbf{0}$$

that is

$$-v_n^* + \frac{\partial F_3(u_{n+1}, \phi_{n+1})}{\partial u} = \mathbf{0}$$

and

$$-(v_0^*)_n + \frac{\partial F_3(u_{n+1}, \phi_{n+1})}{\partial \phi} = \mathbf{0}$$

6. If $\max\{\|u_n - u_{n+1}\|_\infty, \|\phi_{n+1} - \phi_n\|_\infty\} \leq \varepsilon$, then stop, else set $n := n+1$ and go to item 4.

For the case in which $f(x) = 0$, we have obtained numerical results for $K = 1500$ and $K_1 = K/20$. For such a concerning solution u_0 obtained, please see Figure 11.1. For the case in which $f(x) = \sin(\pi x)/2$, we have obtained numerical results for $K = 100$ and $K_1 = K/20$. For such a concerning solution u_0 obtained, please see Figure 11.2.

Remark 11.3.1 *Observe that the solutions obtained are approximate critical points. They are not, in a classical sense, the global solutions for the related optimization problems. Indeed, such solutions reflect the average behavior of weak cluster points for concerning minimizing sequences.*

11.4 A convex dual variational formulation for a third similar model

In this section, we present another duality principle for a third related model in phase transition.

Let $\Omega = [0, 1] \subset \mathbb{R}$ and consider a functional $J : V \to \mathbb{R}$ where

$$J(u) = \frac{1}{2} \int_\Omega \min\{(u'-1)^2, (u'+1)^2\} \, dx + \frac{1}{2} \int_\Omega u^2 \, dx - \langle u, f \rangle_{L^2},$$

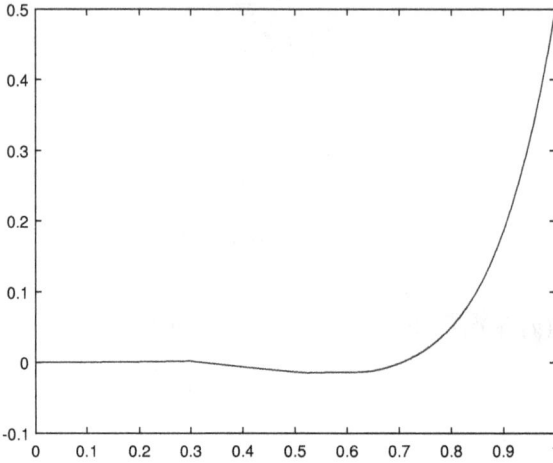

Figure 11.1: Solution $u_0(x)$ for the case $f(x) = 0$.

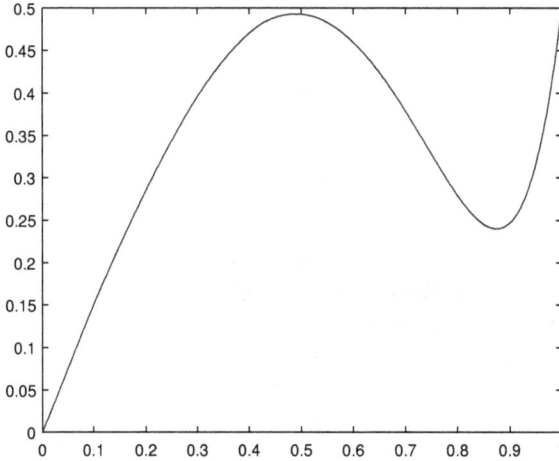

Figure 11.2: Solution $u_0(x)$ for the case $f(x) = \sin(\pi x)/2$.

and where
$$V = \{u \in W^{1,2}(\Omega) \; : \; u(0) = 0 \text{ and } u(1) = 1/2\}$$
and $f \in L^2(\Omega)$.

A global optimum point is not attained for J so that the problem of finding a global minimum for J has no solution.

Anyway, one question remains, how the minimizing sequences behave close to the infimum of J.

We intend to use the duality theory to solve such a global optimization problem in an appropriate sense to be specified.

At this point we define, $F : V \to \mathbb{R}$ and $G : V \to \mathbb{R}$ by

$$
\begin{aligned}
F(u) &= \frac{1}{2} \int_{\Omega} \min\{(u'-1)^2, (u'+1)^2\}\, dx \\
&= \frac{1}{2} \int_{\Omega} (u')^2\, dx - \int_{\Omega} |u'|\, dx + 1/2 \\
&\equiv F_1(u'),
\end{aligned}
\tag{11.11}
$$

and

$$
G(u) = \frac{1}{2} \int_{\Omega} u^2\, dx - \langle u, f \rangle_{L^2}.
$$

Denoting $Y = Y^* = L^2(\Omega)$ we also define the polar functional $F_1^* : Y^* \to \mathbb{R}$ and $G^* : Y^* \to \mathbb{R}$ by

$$
\begin{aligned}
F_1^*(v^*) &= \sup_{v \in Y}\{\langle v, v^* \rangle_{L^2} - F_1(v)\} \\
&= \frac{1}{2} \int_{\Omega} (v^*)^2\, dx + \int_{\Omega} |v^*|\, dx,
\end{aligned}
\tag{11.12}
$$

and

$$
\begin{aligned}
G^*((v^*)') &= \sup_{u \in V}\{-\langle u', v^* \rangle_{L^2} - G(u)\} \\
&= \frac{1}{2} \int_{\Omega} ((v^*)' + f)^2\, dx - \frac{1}{2} v^*(1).
\end{aligned}
\tag{11.13}
$$

Observe this is the scalar case of the calculus of variations, so that from the standard results on convex analysis, we have

$$
\inf_{u \in V} J(u) = \max_{v^* \in Y^*}\{-F_1^*(v^*) - G^*(-(v^*)')\}.
$$

Indeed, from the direct method of the calculus of variations, the maximum for the dual formulation is attained at some $\hat{v}^* \in Y^*$.

Moreover, the corresponding solution $u_0 \in V$ is obtained from the equation

$$
u_0 = \frac{\partial G((\hat{v}^*)')}{\partial (v^*)'} = (\hat{v}^*)' + f.
$$

Finally, the Euler-Lagrange equations for the dual problem stands for

$$
\begin{cases}
(v^*)'' + f' - v^* - \operatorname{sign}(v^*) = 0, & \text{in } \Omega, \\
(v^*)'(0) + f(0) = 0,\ (v^*)'(1) + f(1) = 1/2,
\end{cases}
\tag{11.14}
$$

where $\operatorname{sign}(v^*(x)) = 1$ if $v^*(x) > 0$, $\operatorname{sign}(v^*(x)) = -1$, if $v^*(x) < 0$ and

$$
-1 \le \operatorname{sign}(v^*(x)) \le 1,
$$

if $v^*(x) = 0$.

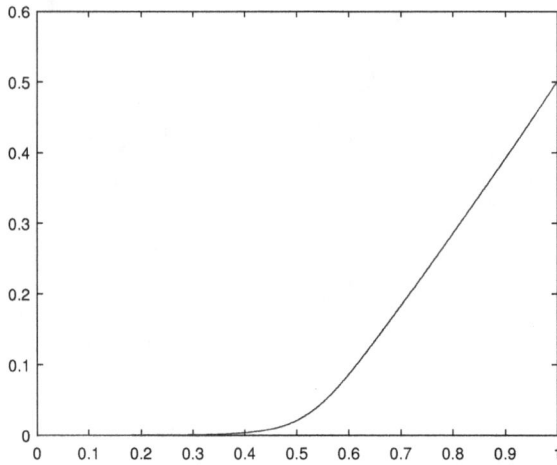

Figure 11.3: Solution $u_0(x)$ for the case $f(x) = 0$.

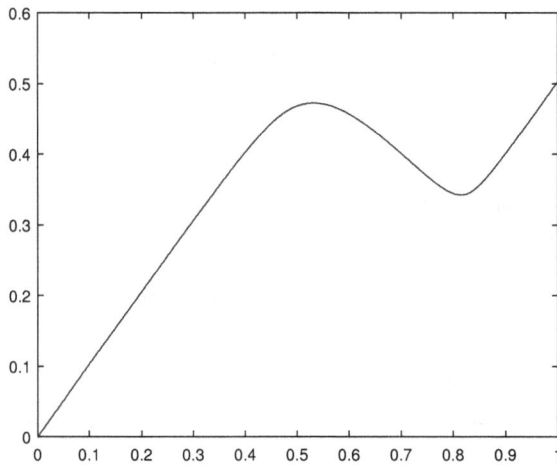

Figure 11.4: Solution $u_0(x)$ for the case $f(x) = \sin(\pi x)/2$.

We have computed the solutions v^* and corresponding solutions $u_0 \in V$ for the cases in which $f(x) = 0$ and $f(x) = \sin(\pi x)/2$.

For the solution $u_0(x)$ for the case in which $f(x) = 0$, please see Figure 11.3.

For the solution $u_0(x)$ for the case in which $f(x) = \sin(\pi x)/2$, please see Figure 11.4.

Remark 11.4.1 *Observe that such solutions u_0 obtained are not the global solutions for the related primal optimization problems. Indeed, such solutions reflect the average behavior of weak cluster points for concerning minimizing sequences.*

11.4.1 The algorithm through which we have obtained the numerical results

In this subsection we present the software in MATLAB® through which we have obtained the last numerical results.

This algorithm is for solving the concerning Euler-Lagrange equations for the dual problem, that is, for solving the equation

$$\begin{cases} (v^*)'' + f' - v^* - \text{sign}(v^*) = 0, & \text{in } \Omega, \\ (v^*)'(0) = 0, \ (v^*)'(1) = 1/2. \end{cases} \tag{11.15}$$

Here the concerning software in MATLAB. We emphasize to have used the smooth approximation

$$|v^*| \approx \sqrt{(v^*)^2 + e_1},$$

where a small value for e_1 is specified in the next lines.

****

1. clear all

2. $m_8 = 800$; (number of nodes)

3. $d = 1/m_8$;

4. $e_1 = 0.00001$;

5. $for \ i = 1 : m_8$
   $yo(i, 1) = 0.01$;
   $y_1(i, 1) = \sin(\pi * i / m_8)/2$;
   end;

6. $for \ i = 1 : m_8 - 1$
   $dy_1(i, 1) = (y_1(i+1, 1) - y_1(i, 1))/d$;
   end;

7. $for \ k = 1 : 3000$ (we have fixed the number of iterations)
   $i = 1$;
   $h_3 = 1/\sqrt{vo(i, 1)^2 + e_1}$;
   $m_{12} = 1 + d^2 * h_3 + d^2$;
   $m_{50}(i) = 1/m_{12}$;
   $z(i) = m_{50}(i) * (dy_1(i, 1) * d^2)$;

8. *for* $i = 2 : m_8 - 1$

$$h_3 = 1/\sqrt{vo(i,1)^2 + e_1};$$
$$m_{12} = 2 + h_3 * d^2 + d^2 - m50(i-1);$$
$$m50(i) = 1/m_{12};$$
$$z(i) = m_{50}(i) * (z(i-1) + dy_1(i,1) * d^2);$$

end;

9. $v(m_8, 1) = (d/2 + z(m_8 - 1))/(1 - m_{50}(m_8 - 1));$

10. *for* $i = 1 : m_8 - 1$

$$v(m_8 - i, 1) = m_{50}(m_8 - i) * v(m_8 - i + 1) + z(m_8 - i);$$

end;

11. $v(m_8/2, 1)$

12. $vo = v;$

end;

13. *for* $i = 1 : m_8 - 1$

$$u(i,1) = (v(i+1,1) - v(i,1))/d + y_1(i,1);$$

end;

14. *for* $i = 1 : m_8 - 1$

$$x(i) = i * d;$$

end;

$$plot(x, u(:, 1))$$

****

11.5 An improvement of the convexity conditions for a non-convex related model through an approximate primal formulation

In this section, we develop an approximate primal dual formulation suitable for a large class of variational models.

Here, the applications are for the Kirchhoff-Love plate model, which may be found in Ciarlet [31].

At this point, we start to describe the primal variational formulation.

Let $\Omega \subset \mathbb{R}^2$ be an open, bounded, connected set which represents the middle surface of a plate of thickness h. The boundary of Ω, which is assumed to be regular (Lipschitzian), is denoted by $\partial \Omega$. The vectorial basis related to the cartesian system $\{x_1, x_2, x_3\}$ is denoted by $(\mathbf{a}_\alpha, \mathbf{a}_3)$, where $\alpha = 1, 2$ (in general Greek indices stand for

1 or 2), and where $\mathbf{a}_3$ is the vector normal to Ω, whereas $\mathbf{a}_1$ and $\mathbf{a}_2$ are orthogonal vectors parallel to Ω. Also, $\mathbf{n}$ is the outward normal to the plate surface.

The displacements will be denoted by

$$\hat{\mathbf{u}} = \{\hat{u}_\alpha, \hat{u}_3\} = \hat{u}_\alpha \mathbf{a}_\alpha + \hat{u}_3 \mathbf{a}_3.$$

The Kirchhoff-Love relations are

$$\hat{u}_\alpha(x_1, x_2, x_3) = u_\alpha(x_1, x_2) - x_3 w(x_1, x_2),_\alpha$$
$$\text{and } \hat{u}_3(x_1, x_2, x_3) = w(x_1, x_2). \tag{11.16}$$

Here $-h/2 \le x_3 \le h/2$ so that we have $u = (u_\alpha, w) \in U$ where

$$
\begin{aligned}
U &= \{u = (u_\alpha, w) \in W^{1,2}(\Omega; \mathbb{R}^2) \times W^{2,2}(\Omega), \\
&\qquad u_\alpha = w = \frac{\partial w}{\partial \mathbf{n}} = 0 \text{ on } \partial\Omega \} \\
&= W_0^{1,2}(\Omega; \mathbb{R}^2) \times W_0^{2,2}(\Omega).
\end{aligned}
$$

It is worth emphasizing that the boundary conditions specified here refer to a clamped plate.

We also define the operator $\Lambda : U \to Y \times Y$, where $Y = Y^* = L^2(\Omega; \mathbb{R}^{2\times2})$, by

$$\Lambda(u) = \{\gamma(u), \kappa(u)\},$$

$$\gamma_{\alpha\beta}(u) = \frac{u_{\alpha,\beta} + u_{\beta,\alpha}}{2} + \frac{w_{,\alpha} w_{,\beta}}{2},$$

$$\kappa_{\alpha\beta}(u) = -w_{,\alpha\beta}.$$

The constitutive relations are given by

$$N_{\alpha\beta}(u) = H_{\alpha\beta\lambda\mu} \gamma_{\lambda\mu}(u), \tag{11.17}$$

$$M_{\alpha\beta}(u) = h_{\alpha\beta\lambda\mu} \kappa_{\lambda\mu}(u), \tag{11.18}$$

where: $\{H_{\alpha\beta\lambda\mu}\}$ and $\left\{h_{\alpha\beta\lambda\mu} = \frac{h^2}{12} H_{\alpha\beta\lambda\mu}\right\}$, are symmetric positive definite fourth order tensors. From now on, we denote $\{\overline{H}_{\alpha\beta\lambda\mu}\} = \{H_{\alpha\beta\lambda\mu}\}^{-1}$ and $\{\overline{h}_{\alpha\beta\lambda\mu}\} = \{h_{\alpha\beta\lambda\mu}\}^{-1}$.

Furthermore $\{N_{\alpha\beta}\}$ denote the membrane force tensor and $\{M_{\alpha\beta}\}$ the moment one. The plate stored energy, represented by $(G \circ \Lambda) : U \to \mathbb{R}$ is expressed by

$$(G \circ \Lambda)(u) = \frac{1}{2} \int_\Omega N_{\alpha\beta}(u) \gamma_{\alpha\beta}(u) \, dx + \frac{1}{2} \int_\Omega M_{\alpha\beta}(u) \kappa_{\alpha\beta}(u) \, dx \tag{11.19}$$

and the external work, represented by $F : U \to \mathbb{R}$, is given by

$$F(u) = \langle w, P \rangle_{L^2} + \langle u_\alpha, P_\alpha \rangle_{L^2}, \tag{11.20}$$

where $P, P_1, P_2 \in L^2(\Omega)$ are external loads in the directions $\mathbf{a}_3$, $\mathbf{a}_1$ and $\mathbf{a}_2$ respectively. The potential energy, denoted by $J : U \to \mathbb{R}$ is expressed by:

$$J(u) = (G \circ \Lambda)(u) - F(u)$$

Define now $J_3 : \tilde{U} \to \mathbb{R}$ by

$$J_3(u) = J(u) + J_5(w).$$

where

$$J_5(w) = 10 \int_\Omega \frac{a^{K b w}}{\ln(a) \, K^{3/2}} \, dx + 10 \int_\Omega \frac{a^{-K(b w - 1/100)}}{\ln(a) \, K^{3/2}} \, dx.$$

In such a case for $a = 2.71$, $K = 185$, $b = P/|P|$ in Ω and

$$\tilde{U} = \{u \in U \,:\, \|w\|_\infty \le 0.01 \text{ and } P\, w \ge 0 \text{ a.e. in } \Omega\},$$

we get

$$\frac{\partial J_3(u)}{\partial w} = \frac{\partial J(u)}{\partial w} + \frac{\partial J_5(u)}{\partial w}$$

$$\approx \frac{\partial J(u)}{\partial w} + \mathscr{O}(\pm 3.0), \qquad (11.21)$$

and

$$\frac{\partial^2 J_3(u)}{\partial w^2} = \frac{\partial^2 J(u)}{\partial w^2} + \frac{\partial^2 J_5(u)}{\partial w^2}$$

$$\approx \frac{\partial^2 J(u)}{\partial w^2} + \mathscr{O}(850). \qquad (11.22)$$

This new functional J_3 has a relevant improvement in the convexity conditions concerning the previous functional J.

Indeed, we have obtained a gain in positiveness for the second variation $\frac{\partial^2 J(u)}{\partial w^2}$, which has increased of order $\mathscr{O}(700 - 1000)$.

Moreover, the difference between the approximate and exact equation

$$\frac{\partial J(u)}{\partial w} = \mathbf{0}$$

is of order $\mathscr{O}(\pm 3.0)$ which corresponds to a small perturbation in the original equation for a load of $P = 1500 \, N/m^2$, for example. Summarizing, the exact equation may be approximately solved in an appropriate sense.

11.6 An exact convex dual variational formulation for a non-convex primal one

In this section, we develop a convex dual variational formulation suitable to compute a critical point for the corresponding primal one.

Let $\Omega \subset \mathbb{R}^2$ be an open, bounded, connected set with a regular (Lipschitzian) boundary denoted by $\partial\Omega$.

Consider a functional $J : V \to \mathbb{R}$ where

$$J(u) = F(u_x, u_y) - \langle u, f\rangle_{L^2},$$

$V = W_0^{1,2}(\Omega)$ and $f \in L^2(\Omega)$.

Here we denote $Y = Y^* = L^2(\Omega)$ and $Y_1 = Y_1^* = L^2(\Omega) \times L^2(\Omega)$.

Defining

$$V_1 = \{u \in V \ : \ \|u\|_{1,\infty} \le K_1\}$$

for some appropriate $K_1 > 0$, suppose also F is twice Fréchet differentiable and

$$\det\left\{\frac{\partial^2 F(u_x, u_y)}{\partial v_1 \partial v_2}\right\} \ne 0,$$

$\forall u \in V_1$.

Define now $F_1 : V \to \mathbb{R}$ and $F_2 : V \to \mathbb{R}$ by

$$F_1(u_x, u_y) = F(u_x, u_y) + \frac{\varepsilon}{2}\int_\Omega u_x^2\, dx + \frac{\varepsilon}{2}\int_\Omega u_y^2\, dx,$$

and

$$F_2(u_x, u_y) = \frac{\varepsilon}{2}\int_\Omega u_x^2\, dx + \frac{\varepsilon}{2}\int_\Omega u_y^2\, dx,$$

where here we denote $dx = dx_1 dx_2$.

Moreover, we define the respective Legendre transform functionals F_1^* and F_2^* as

$$F_1^*(v^*) = \langle v_1, v_1^*\rangle_{L^2} + \langle v_2, v_2^*\rangle_{L^2} - F_1(v_1, v_2),$$

where $v_1, v_2 \in Y$ are such that

$$v_1^* = \frac{\partial F_1(v_1, v_2)}{\partial v_1},$$

$$v_2^* = \frac{\partial F_1(v_1, v_2)}{\partial v_2},$$

and

$$F_2^*(v^*) = \langle v_1, v_1^* + f_1\rangle_{L^2} + \langle v_2, v_2^*\rangle_{L^2} - F_2(v_1, v_2),$$

where $v_1, v_2 \in Y$ are such that

$$v_1^* + f_1 = \frac{\partial F_2(v_1, v_2)}{\partial v_1},$$

$$v_2^* = \frac{\partial F_2(v_1, v_2)}{\partial v_2}.$$

Here f_1 is any function such that

$$(f_1)_x = f, \text{ in } \Omega.$$

Furthermore, we define

$$
\begin{aligned}
J^*(v^*) &= -F_1^*(v^*) + F_2^*(v^*) \\
&= -F_1^*(v^*) + \frac{1}{2\varepsilon}\int_\Omega (v_1^* + f_1)^2 \, dx + \frac{1}{2\varepsilon}\int_\Omega (v_2^*)^2 \, dx. \quad (11.23)
\end{aligned}
$$

Observe that through the target conditions

$$v_1^* + f_1 = \varepsilon u_x,$$

$$v_2^* = \varepsilon u_y,$$

we may obtain the compatibility condition

$$(v_1^* + f_1)_y - (v_2^*)_x = 0.$$

Define now

$$A^* = \{v^* = (v_1^*, v_2^*) \in B_r(0,0) \subset Y_1^* \; : \; (v_1^* + f_1)_y - (v_2^*)_x = 0, \text{ in } \Omega\},$$

for some appropriate $r > 0$ such that J^* is convex in $B_r(0,0)$.

Consider the problem of minimizing J^* subject to $v^* \in A^*$.

Assuming $r > 0$ is large enough so that the restriction in r is not active, at this point we define the associated Lagrangian

$$J_1^*(v^*, \varphi) = J^*(v^*) + \langle \varphi, (v_1^* + f)_y - (v_2^*)_x \rangle_{L^2},$$

where φ is an appropriate Lagrange multiplier.

Therefore

$$
\begin{aligned}
J_1^*(v^*) &= -F_1^*(v^*) + \frac{1}{2\varepsilon}\int_\Omega (v_1^* + f_1)^2 \, dx + \frac{1}{2\varepsilon}\int_\Omega (v_2^*)^2 \, dx \\
&\quad + \langle \varphi, (v_1^* + f)_y - (v_2^*)_x \rangle_{L^2}. \quad (11.24)
\end{aligned}
$$

The optimal point in question will be a solution of the corresponding Euler-Lagrange equations for J_1^*.

From the variation of J_1^* in v_1^* we obtain

$$-\frac{\partial F_1^*(v^*)}{\partial v_1^*} + \frac{v_1^* + f}{\varepsilon} - \frac{\partial \varphi}{\partial y} = 0. \quad (11.25)$$

From the variation of J_1^* in v_2^* we obtain

$$-\frac{\partial F_1^*(v^*)}{\partial v_2^*} + \frac{v_2^*}{\varepsilon} + \frac{\partial \varphi}{\partial x} = 0. \quad (11.26)$$

From the variation of J_1^* in φ we have

$$(v_1^* + f)_y - (v_2^*)_x = 0.$$

From this last equation, we may obtain $u \in V$ such that

$$v_1^* + f = \varepsilon u_x,$$

and

$$v_2^* = \varepsilon u_y.$$

From this and the previous extremal equations indicated we have

$$-\frac{\partial F_1^*(v^*)}{\partial v_1^*} + u_x - \frac{\partial \varphi}{\partial y} = 0,$$

and

$$-\frac{\partial F_1^*(v^*)}{\partial v_2^*} + u_y + \frac{\partial \varphi}{\partial x} = 0.$$

so that

$$v_1^* + f = \frac{\partial F_1(u_x - \varphi_y, u_y + \varphi_x)}{\partial v_1},$$

and

$$v_2^* = \frac{\partial F_1(u_x - \varphi_y, u_y + \varphi_x)}{\partial v_2}.$$

From this and equation (11.25) and (11.26) we have

$$-\varepsilon\left(\frac{\partial F_1^*(v^*)}{\partial v_1^*}\right)_x - \varepsilon\left(\frac{\partial F_1^*(v^*)}{\partial v_2^*}\right)_y$$
$$+ (v_1^* + f_1)_x + (v_2^*)_y$$
$$= -\varepsilon u_{xx} - \varepsilon u_{yy} + (v_1^*)_x + (v_2^*)_y + f = 0. \tag{11.27}$$

Replacing the expressions of v_1^* and v_2^* into this last equation, we have

$$-\varepsilon u_{xx} - \varepsilon u_{yy} + \left(\frac{\partial F_1(u_x - \varphi_y, u_y + \varphi_x)}{\partial v_1}\right)_x + \left(\frac{\partial F_1(u_x - \varphi_y, u_y + \varphi_x)}{\partial v_2}\right)_y + f = 0,$$

so that

$$\left(\frac{\partial F(u_x - \varphi_y, u_y + \varphi_x)}{\partial v_1}\right)_x + \left(\frac{\partial F(u_x - \varphi_y, u_y + \varphi_x)}{\partial v_2}\right)_y + f = 0, \text{ in } \Omega. \tag{11.28}$$

Observe that if

$$\nabla^2 \varphi = 0$$

then there exists $\hat{u}$ such that u and φ are also such that

$$u_x - \varphi_y = \hat{u}_x$$

and

$$u_y + \varphi_x = \hat{u}_y.$$

The boundary conditions for φ must be such that $\hat{u} \in W_0^{1,2}$.
From this and equation (11.28) we obtain

$$\delta J(\hat{u}) = \mathbf{0}.$$

Summarizing, we may obtain a solution $\hat{u} \in W_0^{1,2}$ of equation $\delta J(\hat{u}) = \mathbf{0}$ by minimizing J^* on A^*.

Finally, observe that clearly J^* is convex in an appropriate large ball $B_r(0,0)$ for some appropriate $r > 0$.

11.7 Another primal dual formulation for a related model

Let $\Omega \subset \mathbb{R}^3$ be an open, bounded and connected set with a regular boundary denoted by $\partial \Omega$.

Consider the functional $J : V \to \mathbb{R}$ where

$$
\begin{aligned}
J(u) &= \frac{\gamma}{2} \int_\Omega \nabla u \cdot \nabla u \, dx + \frac{\alpha}{2} \int_\Omega (u^2 - \beta)^2 \, dx \\
&\quad - \langle u, f \rangle_{L^2},
\end{aligned}
\tag{11.29}
$$

$\alpha > 0$, $\beta > 0$, $\gamma > 0$, $V = W_0^{1,2}(\Omega)$ and $f \in L^2(\Omega)$.

Denoting $Y = Y^* = L^2(\Omega)$, define now $J_1^* : V \times Y^* \to \mathbb{R}$ by

$$
\begin{aligned}
J_1^*(u, v_0^*) &= -\frac{\gamma}{2} \int_\Omega \nabla u \cdot \nabla u \, dx - \langle u^2, v_0^* \rangle_{L^2} \\
&\quad + \frac{K_1}{2} \int_\Omega (-\gamma \nabla^2 u + 2 v_0^* u - f)^2 \, dx + \langle u, f \rangle_{L^2} \\
&\quad + \frac{1}{2\alpha} \int_\Omega (v_0^*)^2 \, dx + \beta \int_\Omega v_0^* \, dx,
\end{aligned}
\tag{11.30}
$$

Define also

$$A^+ = \{u \in V \ : \ u f \geq 0, \text{ a.e. in } \Omega\},$$

$$V_2 = \{u \in V \ : \ \|u\|_\infty \leq K_3\},$$

and

$$V_1 = V_2 \cap A^+$$

for some appropriate $K_3 > 0$ to be specified.

Moreover define

$$B^* = \{v_0^* \in Y^* \ : \ \|v_0^*\|_\infty \leq K\}$$

for some appropriate $K > 0$ to be specified.

Observe that, denoting

$$\varphi = -\gamma\nabla^2 u + 2v_0^* u - f$$

we have

$$\frac{\partial^2 J_1^*(u,v_0^*)}{\partial(v_0^*)^2} = \frac{1}{\alpha} + 4K_1 u^2$$

$$\frac{\partial^2 J_1^*(u,v_0^*)}{\partial u^2} = \gamma\nabla^2 - 2v_0^* + K_1(-\gamma\nabla^2 + 2v_0^*)^2$$

and

$$\frac{\partial^2 J_1^*(u,v_0^*)}{\partial u \partial v_0^*} = K_1(2\varphi + 2(-\gamma\nabla^2 u + 2v_0^* u)) - 2u$$

so that

$$
\begin{aligned}
&\det\{\delta^2 J_1^*(u,v_0^*)\} \\
&= \frac{\partial^2 J_1^*(u,v_0^*)}{\partial(v_0^*)^2}\frac{\partial^2 J_1^*(u,v_0^*)}{\partial u^2} - \left(\frac{\partial^2 J_1^*(u,v_0^*)}{\partial u \partial v_0^*}\right)^2 \\
&= \frac{K_1(-\gamma\nabla^2 + 2v_0^*)^2}{\alpha} - \frac{\gamma\nabla^2 + 2v_0^* + 4\alpha u^2}{\alpha} \\
&\quad -4K_1^2\varphi^2 - 8K_1\varphi(-\gamma\nabla^2 + 2v_0^*)u + 8K_1\varphi u \\
&\quad +4K_1(-\gamma\nabla^2 u + 2v_0^* u)u.
\end{aligned}
\tag{11.31}
$$

Observe now that a critical point $\varphi = 0$ and $(-\gamma\nabla^2 u + 2v_0^* u)u = fu \geq 0$ in Ω. Therefore, for an appropriate large $K_1 > 0$, also at a critical point, we have

$$
\begin{aligned}
&\det\{\delta^2 J_1^*(u,v_0^*)\} \\
&= 4K_1 fu - \frac{\delta^2 J(u)}{\alpha} + K_1\frac{(-\gamma\nabla^2 + 2v_0^*)^2}{\alpha} > 0.
\end{aligned}
\tag{11.32}
$$

Remark 11.7.1 *From this last equation we may observe that J_1^* has a large region of convexity about any critical point $(u_0,\hat{v}_0^*)$, that is, there exists a large $r > 0$ such that J_1^* is convex on $B_r(u_0,\hat{v}_0^*)$.*

With such results in mind, we may easily prove the following theorem.

Theorem 11.7.2 *Assume $K_1 \gg \max\{1, K, K_3\}$ and suppose $(u_0,\hat{v}_0^*) \in V_1 \times B^*$ is such that*

$$\delta J_1^*(u_0,\hat{v}_0^*) = \mathbf{0}.$$

Under such hypotheses, there exists $r > 0$ such that J_1^* is convex in $E^* = B_r(u_0,\hat{v}_0^*) \cap (V_1 \times B^*)$,

$$\delta J(u_0) = \mathbf{0},$$

and

$$-J(u_0) = J_1(u_0,\hat{v}_0^*) = \inf_{(u,v_0^*)\in E^*} J_1^*(u,v_0^*).$$

11.8 A third primal dual formulation for a related model

Let $\Omega \subset \mathbb{R}^3$ be an open, bounded and connected set with a regular boundary denoted by $\partial \Omega$.

Consider the functional $J : V \to \mathbb{R}$ where

$$J(u) = \frac{\gamma}{2} \int_\Omega \nabla u \cdot \nabla u \, dx + \frac{\alpha}{2} \int_\Omega (u^2 - \beta)^2 \, dx$$
$$- \langle u, f \rangle_{L^2}, \tag{11.33}$$

$\alpha > 0$, $\beta > 0$, $\gamma > 0$, $V = W_0^{1,2}(\Omega)$ and $f \in L^2(\Omega)$.

Denoting $Y = Y^* = L^2(\Omega)$, define now $J_1^* : V \times Y^* \times Y^* \to \mathbb{R}$ by

$$J_1^*(u, v_0^*, v_1^*) = \frac{\gamma}{2} \int_\Omega \nabla u \cdot \nabla u \, dx + \frac{1}{2} \int_\Omega K u^2 \, dx$$
$$- \langle u, v_1^* \rangle_{L^2} + \frac{1}{2} \int_\Omega \frac{(v_1^*)^2}{(-2v_0^* + K)} \, dx$$
$$+ \frac{1}{2(\alpha + \varepsilon)} \int_\Omega (v_0^* - \alpha(u^2 - \beta))^2 \, dx + \langle u, f \rangle_{L^2}$$
$$- \frac{1}{2\alpha} \int_\Omega (v_0^*)^2 \, dx - \beta \int_\Omega v_0^* \, dx, \tag{11.34}$$

where $\varepsilon > 0$ is a small real constant.

Define also

$$A^+ = \{u \in V : u f \geq 0, \text{ a.e. in } \Omega\},$$

$$V_2 = \{u \in V : \|u\|_\infty \leq K_3\},$$

and

$$V_1 = V_2 \cap A^+$$

for some appropriate $K_3 > 0$ to be specified.

Moreover, define

$$B^* = \{v_0^* \in Y^* : \|v_0^*\|_\infty \leq K_4\}$$

and

$$D^* = \{v_1^* \in Y^* : \|v_1^*\| \leq K_5\},$$

for some appropriate real constants $K_4, K_5 > 0$ to be specified.

Remark 11.8.1 *Define now*

$$H_1(u, v_0^*) = -\gamma \nabla^2 + 2v_0^* + 4\alpha u^2$$

and

$$\hat{E}_{v_0^*} = \{u \in V : H_1(u, v_0^*) \geq \mathbf{0}\}.$$

For a fixed $v_0^* \in B^*$, we are going to prove that $C^* = \hat{E}_{v_0^*} \cap V_1$ is a convex set.

Assume, for a finite dimensional problem version, in a finite differences or finite element context, that

$$-\gamma \nabla^2 - 2v_0^* \leq 0,$$

so that for $K_1 > 0$ be sufficiently large, we have

$$-\gamma \nabla^2 + 2v_0^* - K_1 u^2 \leq 0.$$

Observe now that

$$H_1(u, v_0^*) = -\gamma \nabla^2 + 2v_0^* - K_1 u^2 + 4\alpha u^2 + K_1 u^2.$$

Let $u_1, u_2 \in C^*$ and $\lambda \in [0, 1]$.
Thus

$$sign\,(u_1) = sign\,(u_2)\ in\ \Omega$$

so that

$$\lambda |u_1| + (1 - \lambda)|u_2| = |\lambda u_1 + (1 - \lambda)u_2|\ in\ \Omega.$$

Observe now that

$$H_1(u_1, v_0^*) \geq 0$$

and

$$H_1(u_2, v_0^*) \geq 0$$

so that

$$4\alpha u_1^2 + K_1 u_1^2 \geq \gamma \nabla^2 - 2v_0^* + K_1 u_1^2 \geq 0,$$

and

$$4\alpha u_2^2 + K_1 u_1^2 \geq \gamma \nabla^2 - 2v_0^* + K_1 u_2^2 \geq 0,$$

so that

$$\sqrt{4\alpha + K_1}\,|u_1| \geq \sqrt{\gamma \nabla^2 - 2v_0^* + K_1 u_1^2}$$

and

$$\sqrt{4\alpha + K_1}\,|u_2| \geq \sqrt{\gamma \nabla^2 - 2v_0^* + K_1 u_2^2}.$$

From such results we obtain

$$
\begin{aligned}
\sqrt{4\alpha + K_1}\,|\lambda u_1 + (1 - \lambda)u_2| &= \sqrt{4\alpha + K_1}(\lambda |u_1| + (1 - \lambda)|u_2|) \\
&\geq \lambda \sqrt{\gamma \nabla^2 - 2v_0^* + K_1 u_1^2} + (1 - \lambda)\sqrt{\gamma \nabla^2 - 2v_0^* + K_1 u_2^2} \\
&\geq \sqrt{\gamma \nabla^2 - 2v_0^* + K_1 (\lambda u_1 + (1 - \lambda)u_2)^2}. \quad (11.35)
\end{aligned}
$$

From this we obtain

$$(4\alpha + K_1)(\lambda u_1 + (1 - \lambda)u_2)^2 \geq \gamma \nabla^2 - 2v_0^* + K_1 (\lambda u_1 + (1 - \lambda)u_2)^2,$$

so that

$$H_1(\lambda u_1 + (1 - \lambda)u_2, v_0^*) \geq 0.$$

Hence $\hat{E}_{v_0^*}$ is convex. Since V_1 is also clearly convex, we have obtained that $C^* = \hat{E}_{v_0^*} \cap V_1$ is convex.
Such a result we will be used many times in the next sections.

Observe that, defining

$$\varphi = v_0^* - \alpha(u^2 - \beta)$$

we may obtain

$$\frac{\partial^2 J_1^*(u, v_0^*, v_1^*)}{\partial u^2} = -\gamma \nabla^2 + K + \frac{\alpha}{\alpha + \varepsilon} 4u^2 - 2\varphi \frac{\alpha}{\alpha + \varepsilon}$$

$$\frac{\partial^2 J_1^*(u, v_0^*, v_1^*)}{\partial (v_1^*)^2} = \frac{1}{-2v_0^* + K}$$

and

$$\frac{\partial^2 J_1^*(u, v_0^*, v_1^*)}{\partial u \partial v_1^*} = -1$$

so that

$$\det\left\{ \frac{\partial^2 J_1^*(u, v_0^*, v_1^*)}{\partial u \partial v_1^*} \right\}$$

$$= \frac{\partial^2 J_1^*(u, v_1^*, v_0^*)}{\partial (v_1^*)^2} \frac{\partial^2 J_1^*(u, v_1^*, v_0^*)}{\partial u^2} - \left(\frac{\partial^2 J_1^*(u, v_1^*, v_0^*)}{\partial u \partial v_1^*} \right)^2$$

$$= \frac{-\gamma \nabla^2 + 2v_0^* + 4\frac{\alpha^2}{\alpha + \varepsilon}u^2 - 2\frac{\alpha}{\alpha + \varepsilon}\varphi}{-2v_0^* + K}$$

$$\equiv H(u, v_0^*). \tag{11.36}$$

However, at a critical point, we have $\varphi = \mathbf{0}$ so that, we define

$$C_{v_0^*}^* = \{u \in V \ : \ \varphi \le \mathbf{0}\}.$$

From such results, assuming $K \gg \max\{K_3, K_4, K_5\}$, define now

$$E_{v_0^*} = \{u \in V \ : \ H(u, v_0^*) > \mathbf{0}\}.$$

Observe that similarly as it was develop in remark 11.8.1, we may prove that $E_{v_0^*}$ is a convex set.

With such results in mind, we may easily prove the following theorem.

Theorem 11.8.2 *Suppose* $(u_0, \hat{v}_1^*, \hat{v}_0^*) \in E^* = (V_1 \cap C_{\hat{v}_0^*}^* \cap E_{\hat{v}_0^*}) \times D^* \times B^*$ *and* $-\gamma \nabla^2 - 2\hat{v}_0^* \le \mathbf{0}$ *is such that*

$$\delta J_1^*(u_0, \hat{v}_0^*, \hat{v}_1^*) = \mathbf{0}.$$

Under such hypotheses, we have that

$$\delta J(u_0) = \mathbf{0}$$

and

$$
\begin{aligned}
J(u_0) &= \inf_{u \in V_1} J(u) \\
&= J_1^*(u_0, \hat{v}_1^*, \hat{v}_0^*) \\
&= \inf_{(u,v_1^*) \in V_1 \times D^*} \left\{ \sup_{v_0^* \in B^*} J_1^*(u, v_1^*, v_0^*) \right\} \\
&= \sup_{v_0^* \in B^*} \left\{ \inf_{(u,v_1^*) \in V_1 \times D^*} J_1^*(u, v_1^*, v_0^*) \right\}.
\end{aligned}
\tag{11.37}
$$

Proof 11.1   The proof that
$$
\delta J(u_0) = \mathbf{0}
$$
and
$$
J(u_0) = J_1^*(u_0, \hat{v}_1^*, \hat{v}_0^*)
$$
may be easily made similarly as in the previous sections.

Moreover, from the hypotheses, we have
$$
J_1^*(u_0, \hat{v}_1^*, \hat{v}_0^*) = \inf_{(u,v_1^*) \in V_1 \times D^*} J_1^*(u, v_1^*, \hat{v}_0^*)
$$

and
$$
J_1^*(u_0, \hat{v}_1^*, \hat{v}_0^*) = \sup_{v_0^* \in B^*} J_1^*(u_0, \hat{v}_1^*, v_0^*).
$$

From this, from a standard saddle point theorem and the remaining hypotheses, we may infer that

$$
\begin{aligned}
J(u_0) &= J_1^*(u_0, \hat{v}_1^*, \hat{v}_0^*) \\
&= \inf_{(u,v_1^*) \in V_1 \times D^*} \left\{ \sup_{v_0^* \in B^*} J_1^*(u, v_1^*, v_0^*) \right\} \\
&= \sup_{v_0^* \in B^*} \left\{ \inf_{(u,v_1^*) \in V_1 \times D^*} J_1^*(u, v_1^*, v_0^*) \right\}.
\end{aligned}
\tag{11.38}
$$

Moreover, observe that

$$
\begin{aligned}
J_1^*(u_0, \hat{v}_1^*, \hat{v}_0^*) &= \inf_{(u, v_1^*) \in V_1 \times D^*} J_1^*(u, v_1^*, \hat{v}_0^*) \\
&\leq \frac{\gamma}{2} \int_\Omega \nabla u \cdot \nabla u \, dx + \frac{K}{2} \int_\Omega u^2 \, dx \\
&\quad + \langle u^2, \hat{v}_0^* \rangle_{L^2} - \frac{K}{2} \int_\Omega u^2 \, dx \\
&\quad - \frac{1}{2\alpha} \int_\Omega (\hat{v}_0^*)^2 \, dx - \beta \int_\Omega \hat{v}_0^* \, dx \\
&\quad + \frac{1}{2(\alpha + \varepsilon)} \int_\Omega (\hat{v}_0^* - \alpha(u^2 - \beta))^2 \, dx - \langle u, f \rangle_{L^2} \\
&\leq \sup_{v_0^* \in Y^*} \left\{ \frac{\gamma}{2} \int_\Omega \nabla u \cdot \nabla u \, dx + \langle u^2, v_0^* \rangle \right. \\
&\quad - \frac{1}{2\alpha} \int_\Omega (v_0^*)^2 \, dx - \beta \int_\Omega v_0^* \, dx \\
&\quad \left. + \frac{1}{2(\alpha + \varepsilon)} \int_\Omega (v_0^* - \alpha(u^2 - \beta))^2 \, dx - \langle u, f \rangle_{L^2} \right\} \\
&= \frac{\gamma}{2} \int_\Omega \nabla u \cdot \nabla u \, dx + \frac{\alpha}{2} \int_\Omega (u^2 - \beta)^2 \, dx \\
&\quad - \langle u, f \rangle_{L^2}, \ \forall u \in V_1.
\end{aligned}
\tag{11.39}
$$

Summarizing, we have got

$$
J(u_0) = J_1^*(u_0, \hat{v}_1^*, \hat{v}_0^*) \leq \inf_{u \in V_1} J(u).
$$

From such results, we may infer that

$$
\begin{aligned}
J(u_0) &= \inf_{u \in V_1} J(u) \\
&= J_1^*(u_0, \hat{v}_1^*, \hat{v}_0^*) \\
&= \inf_{(u, v_1^*) \in V_1 \times D^*} \left\{ \sup_{v_0^* \in B^*} J_1^*(u, v_1^*, v_0^*) \right\} \\
&= \sup_{v_0^* \in B^*} \left\{ \inf_{(u, v_1^*) \in V_1 \times D^*} J_1^*(u, v_1^*, v_0^*) \right\}.
\end{aligned}
\tag{11.40}
$$

The proof is complete.

11.9   A fourth primal dual formulation for a related model

Let $\Omega \subset \mathbb{R}^3$ be an open, bounded and connected set with a regular boundary denoted by $\partial \Omega$.

Consider the functional $J : V \to \mathbb{R}$ where

$$J(u) = \frac{\gamma}{2} \int_\Omega \nabla u \cdot \nabla u \, dx + \frac{\alpha}{2} \int_\Omega (u^2 - \beta)^2 \, dx$$
$$- \langle u, f \rangle_{L^2}, \tag{11.41}$$

$\alpha > 0$, $\beta > 0$, $\gamma > 0$, $V = W_0^{1,2}(\Omega)$ and $f \in L^2(\Omega)$.

Denoting $Y = Y^* = L^2(\Omega)$, define now $J_1^* : V \times Y^* \to \mathbb{R}$ by

$$J_1^*(u, v_0^*) = \frac{\gamma}{2} \int_\Omega \nabla u \cdot \nabla u \, dx - \langle u^2, v_0^* \rangle_{L^2}$$
$$+ \frac{1}{2(\alpha + \varepsilon)} \int_\Omega (v_0^* - \alpha(u^2 - \beta))^2 \, dx - \langle u, f \rangle_{L^2}$$
$$- \frac{1}{2\alpha} \int_\Omega (v_0^*)^2 \, dx - \beta \int_\Omega v_0^* \, dx, \tag{11.42}$$

where $\varepsilon > 0$ is a small real constant.

Define also

$$A^+ = \{u \in V \ : \ u f \geq 0, \text{ a.e. in } \Omega\},$$
$$V_2 = \{u \in V \ : \ \|u\|_\infty \leq K_3\},$$

and

$$V_1 = V_2 \cap A^+$$

for some appropriate real constant $K_3 > 0$.

Moreover define

$$B^* = \{v_0^* \in Y^* \ : \ \|v_0^*\|_\infty \leq K_4\}$$

for some appropriate real constant $K_4 > 0$.

Observe that, denoting $\varphi = v_0^* - \alpha(u^2 - \beta)$, we may obtain

$$\frac{\partial^2 J_1^*(u, v_0^*)}{\partial u^2} = -\gamma \nabla^2 + 2v_0$$
$$+ \frac{\alpha^2}{\alpha + \varepsilon} 4u^2 - 2 \frac{\varphi}{\alpha + \varepsilon} \alpha$$
$$\equiv H(u, v_0^*), \tag{11.43}$$

and

$$\frac{\partial^2 J_1^*(u, v_0^*)}{\partial (v_0^*)^2} = -\frac{1}{\alpha} + \frac{1}{\alpha + \varepsilon} < 0$$

However, at a critical point, we have $\varphi = 0$ so that, we define

$$C_{v_0^*}^* = \{u \in V \ : \ \varphi \leq 0\}.$$

Define also,

$$E_{v_0^*} = \{u \in V \ : \ H(u, v_0^*) > 0\}.$$

Remark 11.9.1 *Similarly as it was developed in remark 11.8.1 we may prove that such a $E_{v_0^*}$ is a convex set.*

With such results in mind, we may easily prove the following theorem.

Theorem 11.9.2 *Suppose $(u_0, \hat{v}_0^*) \in E^* = (V_1 \cap C_{\hat{v}_0^*} \cap E_{\hat{v}_0^*}) \times B^*$ is such that*

$$\delta J_1^*(u_0, \hat{v}_0^*) = \mathbf{0}.$$

Under such hypotheses, we have that

$$\delta J(u_0) = \mathbf{0}$$

and

$$
\begin{aligned}
J(u_0) &= \inf_{u \in V_1} J(u) \\
&= J_1^*(u_0, \hat{v}_0^*) \\
&= \inf_{u \in V_1} \left\{ \sup_{v_0^* \in B^*} J_1^*(u, v_0^*) \right\} \\
&= \sup_{v_0^* \in B^*} \left\{ \inf_{u \in V_1} J_1^*(u, v_0^*) \right\}.
\end{aligned}
\tag{11.44}
$$

Proof 11.2 The proof that

$$\delta J(u_0) = \mathbf{0}$$

and

$$J(u_0) = J_1^*(u_0, \hat{v}_0^*)$$

may be easily made similarly as in the previous sections.

Moreover, from the hypotheses, we have

$$J_1^*(u_0, \hat{v}_0^*) = \inf_{u \in V_1} J_1^*(u, \hat{v}_0^*)$$

and

$$J_1^*(u_0, \hat{v}_0^*) = \sup_{v_0^* \in B^*} J_1^*(u_0, v_0^*).$$

From this, from a standard saddle point theorem and the remaining hypotheses, we may infer that

$$
\begin{aligned}
J(u_0) &= J_1^*(u_0, \hat{v}_0^*) \\
&= \inf_{u \in V_1} \left\{ \sup_{v_0^* \in B^*} J_1^*(u, v_0^*) \right\} \\
&= \sup_{v_0^* \in B^*} \left\{ \inf_{u \in V_1} J_1^*(u, v_0^*) \right\}.
\end{aligned}
\tag{11.45}
$$

Moreover, observe that

$$
\begin{aligned}
J_1^*(u_0, \hat{v}_0^*) &= \inf_{u \in V_1} J_1^*(u, \hat{v}_0^*) \\
&\leq \frac{\gamma}{2} \int_\Omega \nabla u \cdot \nabla u \, dx + \langle u^2, \hat{v}_0^* \rangle_{L^2} \\
&\quad - \frac{1}{2\alpha} \int_\Omega (\hat{v}_0^*)^2 \, dx - \beta \int_\Omega \hat{v}_0^* \, dx \\
&\quad + \frac{1}{2(\alpha + \varepsilon)} \int_\Omega (\hat{v}_0^* - \alpha(u^2 - \beta))^2 \, dx - \langle u, f \rangle_{L^2} \\
&\leq \sup_{v_0^* \in Y^*} \left\{ \frac{\gamma}{2} \int_\Omega \nabla u \cdot \nabla u \, dx + \langle u^2, v_0^* \rangle \right. \\
&\quad - \frac{1}{2\alpha} \int_\Omega (v_0^*)^2 \, dx - \beta \int_\Omega v_0^* \, dx \\
&\quad \left. + \frac{1}{2(\alpha + \varepsilon)} \int_\Omega (v_0^* - \alpha(u^2 - \beta))^2 \, dx - \langle u, f \rangle_{L^2} \right\} \\
&= \frac{\gamma}{2} \int_\Omega \nabla u \cdot \nabla u \, dx + \frac{\alpha}{2} \int_\Omega (u^2 - \beta)^2 \, dx \\
&\quad - \langle u, f \rangle_{L^2}, \ \forall u \in V_1.
\end{aligned}
\tag{11.46}
$$

Summarizing, we have got

$$
J(u_0) = J_1^*(u_0, \hat{v}_0^*) \leq \inf_{u \in V_1} J(u).
$$

From such results, we may infer that

$$
\begin{aligned}
J(u_0) &= \inf_{u \in V_1} J(u) \\
&= J_1^*(u_0, \hat{v}_0^*) \\
&= \inf_{u \in V_1} \left\{ \sup_{v_0^* \in B^*} J_1^*(u, v_0^*) \right\} \\
&= \sup_{v_0^* \in B^*} \left\{ \inf_{u \in V_1} J_1^*(u, v_0^*) \right\}.
\end{aligned}
\tag{11.47}
$$

The proof is complete.

11.10 One more primal dual formulation for a related model

Let $\Omega \subset \mathbb{R}^3$ be an open, bounded and connected set with a regular boundary denoted by $\partial\Omega$.

Consider the functional $J : V \to \mathbb{R}$ where

$$
\begin{aligned}
J(u) &= \frac{\gamma}{2} \int_\Omega \nabla u \cdot \nabla u \, dx + \frac{\alpha}{2} \int_\Omega (u^2 - \beta)^2 \, dx \\
&\quad - \langle u, f \rangle_{L^2},
\end{aligned}
\tag{11.48}
$$

$\alpha > 0$, $\beta > 0$, $\gamma > 0$, $V = W_0^{1,2}(\Omega)$ and $f \in L^2(\Omega)$.

Denoting $Y = Y^* = L^2(\Omega)$, define now $J_1^* : V \times Y^* \times Y^* \to \mathbb{R}$ by

$$
\begin{aligned}
J_1^*(u, v_1^*, v_0^*) &= \frac{\gamma}{2} \int_\Omega \nabla u \cdot \nabla u \, dx + \frac{K}{2} \int_\Omega u^2 \, dx - \langle u, v_1^* \rangle_{L^2} \\
&\quad + \frac{1}{2} \int_\Omega \frac{(v_1^*)^2}{-2v_0^* + K} \, dx - \langle u, f \rangle_{L^2} \\
&\quad + \frac{K^2}{2} \int_\Omega \left(\frac{v_1^* + f}{-\gamma \nabla^2 + K} - \frac{v_1^*}{-2v_0^* + K} \right)^2 dx \\
&\quad - \frac{1}{2\alpha} \int_\Omega (v_0^*)^2 \, dx - \beta \int_\Omega v_0^* \, dx,
\end{aligned}
\tag{11.49}
$$

Define also

$$
A^+ = \{ u \in V \ : \ u f \geq 0, \text{ a.e. in } \Omega \},
$$
$$
V_2 = \{ u \in V \ : \ \|u\|_\infty \leq K_3 \},
$$

and

$$
V_1 = V_2 \cap A^+
$$

specifically for a constant $K_3 = \sqrt{\frac{1}{5\alpha}}$.

Moreover define

$$
B^* = \{ v_0^* \in Y^* \ : \ \|v_0^*\|_\infty \leq K_4 \}
$$

and

$$
D^* = \{ v_1^* \in Y^* \ : \ \|v_1^*\|_\infty \leq K_5 \}
$$

for some appropriate real constants $K_4 > 0$ and $K_5 > 0$.

Observe that

$$
\frac{\partial^2 J_1^*(u, v_1^*, v_0^*)}{\partial u^2} = -\gamma \nabla^2 + K,
$$

$$
\frac{\partial^2 J_1^*(u, v_1^*, v_0^*)}{\partial (v_1^*)^2} = \frac{1}{-2v_0^* + K} + \frac{K^2(-\gamma \nabla^2 + 2v_0^*)^2}{[(-\gamma \nabla^2 + K)(-2v_0^* + K)]^2},
$$

$$
\frac{\partial^2 J_1^*(u, v_1^*, v_0^*)}{\partial u \, \partial v_1^*} = -1,
$$

so that

$$
\begin{aligned}
\det \left(\frac{\partial^2 J_1^*(u, v_1^*, v_0^*)}{\partial u \, \partial v_1^*} \right) &= \frac{\partial^2 J_1^*(u, v_1^*, v_0^*)}{\partial (v_1^*)^2} \frac{\partial^2 J_1^*(u, v_1^*, v_0^*)}{\partial u^2} - \left(\frac{\partial^2 J_1^*(u, v_1^*, v_0^*)}{\partial u \, \partial v_1^*} \right)^2 \\
&= \mathcal{O} \left(\frac{K^2(2(-\gamma \nabla^2 + 2v_0^*) + 2(-\gamma \nabla^2 + 2v_0^*)^2)}{(-\gamma \nabla^2 + K)(-2v_0^* + K)^2} \right) \\
&\equiv H(v_0^*).
\end{aligned}
\tag{11.50}
$$

With such results in mind, we may easily prove the following theorem.

Theorem 11.10.1 *Assume $K \gg \max\{K_3, K_4, K_5, 1\}$ and suppose $(u_0, \hat{v}_1^*, \hat{v}_0^*) \in V_1 \times D^* \times B^*$ is such that*

$$\delta J_1^*(u_0, \hat{v}_1^*, \hat{v}_0^*) = 0.$$

Suppose also $H(\hat{v}_0^*) > 0$.
Under such hypotheses, we have that

$$\delta J(u_0) = \mathbf{0}$$

and

$$
\begin{aligned}
J(u_0) &= \inf_{u \in V_1} \left\{ J(u) + \frac{K^2}{2} \int_\Omega \left(\frac{(-\gamma \nabla^2 u + 2\hat{v}_0^* u - f)}{-\gamma \nabla^2 + K} \right)^2 dx \right\} \\
&= J_1^*(u_0, \hat{v}_1^*, \hat{v}_0^*) \\
&= \inf_{(u, v_1^*) \in V_1 \times D^*} \left\{ \sup_{v_0^* \in B^*} J_1^*(u,, v_1^*, v_0^*) \right\} \\
&= \sup_{v_0^* \in B^*} \left\{ \inf_{(u, v_1^*) \in V_1 \times D^*} J_1^*(u, v_1^*, v_0^*) \right\}.
\end{aligned}
\tag{11.51}
$$

Proof 11.3 The proof that

$$\delta J(u_0) = -\gamma \nabla^2 u_0 + 2\alpha(u^2 - \beta)u_0 - f = \mathbf{0},$$
$$\hat{v}_0^* = \alpha(u_0^2 - \beta)$$

and

$$J(u_0) = J(u_0) + \frac{K^2}{2} \int_\Omega \left(\frac{(-\gamma \nabla^2 u_0 + 2\hat{v}_0^* u_0 - f)}{-\gamma \nabla^2 + K} \right)^2 dx = J_1^*(u_0, \hat{v}_1^*, \hat{v}_0^*)$$

may be easily made similarly as in the previous sections.
Moreover, from the hypotheses, we have

$$J_1^*(u_0, \hat{v}_1^*, \hat{v}_0^*) = \inf_{(u, v_1^*) \in V_1 \times D^*} J_1^*(u, v_1^*, \hat{v}_0^*)$$

and

$$J_1^*(u_0, \hat{v}_1^*, \hat{v}_0^*) = \sup_{v_0^* \in B^*} J_1^*(u_0, \hat{v}_1^*, v_0^*).$$

From this, from a standard saddle point theorem and the remaining hypotheses, we may infer that

$$
\begin{aligned}
J(u_0) &= J_1^*(u_0, \hat{v}_1^*, \hat{v}_0^*) \\
&= \inf_{(u, v_1^*) \in V_1 \times D^*} \left\{ \sup_{v_0^* \in B^*} J_1^*(u, v_1^*, v_0^*) \right\} \\
&= \sup_{v_0^* \in B^*} \left\{ \inf_{(u, v_1^*) \in V_1 \times D^*} J_1^*(u, v_1^*, v_0^*) \right\}.
\end{aligned}
\tag{11.52}
$$

Moreover, observe that

$$
\begin{aligned}
J_1^*(u_0, \hat{v}_1^*, \hat{v}_0^*) &= \inf_{(u,v_1^*)\in V_1 \times D^*} J_1^*(u, v_1^*, \hat{v}_0^*) \\
&\leq \frac{\gamma}{2}\int_\Omega \nabla u \cdot \nabla u \, dx + \langle u^2, \hat{v}_0^*\rangle_{L^2} \\
&\quad -\frac{1}{2\alpha}\int_\Omega (\hat{v}_0^*)^2 \, dx - \beta \int_\Omega \hat{v}_0^* \, dx - \langle u, f\rangle_{L^2} \\
&\quad +\frac{K^2}{2}\int_\Omega \left(\frac{(-\gamma\nabla^2 u + 2\hat{v}_0^* u - f)}{-\gamma\nabla^2 + K}\right)^2 dx \\
&\leq \sup_{v_0^*\in Y^*} \left\{ \frac{\gamma}{2}\int_\Omega \nabla u \cdot \nabla u \, dx + \langle u^2, v_0^*\rangle \right. \\
&\quad -\frac{1}{2\alpha}\int_\Omega (v_0^*)^2 \, dx - \beta \int_\Omega v_0^* \, dx - \langle u, f\rangle_{L^2} \\
&\quad \left. +\frac{K^2}{2}\int_\Omega \left(\frac{(-\gamma\nabla^2 u + 2\hat{v}_0^* u - f)}{-\gamma\nabla^2 + K}\right)^2 dx \right\} \\
&= \frac{\gamma}{2}\int_\Omega \nabla u \cdot \nabla u \, dx + \frac{\alpha}{2}\int_\Omega (u^2 - \beta)^2 \, dx \\
&\quad -\langle u, f\rangle_{L^2} \\
&\quad +\frac{K^2}{2}\int_\Omega \left(\frac{(-\gamma\nabla^2 u + 2\hat{v}_0^* u - f)}{-\gamma\nabla^2 + K}\right)^2 dx, \ \forall u \in V_1. \quad (11.53)
\end{aligned}
$$

From this we have got

$$
\begin{aligned}
J_1^*(u_0, \hat{v}_1^*, \hat{v}_0^*) &\leq \frac{\gamma}{2}\int_\Omega \nabla u \cdot \nabla u \, dx + \frac{\alpha}{2}\int_\Omega (u^2 - \beta)^2 \, dx - \langle u, f\rangle_{L^2} \\
&\quad +\frac{K^2}{2}\int_\Omega \left(\frac{(-\gamma\nabla^2 u + 2\hat{v}_0^* u - f)}{-\gamma\nabla^2 + K}\right)^2 dx, \ \forall u \in V_1. \quad (11.54)
\end{aligned}
$$

Therefore, from such results we may obtain

$$
\begin{aligned}
J(u_0) &= \inf_{u\in V_1} \left\{ J(u) + \frac{K^2}{2}\int_\Omega \left(\frac{(-\gamma\nabla^2 u + 2\hat{v}_0^* u - f)}{-\gamma\nabla^2 + K}\right)^2 dx \right\} \\
&= J_1^*(u_0, \hat{v}_1^*, \hat{v}_0^*) \\
&= \inf_{(u,v_1^*)\in V_1 \times D^*} \left\{ \sup_{v_0^*\in B^*} J_1^*(u,, v_1^*, v_0^*) \right\} \\
&= \sup_{v_0^*\in B^*} \left\{ \inf_{(u,v_1^*)\in V_1 \times D^*} J_1^*(u, v_1^*, v_0^*) \right\}. \quad (11.55)
\end{aligned}
$$

The proof is complete.

11.11 Another primal dual formulation for a related model

In this section, we present another primal dual formulation.

Let $\Omega \subset \mathbb{R}^3$ be an open, bounded and connected set with a regular boundary denoted by $\partial \Omega$.

Consider the functional $J : V \to \mathbb{R}$ where

$$
\begin{aligned}
J(u) &= \frac{\gamma}{2} \int_\Omega \nabla u \cdot \nabla u \, dx + \frac{\alpha}{2} \int_\Omega (u^2 - \beta)^2 \, dx \\
&\quad - \langle u, f \rangle_{L^2},
\end{aligned}
\tag{11.56}
$$

$\alpha > 0$, $\beta > 0$, $\gamma > 0$, $V = W_0^{1,2}(\Omega)$ and $f \in L^2(\Omega)$.

Denoting $Y = Y^* = L^2(\Omega)$, define now $J_1^* : V \times Y^* \to \mathbb{R}$ by

$$
\begin{aligned}
J_1^*(u, v_0^*) &= \frac{\gamma}{2} \int_\Omega \nabla u \cdot \nabla u \, dx + \langle u^2, v_0^* \rangle_{L^2} \\
&\quad + \frac{\alpha - \varepsilon}{2} \int_\Omega u^4 \, dx - \langle u, f \rangle_{L^2} \\
&\quad - \frac{1}{2\varepsilon} \int_\Omega (v_0^* + \alpha \beta)^2 \, dx,
\end{aligned}
\tag{11.57}
$$

and $J_2^* : V \times Y^* \to \mathbb{R}$, by

$$
\begin{aligned}
J_2^*(u, v_0^*) &= \frac{\gamma}{2} \int_\Omega \nabla u \cdot \nabla u \, dx + \langle u^2, v_0^* \rangle_{L^2} \\
&\quad + \frac{K_1}{2} \int_\Omega (-\gamma \nabla^2 u + 2 v_0^* u - 2(\alpha - \varepsilon) u^3 - f)^2 \, dx \\
&\quad + \frac{\alpha - \varepsilon}{2} \int_\Omega u^4 \, dx - \langle u, f \rangle_{L^2} \\
&\quad - \frac{1}{2\varepsilon} \int_\Omega (v_0^* + \alpha \beta)^2 \, dx,
\end{aligned}
\tag{11.58}
$$

Define also

$$
A^+ = \{ u \in V \ : \ u f \geq 0, \text{ a.e. in } \Omega \},
$$

$$
V_2 = \{ u \in V \ : \ \|u\|_\infty \leq K_3 \},
$$

and

$$
V_1 = V_2 \cap A^+.
$$

Moreover define

$$
B^* = \{ v_0^* \in Y^* \ : \ \|v_0^*\|_\infty \leq K_4 \}
$$

for some appropriate constants $K_3 > 0$ and $K_4 > 0$.

Observe that, for $K_1 = 1/\sqrt{\varepsilon}$, we have

$$
\begin{aligned}
\frac{\partial^2 J_2^*(u, v_0^*)}{\partial u^2} &= (-\gamma \nabla^2 + 2v_0^* + 6(\alpha - \varepsilon)u^2) + K_1(-\gamma \nabla^2 + 2v_0^* + 6(\alpha - \varepsilon)u^2)^2 \\
&\quad + K_1(-\gamma \nabla^2 u + 2v_0^* u + 2(\alpha - \varepsilon)u^3 - f)12(\alpha - \varepsilon)u \, dx, \quad (11.59)
\end{aligned}
$$

$$
\begin{aligned}
\frac{\partial^2 J_2^*(u, v_0^*)}{\partial (v_0^*)^2} &= K_1 4u^2 - \frac{1}{\varepsilon} \\
&< \mathbf{0}, \ \forall u \in V_1, \ v_0^* \in B^*. \quad (11.60)
\end{aligned}
$$

Define now

$$
A_2(u, v_0^*) = (-\gamma \nabla^2 u + 2v_0^* u + 2(\alpha - \varepsilon)u^3 - f)12(\alpha - \varepsilon)u,
$$

$$
C^* = \{(u, v_0^*) \in V \times B^* \ : \ \|A_2(u, v_0^*)\|_\infty \leq \varepsilon_1\}
$$

for a small real parameter $\varepsilon_1 > 0$.

Finally, define

$$
E_{v_0^*} = \left\{ u \in V \ : \ \frac{\partial^2 A_2(u, v_0^*)}{\partial u^2} > \mathbf{0} \right\}.
$$

Remark 11.11.1 *Similarly as it was developed in remark 11.8.1, we may prove that such a $E_{v_0^*}$ is a convex set.*

Thus,

$$
E_{v_0^*} \cap V_1
$$

is a convex set, $\forall v_0^* \in B^*$ (for the proof of a similar result please see Theorem 8.7.1 at pages 297, 298 and 299 in [13]).)

With such results in mind, we may easily prove the following theorem.

Theorem 11.11.2 *Assume $K_1 \gg 1 \gg \varepsilon_1$ and suppose $(u_0, \hat{v}_0^*) \in V_1 \times B^*$ is such that*

$$
\delta J_2^*(u_0, \hat{v}_0^*) = \mathbf{0}
$$

and $u_0 \in E_{\hat{v}_0^*}$.

Under such hypotheses, we have that

$$
\delta J(u_0) = \mathbf{0}
$$

and

$$
\begin{aligned}
J(u_0) &= \inf_{u \in V_1} \left\{ J(u) + \frac{K_1}{2} \int_\Omega (-\gamma \nabla^2 u + 2\hat{v}_0^* u + 2(\alpha - \varepsilon)u^3 - f)^2 \, dx \right\} \\
&= J_2^*(u_0, \hat{v}_0^*) \\
&= \sup_{v_0^* \in B^*} \left\{ \inf_{u \in V_1} J_2^*(u, v_0^*) \right\}. \quad (11.61)
\end{aligned}
$$

Proof 11.4 The proof that

$$\delta J(u_0) = -\gamma \nabla^2 u_0 + 2\alpha(u^2 - \beta)u_0 - f = \mathbf{0},$$

and

$$J(u_0) = J(u_0) + \frac{K_1}{2} \int_\Omega (-\gamma \nabla^2 u_0 + 2\hat{v}_0^* u_0 + 2(\alpha - \varepsilon)u_0^3 - f)^2 \, dx = J_2^*(u_0, \hat{v}_0^*)$$

may be easily made similarly as in the previous sections.

Moreover, from the hypotheses and from the above lines, since J_2^* is concave in v_0^* on $V_1 \times B^*$ and $u_0 \in E_{\hat{v}_0^*}$, we have that

$$J_2^*(u_0, \hat{v}_0^*) = \inf_{u \in V_1} J_2^*(u, \hat{v}_0^*)$$

and

$$J_2^*(u_0, \hat{v}_0^*) = \sup_{v_0^* \in B^*} J_2^*(u_0, v_0^*).$$

From this, from the standard Saddle Point Theorem and the remaining hypotheses, we may infer that

$$
\begin{aligned}
J(u_0) &= J_2^*(u_0, \hat{v}_0^*) \\
&= \inf_{u \in V_1} \left\{ \sup_{v_0^* \in B^*} J_2^*(u, v_1^*, v_0^*) \right\} \\
&= \sup_{v_0^* \in B^*} \left\{ \inf_{u \in V_1} J_2^*(u, v_0^*) \right\}.
\end{aligned}
\tag{11.62}
$$

Moreover, observe that

$$
\begin{aligned}
J_2^*(u_0, \hat{v}_0^*) &= \inf_{u \in V_1} J_2^*(u, \hat{v}_0^*) \\
&\leq \frac{\gamma}{2} \int_\Omega \nabla u \cdot \nabla u \, dx + \langle u^2, \hat{v}_0^* \rangle_{L^2} + \frac{\alpha - \varepsilon}{2} \int_\Omega u^4 \, dx \\
&\quad - \frac{1}{2\varepsilon} \int_\Omega (\hat{v}_0^* + \alpha\beta)^2 \, dx - \langle u, f \rangle_{L^2} \\
&\quad + \frac{K_1}{2} \int_\Omega (-\gamma \nabla^2 u + 2\hat{v}_0^* u + 2(\alpha - \varepsilon)u^3 - f)^2 \, dx \\
&\leq \sup_{v_0^* \in Y^*} \left\{ \frac{\gamma}{2} \int_\Omega \nabla u \cdot \nabla u \, dx + \langle u^2, v_0^* \rangle_{L^2} + \frac{\alpha - \varepsilon}{2} \int_\Omega u^4 \, dx \right. \\
&\quad - \frac{1}{2\varepsilon} \int_\Omega (v_0^* + \alpha\beta)^2 \, dx - \langle u, f \rangle_{L^2} \\
&\quad \left. + \frac{K_1}{2} \int_\Omega (-\gamma \nabla^2 u + 2\hat{v}_0^* u + 2(\alpha - \varepsilon)u^3 - f)^2 \, dx \right\} \\
&= J(u) + \frac{K_1}{2} \int_\Omega (-\gamma \nabla^2 u + 2\hat{v}_0^* u + 2(\alpha - \varepsilon)u^3 - f)^2 \, dx, \ \forall u \in V_1. \tag{11.63}
\end{aligned}
$$

From this we have got

$$J_2^*(u_0, \hat{v}_0^*) \leq J(u) + \frac{K_1}{2} \int_\Omega (-\gamma \nabla^2 u + 2\hat{v}_0^* u + 2(\alpha - \varepsilon)u^3 - f)^2 \, dx, \ \forall u \in V_1. \quad (11.64)$$

Therefore, from such results we may obtain

$$
\begin{aligned}
J(u_0) &= \inf_{u \in V_1} \left\{ J(u) + \frac{K_1}{2} \int_\Omega (-\gamma \nabla^2 u + 2\hat{v}_0^* u + 2(\alpha - \varepsilon)u^3 - f)^2 \, dx \right\} \\
&= J_2^*(u_0, \hat{v}_0^*) \\
&= \sup_{v_0^* \in B^*} \left\{ \inf_{u \in V_1} J_2^*(u, v_0^*) \right\}. \quad (11.65)
\end{aligned}
$$

The proof is complete.

Chapter 12

Dual Variational Formulations for a Large Class of Non-Convex Models in the Calculus of Variations

12.1 Introduction

This article develops dual variational formulations for a large class of models in variational optimization. The results are established through basic tools of functional analysis, convex analysis, and the duality theory. The main duality principle is developed as an application to a Ginzburg-Landau type system in superconductivity in the absence of a magnetic field. In the first sections, we develop new general dual convex variational formulations, more specifically, dual formulations with a large region of convexity around the critical points which are suitable for the non-convex optimization for a large class of models in physics and engineering. Finally, in the last section we present some numerical results concerning the generalized method of lines applied to a Ginzburg-Landau type equation.

Such results are based on the works of J.J. Telega and W.R. Bielski [10, 11, 74, 75] and on a D.C. optimization approach developed in Toland [81].

Remark 12.1.1 *This chapter has been published in a similar article format by the MDPI Journal Mathematics, reference [20]:*

F.S. Botelho, Dual Variational Formulations for a Large Class of Non-Convex Models in the Calculus of Variations, Mathematics 2023, 11(1), 63; https://doi.org/10.3390/math11010063 - 24 Dec 2022.

About the other references, details on the Sobolev spaces involved are found in [1]. Related results on convex analysis and duality theory are addressed in [12, 13, 14, 22, 62]. Finally, similar models on the superconductivity physics may be found in [4, 52].

Remark 12.1.2 *It is worth highlighting, we may generically denote*

$$\int_\Omega [(-\gamma \nabla^2 + K I_d)^{-1} v^*] v^* \, dx$$

simply by

$$\int_\Omega \frac{(v^*)^2}{-\gamma \nabla^2 + K} \, dx,$$

where I_d denotes a concerning identity operator.

Other similar notations may be used along this text as their indicated meaning are sufficiently clear.

Also, ∇^2 denotes the Laplace operator and for real constants $K_2 > 0$ and $K_1 > 0$, the notation $K_2 \gg K_1$ means that $K_2 > 0$ is much larger than $K_1 > 0$.

Finally, we adopt the standard Einstein convention of summing up repeated indices, unless otherwise indicated.

In order to clarify the notation, here we introduce the definition of topological dual space.

Definition 12.1.3 (Topological dual spaces) *Let U be a Banach space. We shall define its dual topological space, as the set of all linear continuous functionals defined on U. We suppose such a dual space of U, may be represented by another Banach space U^*, through a bilinear form $\langle \cdot, \cdot \rangle_U : U \times U^* \to \mathbb{R}$ (here we are referring to standard representations of dual spaces of Sobolev and Lebesgue spaces). Thus, given $f : U \to \mathbb{R}$ linear and continuous, we assume the existence of a unique $u^* \in U^*$ such that*

$$f(u) = \langle u, u^* \rangle_U, \forall u \in U. \tag{12.1}$$

The norm of f, denoted by $\|f\|_{U^*}$, is defined as

$$\|f\|_{U^*} = \sup_{u \in U} \{ |\langle u, u^* \rangle_U| \ : \ \|u\|_U \leq 1 \} \equiv \|u^*\|_{U^*}. \tag{12.2}$$

At this point we start to describe the primal and dual variational formulations.

Let $\Omega \subset \mathbb{R}^3$ be an open, bounded, connected set with a regular (Lipschitzian) boundary denoted by $\partial \Omega$.

Firstly we emphasize that, for the Banach space $Y = Y^* = L^2(\Omega)$, we have

$$\langle v, v^* \rangle_{L^2} = \int_\Omega v\, v^*\, dx, \ \forall v, v^* \in L^2(\Omega).$$

For the primal formulation we consider the functional $J : U \to \mathbb{R}$ where

$$
\begin{aligned}
J(u) \ &= \ \frac{\gamma}{2} \int_\Omega \nabla u \cdot \nabla u\, dx \\
&\quad + \frac{\alpha}{2} \int_\Omega (u^2 - \beta)^2\, dx - \langle u, f \rangle_{L^2}.
\end{aligned}
\tag{12.3}
$$

Here we assume $\alpha > 0, \beta > 0, \gamma > 0$, $U = W_0^{1,2}(\Omega)$, $f \in L^2(\Omega)$. Moreover we denote

$$Y = Y^* = L^2(\Omega).$$

Define also $G_1 : U \to \mathbb{R}$ by

$$G_1(u) = \frac{\gamma}{2} \int_\Omega \nabla u \cdot \nabla u\, dx,$$

$G_2 : U \times Y \to \mathbb{R}$ by

$$G_2(u, v) = \frac{\alpha}{2} \int_\Omega (u^2 - \beta + v)^2\, dx + \frac{K}{2} \int_\Omega u^2\, dx,$$

and $F : U \to \mathbb{R}$ by

$$F(u) = \frac{K}{2} \int_\Omega u^2\, dx,$$

where $K \gg \gamma$.

It is worth highlighting that in such a case

$$J(u) = G_1(u) + G_2(u, 0) - F(u) - \langle u, f \rangle_{L^2}, \ \forall u \in U.$$

Furthermore, define the following specific polar functionals specified, namely, $G_1^* : [Y^*]^2 \to \mathbb{R}$ by

$$
\begin{aligned}
G_1^*(v_1^* + z^*) \ &= \ \sup_{u \in U} \{ \langle u, v_1^* + z^* \rangle_{L^2} - G_1(u) \} \\
&= \ \frac{1}{2} \int_\Omega [(-\gamma \nabla^2)^{-1}(v_1^* + z^*)](v_1^* + z^*)\, dx,
\end{aligned}
\tag{12.4}
$$

$G_2^* : [Y^*]^2 \to \mathbb{R}$ by

$$
\begin{aligned}
G_2^*(v_2^*, v_0^*) \ &= \ \sup_{(u,v) \in U \times Y} \{ \langle u, v_2^* \rangle_{L^2} + \langle v, v_0^* \rangle_{L^2} - G_2(u, v) \} \\
&= \ \frac{1}{2} \int_\Omega \frac{(v_2^*)^2}{2v_0^* + K}\, dx \\
&\quad + \frac{1}{2\alpha} \int_\Omega (v_0^*)^2\, dx + \beta \int_\Omega v_0^*\, dx,
\end{aligned}
\tag{12.5}
$$

if $v_0^* \in B^*$ where
$$B^* = \{v_0^* \in Y^* \ : \ 2v_0^* + K > K/2 \text{ in } \Omega\},$$

and finally, $F^* : Y^* \to \mathbb{R}$ by

$$
\begin{aligned}
F^*(z^*) &= \sup_{u \in U} \{\langle u, z^* \rangle_{L^2} - F(u)\} \\
&= \frac{1}{2K} \int_\Omega (z^*)^2 \, dx.
\end{aligned}
\tag{12.6}
$$

Define also

$$A^* = \{v^* = (v_1^*, v_2^*, v_0^*) \in [Y^*]^2 \times B^* \ : \ v_1^* + v_2^* - f = 0, \text{ in } \Omega\},$$

$J^* : [Y^*]^4 \to \mathbb{R}$ by

$$J^*(v^*, z^*) = -G_1^*(v_1^* + z^*) - G_2^*(v_2^*, v_0^*) + F^*(z^*)$$

and $J_1^* : [Y^*]^4 \times U \to \mathbb{R}$ by

$$J_1^*(v^*, z^*, u) = J^*(v^*, z^*) + \langle u, v_1^* + v_2^* - f \rangle_{L^2}.$$

12.2 The main duality principle, a convex dual formulation and the concerning proximal primal functional

Our main result is summarized by the following theorem.

Theorem 12.2.1 *Considering the definitions and statements in the last section, suppose also $(\hat{v}^*, \hat{z}^*, u_0) \in [Y^*]^2 \times B^* \times Y^* \times U$ is such that*

$$\delta J_1^*(\hat{v}^*, \hat{z}^*, u_0) = \mathbf{0}.$$

Under such hypotheses, we have

$$\delta J(u_0) = \mathbf{0},$$

$$\hat{v}^* \in A^*$$

and

$$
\begin{aligned}
J(u_0) &= \inf_{u \in U} \left\{ J(u) + \frac{K}{2} \int_\Omega |u - u_0|^2 \, dx \right\} \\
&= J^*(\hat{v}^*, \hat{z}^*) \\
&= \sup_{v^* \in A^*} \{J^*(v^*, \hat{z}^*)\}.
\end{aligned}
\tag{12.7}
$$

Proof 12.1 Since

$$\delta J_1^*(\hat{v}^*, \hat{z}^*, u_0) = 0$$

from the variation in v_1^* we obtain

$$-\frac{(\hat{v}_1^* + \hat{z}^*)}{-\gamma\nabla^2} + u_0 = 0 \text{ in } \Omega,$$

so that

$$\hat{v}_1^* + \hat{z}^* = -\gamma\nabla^2 u_0.$$

From the variation in v_2^* we obtain

$$-\frac{\hat{v}_2^*}{2\hat{v}_0^* + K} + u_0 = 0, \text{ in } \Omega.$$

From the variation in v_0^* we also obtain

$$\frac{(\hat{v}_2^*)^2}{(2\hat{v}_0^* + K)^2} - \frac{\hat{v}_0^*}{\alpha} - \beta = 0$$

and therefore,

$$\hat{v}_0^* = \alpha(u_0^2 - \beta).$$

From the variation in u we get

$$\hat{v}_1^* + \hat{v}_2^* - f = 0, \text{ in } \Omega$$

and thus

$$\hat{v}^* \in A^*.$$

Finally, from the variation in z^*, we obtain

$$-\frac{(\hat{v}_1^* + \hat{z}^*)}{-\gamma\nabla^2} + \frac{\hat{z}^*}{K} = 0, \text{ in } \Omega.$$

so that

$$-u_0 + \frac{\hat{z}^*}{K} = 0,$$

that is,

$$\hat{z}^* = Ku_0 \text{ in } \Omega.$$

From such results and $\hat{v}^* \in A^*$ we get

$$
\begin{aligned}
0 &= \hat{v}_1^* + \hat{v}_2^* - f \\
&= -\gamma\nabla^2 u_0 - \hat{z}^* + 2(v_0^*)u_0 + Ku_0 - f \\
&= -\gamma\nabla^2 u_0 + 2\alpha(u_0^2 - \beta)u_0 - f,
\end{aligned}
\tag{12.8}
$$

so that

$$\delta J(u_0) = \mathbf{0}.$$

Also from this and from the Legendre transform proprieties we have

$$G_1^*(\hat{v}_1^* + \hat{z}^*) = \langle u_0, \hat{v}_1^* + \hat{z}^* \rangle_{L^2} - G_1(u_0),$$

$$G_2^*(\hat{v}_2^*, \hat{v}_0^*) = \langle u_0, \hat{v}_2^* \rangle_{L^2} + \langle 0, v_0^* \rangle_{L^2} - G_2(u_0, 0),$$

$$F^*(\hat{z}^*) = \langle u_0, \hat{z}^* \rangle_{L^2} - F(u_0)$$

and thus we obtain

$$
\begin{aligned}
J^*(\hat{v}^*, \hat{z}^*) &= -G_1^*(\hat{v}_1^* + \hat{z}^*) - G_2^*(\hat{v}_2^*, \hat{v}_0^*) + F^*(\hat{z}^*) \\
&= -\langle u_0, \hat{v}_1^* + \hat{v}_2^* \rangle + G_1(u_0) + G_2(u_0, 0) - F(u_0) \\
&= -\langle u_0, f \rangle_{L^2} + G_1(u_0) + G_2(u_0, 0) - F(u_0) \\
&= J(u_0). \tag{12.9}
\end{aligned}
$$

Summarizing, we have got

$$J^*(\hat{v}^*, \hat{z}^*) = J(u_0). \tag{12.10}$$

On the other hand

$$
\begin{aligned}
J^*(\hat{v}^*, \hat{z}^*) &= -G_1^*(\hat{v}_1^* + \hat{z}^*) - G_2^*(\hat{v}_2^*, \hat{v}_0^*) + F^*(\hat{z}^*) \\
&\leq -\langle u, \hat{v}_1^* + \hat{z}^* \rangle_{L^2} - \langle u, \hat{v}_2^* \rangle_{L^2} - \langle 0, v_0^* \rangle_{L^2} + G_1(u) + G_2(u, 0) + F^*(\hat{z}^*) \\
&= -\langle u, f \rangle_{L^2} + G_1(u) + G_2(u, 0) - \langle u, \hat{z}^* \rangle_{L^2} + F^*(\hat{z}^*) \\
&= -\langle u, f \rangle_{L^2} + G_1(u) + G_2(u, 0) - F(u) + F(u) - \langle u, \hat{z}^* \rangle_{L^2} + F^*(\hat{z}^*) \\
&= J(u) + \frac{K}{2} \int_\Omega u^2 \, dx - \langle u, \hat{z}^* \rangle_{L^2} + F^*(\hat{z}^*) \\
&= J(u) + \frac{K}{2} \int_\Omega u^2 \, dx - K \langle u, u_0 \rangle_{L^2} + \frac{K}{2} \int_\Omega u_0^2 \, dx \\
&= J(u) + \frac{K}{2} \int_\Omega |u - u_0|^2 \, dx, \ \forall u \in U. \tag{12.11}
\end{aligned}
$$

Finally by a simple computation we may obtain the Hessian

$$\left\{ \frac{\partial^2 J^*(v^*, z^*)}{\partial (v^*)^2} \right\} < 0$$

in $[Y^*]^2 \times B^* \times Y^*$, so that we may infer that J^* is concave in v^* in $[Y^*]^2 \times B^* \times Y^*$. Therefore, from this, (12.10) and (12.11), we have

$$
\begin{aligned}
J(u_0) &= \inf_{u \in U} \left\{ J(u) + \frac{K}{2} \int_\Omega |u - u_0|^2 \, dx \right\} \\
&= J^*(\hat{v}^*, \hat{z}^*) \\
&= \sup_{v^* \in A^*} \{ J^*(v^*, \hat{z}^*) \}. \tag{12.12}
\end{aligned}
$$

The proof is complete.

12.3   A primal dual variational formulation

In this section, we develop a more general primal dual variational formulation suitable for a large class of models in non-convex optimization.

Consider again $U = W_0^{1,2}(\Omega)$ and let $G : U \to \mathbb{R}$ and $F : U \to \mathbb{R}$ be three times Fréchet differentiable functionals. Let $J : U \to \mathbb{R}$ be defined by

$$J(u) = G(u) - F(u), \ \forall u \in U.$$

Assume $u_0 \in U$ is such that

$$\delta J(u_0) = \mathbf{0}$$

and

$$\delta^2 J(u_0) > \mathbf{0}.$$

Denoting $v^* = (v_1^*, v_2^*)$, define $J^* : U \times Y^* \times Y^* \to \mathbb{R}$ by

$$J^*(u, v^*) = \frac{1}{2}\|v_1^* - G'(u)\|_2^2 + \frac{1}{2}\|v_2^* - F'(u)\|_2^2 + \frac{1}{2}\|v_1^* - v_2^*\|_2^2 \qquad (12.13)$$

Denoting $L_1^*(u, v^*) = v_1^* - G'(u)$ and $L_2^*(u, v^*) = v_2^* - F'(u)$, define also

$$C^* = \left\{ (u, v^*) \in U \times Y^* \times Y^* \ : \ \|L_1^*(u, v_1^*)\|_\infty \le \frac{1}{K} \text{ and } \|L_2^*(u, v_1^*)\|_\infty \le \frac{1}{K} \right\},$$

for an appropriate $K > 0$ to be specified.

Observe that in C^* the Hessian of J^* is given by

$$\{\delta^2 J^*(u, v^*)\} = \left\{ \begin{array}{ccc} G''(u)^2 + F''(u)^2 + \mathscr{O}(1/K) & -G''(u) & -F''(u) \\ -G''(u) & 2 & -1 \\ -F''(u) & -1 & 2 \end{array} \right\}, \quad (12.14)$$

Observe also that

$$\det\left\{ \frac{\partial^2 J^*(u, v^*)}{\partial v_1^* \partial v_2^*} \right\} = 3,$$

and

$$\det\{\delta^2 J^*(u, v^*)\} = (G''(u) - F''(u))^2 + \mathscr{O}(1/K) = (\delta^2 J(u))^2 + \mathscr{O}(1/K).$$

Define now

$$\hat{v}_1^* = G'(u_0),$$
$$\hat{v}_2^* = F'(u_0),$$

so that

$$\hat{v}_1^* - \hat{v}_2^* = \mathbf{0}.$$

From this we may infer that $(u_0, \hat{v}_1^*, \hat{v}_2^*) \in C^*$ and

$$J^*(u_0, \hat{v}^*) = 0 = \min_{(u, v^*) \in C^*} J^*(u, v^*).$$

Moreover, for $K > 0$ sufficiently big, J^* is convex in a neighborhood of $(u_0, \hat{v}^*)$. Therefore, in the last lines, we have proven the following theorem.

Theorem 12.3.1 *Under the statements and definitions of the last lines, there exist* $r_0 > 0$ *and* $r_1 > 0$ *such that*

$$J(u_0) = \min_{u \in B_{r_0}(u_0)} J(u)$$

and $(u_0, \hat{v}_1^*, \hat{v}_2^*) \in C^*$ *is such that*

$$J^*(u_0, \hat{v}^*) = 0 = \min_{(u,v^*) \in U \times [Y^*]^2} J^*(u, v^*).$$

Moreover, J^* is convex in

$$B_{r_1}(u_0, \hat{v}^*).$$

12.4 One more duality principle and a concerning primal dual variational formulation

In this section, we establish a new duality principle and a related primal dual formulation.

The results are based on the approach of Toland [81].

12.4.1 Introduction

Let $\Omega \subset \mathbb{R}^3$ be an open, bounded, connected set with a regular (Lipschitzian) boundary denoted by $\partial \Omega$.

Let $J : V \to \mathbb{R}$ be a functional such that

$$J(u) = G(u) - F(u), \forall u \in V,$$

where $V = W_0^{1,2}(\Omega)$.

Suppose G, F are both three times Fréchet differentiable convex functionals such that

$$\frac{\partial^2 G(u)}{\partial u^2} > 0$$

and

$$\frac{\partial^2 F(u)}{\partial u^2} > 0$$

$\forall u \in V$.

Assume also there exists $\alpha_1 \in \mathbb{R}$ such that

$$\alpha_1 = \inf_{u \in V} J(u).$$

Moreover, suppose that if $\{u_n\} \subset V$ is such that

$$\|u_n\|_V \to \infty$$

then
$$J(u_n) \to +\infty, \text{ as } n \to \infty.$$

At this point we define $J^{**} : V \to \mathbb{R}$ by
$$J^{**}(u) = \sup_{(v^*,\alpha) \in H^*} \{\langle u, v^* \rangle + \alpha\},$$

where
$$H^* = \{(v^*, \alpha) \in V^* \times \mathbb{R} : \langle v, v^* \rangle_V + \alpha \le F(v), \forall v \in V\}.$$

Observe that $(0, \alpha_1) \in H^*$, so that
$$J^{**}(u) \ge \alpha_1 = \inf_{u \in V} J(u).$$

On the other hand, clearly we have
$$J^{**}(u) \le J(u), \forall u \in V,$$

so that we have got
$$\alpha_1 = \inf_{u \in V} J(u) = \inf_{u \in V} J^{**}(u).$$

Let $u \in V$.

Since J is strongly continuous, there exist $\delta > 0$ and $A > 0$ such that,
$$\alpha_1 \le J^{**}(v) \le J(v) \le A, \forall v \in B_\delta(u).$$

From this, considering that J^{**} is convex on V, we may infer that J^{**} is continuous at u, $\forall u \in V$.

Hence J^{**} is strongly lower semi-continuous on V, and since J^{**} is convex we may infer that J^{**} is weakly lower semi-continuous on V.

Let $\{u_n\} \subset V$ be a sequence such that
$$\alpha_1 \le J(u_n) < \alpha_1 + \frac{1}{n}, \forall n \in \mathbb{N}.$$

Hence
$$\alpha_1 = \lim_{n \to \infty} J(u_n) = \inf_{u \in V} J(u) = \inf_{u \in V} J^{**}(u).$$

Suppose there exists a subsequence $\{u_{n_k}\}$ of $\{u_n\}$ such that
$$\|u_{n_k}\|_V \to \infty, \text{ as } k \to \infty.$$

From the hypothesis we have
$$J(u_{n_k}) \to +\infty, \text{ as } k \to \infty,$$

which contradicts
$$\alpha_1 \in \mathbb{R}.$$

Therefore there exists $K > 0$ such that

$$\|u_n\|_V \leq K, \ \forall u \in V.$$

Since V is reflexive, from this and the Katutani Theorem, there exists a subsequence $\{u_{n_k}\}$ of $\{u_n\}$ and $u_0 \in V$ such that

$$u_{n_k} \rightharpoonup u_0, \text{ weakly in } V.$$

Consequently, from this and considering that J^{**} is weakly lower semi-continuous, we have got

$$\alpha_1 = \liminf_{k \to \infty} J^{**}(u_{n_k}) \geq J^{**}(u_0),$$

so that

$$J^{**}(u_0) = \min_{u \in V} J^{**}(u).$$

Define $G^*, F^* : V^* \to \mathbb{R}$ by

$$G^*(v^*) = \sup_{u \in V} \{\langle u, v^* \rangle_V - G(u)\},$$

and

$$F^*(v^*) = \sup_{u \in V} \{\langle u, v^* \rangle_V - F(u)\}.$$

Defining also $J^* : V \to \mathbb{R}$ by

$$J^*(v^*) = F^*(v^*) - G^*(v^*),$$

from the results in [81], we may obtain

$$\inf_{u \in V} J(u) = \inf_{v^* \in V^*} J^*(v^*),$$

so that

$$
\begin{aligned}
J^{**}(u_0) &= \inf_{u \in V} J^{**}(u) \\
&= \inf_{u \in V} J(u) = \inf_{v^* \in V^*} J^*(v^*).
\end{aligned}
\tag{12.15}
$$

Suppose now there exists $\hat{u} \in V$ such that

$$J(\hat{u}) = \inf_{u \in V} J(u).$$

From the standard necessary conditions, we have

$$\delta J(\hat{u}) = \mathbf{0},$$

so that

$$\frac{\partial G(\hat{u})}{\partial u} - \frac{\partial F(\hat{u})}{\partial u} = \mathbf{0}.$$

Define now

$$v_0^* = \frac{\partial F(\hat{u})}{\partial u}.$$

From these last two equations we obtain

$$v_0^* = \frac{\partial G(\hat{u})}{\partial u}.$$

From such results and the Legendre transform properties, we have

$$\hat{u} = \frac{\partial F^*(v_0^*)}{\partial v^*},$$

$$\hat{u} = \frac{\partial G^*(v_0^*)}{\partial v^*},$$

so that

$$\delta J^*(v_0^*) = \frac{\partial F^*(v_0^*)}{\partial v^*} - \frac{\partial G^*(v_0^*)}{\partial v^*} = \hat{u} - \hat{u} = \mathbf{0},$$

$$G^*(v_0^*) = \langle \hat{u}, v_0^* \rangle_V - G(\hat{u})$$

and

$$F^*(v_0^*) = \langle \hat{u}, v_0^* \rangle_V - F(\hat{u})$$

so that

$$\begin{aligned}
\inf_{u \in V} J(u) &= J(\hat{u}) \\
&= G(\hat{u}) - F(\hat{u}) \\
&= \inf_{v^* \in V^*} J^*(v^*) \\
&= F^*(v_0^*) - G^*(v_0^*) \\
&= J^*(v_0^*).
\end{aligned} \tag{12.16}$$

12.4.2 The main duality principle and a related primal dual variational formulation

Considering these last statements and results, we may prove the following theorem.

Theorem 12.4.1 *Let $\Omega \subset \mathbb{R}^3$ be an open, bounded, connected set with a regular (Lipschitzian) boundary denoted by $\partial \Omega$.*

Let $J : V \to \mathbb{R}$ be a functional such that

$$J(u) = G(u) - F(u), \forall u \in V,$$

where $V = W_0^{1,2}(\Omega)$.

Suppose G, F are both three times Fréchet differentiable functionals such that there exists $K > 0$ such that

$$\frac{\partial^2 G(u)}{\partial u^2} + K > 0$$

and

$$\frac{\partial^2 F(u)}{\partial u^2} + K > 0$$

$\forall u \in V$.

Assume also there exists $u_0 \in V$ and $\alpha_1 \in \mathbb{R}$ such that

$$\alpha_1 = \inf_{u \in V} J(u) = J(u_0).$$

Assume $K_3 > 0$ is such that

$$\|u_0\|_\infty < K_3.$$

Define

$$\tilde{V} = \{u \in V \; : \; \|u\|_\infty \leq K_3\}.$$

Assume $K_1 > 0$ is such that if $u \in \tilde{V}$ then

$$\max\left\{\|F'(u)\|_\infty, \|G'(u)\|_\infty, \|F''(u)\|_\infty, \|F'''(u)\|_\infty, \|G''(u)\|_\infty, \|G'''(u)\|_\infty\right\} \leq K_1.$$

Suppose also

$$K \gg \max\{K_1, K_3\}.$$

Define $F_K, G_K : V \to \mathbb{R}$ by

$$F_K(u) = F(u) + \frac{K}{2} \int_\Omega u^2 \, dx,$$

and

$$G_K(u) = G(u) + \frac{K}{2} \int_\Omega u^2 \, dx,$$

$\forall u \in V$.

Define also $G_K^*, F_K^* : V^* \to \mathbb{R}$ by

$$G_K^*(v^*) = \sup_{u \in V}\{\langle u, v^* \rangle_V - G_K(u)\},$$

and

$$F_K^*(v^*) = \sup_{u \in V}\{\langle u, v^* \rangle_V - F_K(u)\}.$$

Observe that since $u_0 \in V$ is such that

$$J(u_0) = \inf_{u \in V} J(u),$$

we have

$$\delta J(u_0) = \mathbf{0}.$$

Let $\varepsilon > 0$ be a small constant.
Define

$$v_0^* = \frac{\partial F_K(u_0)}{\partial u} \in V^*.$$

Under such hypotheses, defining $J_1^* : V \times V^* \to \mathbb{R}$ by

$$
\begin{aligned}
J_1^*(u, v^*) &= F_K^*(v^*) - G_K^*(v^*) \\
&+ \frac{1}{2\varepsilon}\left\| \frac{\partial G_K^*(v^*)}{\partial v^*} - u \right\|_2^2 + \frac{1}{2\varepsilon}\left\| \frac{\partial F_K^*(v^*)}{\partial v^*} - u \right\|_2^2 \\
&+ \frac{1}{2\varepsilon}\left\| \frac{\partial G_K^*(v^*)}{\partial v^*} - \frac{\partial F_K^*(v^*)}{\partial v^*} \right\|_2^2,
\end{aligned}
\tag{12.17}
$$

we have

$$
\begin{aligned}
J(u_0) &= \inf_{u \in V} J(u) \\
&= \inf_{(u,v^*) \in V \times V^*} J_1^*(u, v^*) \\
&= J_1^*(u_0, v_0^*).
\end{aligned}
\tag{12.18}
$$

Proof 12.2 Observe that from the hypotheses and the results and statements of the last subsection

$$J(u_0) = \inf_{u \in V} J(u) = \inf_{v^* \in Y^*} J_K^*(v^*) = J_K^*(v_0^*),$$

where

$$J_K^*(v^*) = F_K^*(v^*) - G_K^*(v^*), \forall v^* \in V^*.$$

Moreover we have

$$J_1^*(u, v^*) \geq J_K^*(v^*), \forall u \in V, \ v^* \in V^*.$$

Also from hypotheses and the last subsection results,

$$u_0 = \frac{\partial F_K^*(v_0^*)}{\partial v^*} = \frac{\partial G_K^*(v_0^*)}{\partial v^*},$$

so that clearly we have

$$J_1^*(u_0, v_0^*) = J_K^*(v_0^*).$$

From these last results, we may infer that

$$
\begin{aligned}
J(u_0) &= \inf_{u \in V} J(u) \\
&= \inf_{v^* \in V^*} J_K^*(v^*) \\
&= J_K^*(v_0^*) \\
&= \inf_{(u,v^*) \in V \times V^*} J_1^*(u,v^*) \\
&= J_1^*(u_0, v_0^*).
\end{aligned}
\tag{12.19}
$$

The proof is complete.

Remark 12.4.2 *At this point we highlight that J_1^* has a large region of convexity around the optimal point (u_0, v_0^*), for $K > 0$ sufficiently large and corresponding $\varepsilon > 0$ sufficiently small.*

Indeed, observe that for $v^* \in V^*$,

$$
G_K^*(v^*) = \sup_{u \in V} \{ \langle u, v^* \rangle_V - G_K(u) \} = \langle \hat{u}, v^* \rangle_V - G_K(\hat{u})
$$

where $\hat{u} \in V$ is such that

$$
v^* = \frac{\partial G_K(\hat{u})}{\partial u} = G'(\hat{u}) + K\hat{u}.
$$

Taking the variation in v^* in this last equation, we obtain

$$
1 = G''(u) \frac{\partial \hat{u}}{\partial v^*} + K \frac{\partial \hat{u}}{\partial v^*},
$$

so that

$$
\frac{\partial \hat{u}}{\partial v^*} = \frac{1}{G''(u) + K} = \mathcal{O}\left(\frac{1}{K} \right).
$$

From this we get

$$
\begin{aligned}
\frac{\partial^2 \hat{u}}{\partial (v^*)^2} &= -\frac{1}{(G''(u)+K)^2} G'''(u) \frac{\partial \hat{u}}{\partial v^*} \\
&= -\frac{1}{(G''(u)+K)^3} G'''(u) \\
&= \mathcal{O}\left(\frac{1}{K^3} \right).
\end{aligned}
\tag{12.20}
$$

On the other hand, from the implicit function theorem

$$
\frac{\partial G_K^*(v^*)}{\partial v^*} = u + [v^* - G_K'(\hat{u})] \frac{\partial \hat{u}}{\partial v^*} = u,
$$

so that

$$
\frac{\partial^2 G_K^*(v^*)}{\partial (v^*)^2} = \frac{\partial \hat{u}}{\partial v^*} = \mathcal{O}\left(\frac{1}{K} \right)
$$

and

$$\frac{\partial^3 G_K^*(v^*)}{\partial (v^*)^3} = \frac{\partial^2 \hat{u}}{\partial (v^*)^2} = \mathcal{O}\left(\frac{1}{K^3}\right).$$

Similarly, we may obtain

$$\frac{\partial^2 F_K^*(v^*)}{\partial (v^*)^2} = \mathcal{O}\left(\frac{1}{K}\right)$$

and

$$\frac{\partial^3 F_K^*(v^*)}{\partial (v^*)^3} = \mathcal{O}\left(\frac{1}{K^3}\right).$$

Denoting

$$A = \frac{\partial^2 F_K^*(v_0^*)}{\partial (v^*)^2}$$

and

$$B = \frac{\partial^2 G_K^*(v_0^*)}{\partial (v^*)^2},$$

we have

$$\frac{\partial^2 J_1^*(u_0, v_0^*)}{\partial (v^*)^2} = A - B + \frac{1}{\varepsilon}\left(2A^2 + 2B^2 - 2AB\right),$$

$$\frac{\partial^2 J_1^*(u_0, v_0^*)}{\partial u^2} = \frac{2}{\varepsilon},$$

and

$$\frac{\partial^2 J_1^*(u_0, v_0^*)}{\partial (v^*)\partial u} = -\frac{1}{\varepsilon}(A + B).$$

From this we get

$$\begin{aligned}
\det(\delta^2 J^*(v_0^*, u_0)) &= \frac{\partial^2 J_1^*(u_0, v_0^*)}{\partial (v^*)^2} \frac{\partial^2 J_1^*(u_0, v_0^*)}{\partial u^2} - \left[\frac{\partial^2 J_1^*(u_0, v_0^*)}{\partial (v^*)\partial u}\right]^2 \\
&= 2\frac{A - B}{\varepsilon} + 2\frac{(A - B)^2}{\varepsilon^2} \\
&= \mathcal{O}\left(\frac{1}{\varepsilon^2}\right) \\
&\gg 0
\end{aligned}$$

(12.21)

about the optimal point (u_0, v_0^*).

12.5 One more dual variational formulation

In this section, again for $\Omega \subset \mathbb{R}^3$ an open, bounded, connected set with a regular (Lipschitzian) boundary $\partial\Omega$, $\gamma > 0$, $\alpha > 0$, $\beta > 0$ and $f \in L^2(\Omega)$, we denote $F_1 : V \times Y \to \mathbb{R}$, $F_2 : V \to \mathbb{R}$ and $G : V \times Y \to \mathbb{R}$ by

$$
\begin{aligned}
F_1(u, v_0^*) &= \frac{\gamma}{2} \int_\Omega \nabla u \cdot \nabla u \, dx - \frac{K}{2} \int_\Omega u^2 \, dx \\
&\quad + \frac{K_1}{2} \int_\Omega (-\gamma \nabla^2 u + 2v_0^* u - f)^2 \, dx + \frac{K_2}{2} \int_\Omega u^2 \, dx, \quad (12.22)
\end{aligned}
$$

$$
F_2(u) = \frac{K_2}{2} \int_\Omega u^2 \, dx + \langle u, f \rangle_{L^2},
$$

and

$$
G(u, v) = \frac{\alpha}{2} \int_\Omega (u^2 - \beta + v)^2 \, dx + \frac{K}{2} \int_\Omega u^2 \, dx.
$$

We define also

$$
J_1(u, v_0^*) = F_1(u, v_0^*) - F_2(u) + G(u, 0),
$$

$$
J(u) = \frac{\gamma}{2} \int_\Omega \nabla u \cdot \nabla u \, dx + \frac{\alpha}{2} \int_\Omega (u^2 - \beta)^2 \, dx - \langle u, f \rangle_{L^2},
$$

and $F_1^* : [Y^*]^3 \to \mathbb{R}$, $F_2^* : Y^* \to \mathbb{R}$, and $G^* : [Y^*]^2 \to \mathbb{R}$, by

$$
\begin{aligned}
&F_1^*(v_2^*, v_1^*, v_0^*) \\
&= \sup_{u \in V} \{ \langle u, v_1^* + v_2^* \rangle_{L^2} - F_1(u, v_0^*) \} \\
&= \frac{1}{2} \int_\Omega \frac{(v_1^* + v_2^* + K_1(-\gamma \nabla^2 + 2v_0^*)f)^2}{(-\gamma \nabla^2 - K + K_2 + K_1(-\gamma \nabla^2 + 2v_0^*)^2)} \, dx \\
&\quad - \frac{K_1}{2} \int_\Omega f^2 \, dx, \quad (12.23)
\end{aligned}
$$

$$
\begin{aligned}
F_2^*(v_2^*) &= \sup_{u \in V} \{ \langle u, v_2^* \rangle_{L^2} - F_2(u) \} \\
&= \frac{1}{2K_2} \int_\Omega (v_2^* - f)^2 \, dx, \quad (12.24)
\end{aligned}
$$

and

$$
\begin{aligned}
G^*(v_1^*, v_0^*) &= \sup_{(u,v) \in V \times Y} \{ \langle u, v_1^* \rangle_{L^2} - \langle v, v_0^* \rangle_{L^2} - G(u, v) \} \\
&= \frac{1}{2} \int_\Omega \frac{(v_1^*)^2}{2v_0^* + K} \, dx + \frac{1}{2\alpha} \int_\Omega (v_0^*)^2 \, dx \\
&\quad + \beta \int_\Omega v_0^* \, dx \quad (12.25)
\end{aligned}
$$

if $v_0^* \in B^*$ where
$$B^* = \{v_0^* \in Y^* : \|v_0^*\|_\infty \leq K/2\}.$$

Define also
$$V_2 = \{u \in V : \|u\|_\infty \leq K_3\},$$
$$A^+ = \{u \in V : u f \geq 0 \text{ a.e. in } \Omega\},$$
$$V_1 = V_2 \cap A^+,$$

$$B_2^* = \{v_0^* \in Y^* : -\gamma \nabla^2 - K + K_1(-\gamma \nabla^2 + 2v_0^*)^2 > 0\},$$

$$D_3^* = \{(v_1^*, v_2^*) \in Y^* \times Y^* : -1/\alpha + 4K_1[u(v_1^*, v_2^*, v_0^*)^2] + 100/K_2 \leq 0, \forall v_0^* \in B^*\},$$

where
$$u(v_2^*, v_0^*) = \frac{\varphi_1}{\varphi},$$

$$\varphi_1 = (v_1^* + v_2^* + K_1(-\gamma \nabla^2 + 2v_0^*)f)$$

and
$$\varphi = (-\gamma \nabla^2 - K + K_1(-\gamma \nabla^2 + 2v_0^*)^2 + K_2),$$

$$D^* = \{v_2^* \in Y^*; \|v_2^*\|_\infty < K_4\}$$
$$E^* = \{v_1^* \in Y^* : \|v_1^*\|_\infty \leq K_5\},$$

for some $K_3, K_4, K_5 > 0$ to be specified,

Finally, we also define $J_1^* : [Y^*]^2 \times B^* \to \mathbb{R}$,

$$J_1^*(v_2^*, v_1^*, v_0^*) = -F_1^*(v_2^*, v_1^*, v_0^*) + F_2^*(v_2^*) - G^*(v_1^*, v_0^*).$$

Assume now $K_1 = 1/[4(\alpha + \varepsilon)K_3^2]$,

$$K_2 \gg K_1 \gg \max\{K_3, K_4, K_5, 1, \gamma, \alpha, \beta\}.$$

Observe that, by direct computation, we may obtain

$$\frac{\partial^2 J_1^*(v_2^*, v_1^*, v_0^*)}{\partial (v_0^*)^2} = -\frac{1}{\alpha} + 4K_1 u(v^*)^2 + \mathcal{O}(1/K_2) < 0,$$

for $v_0^* \in B_3^*$.

Considering such statements and definitions, we may prove the following theorem.

Theorem 12.5.1 *Let* $(\hat{v}_2^*, \hat{v}_1^*, \hat{v}_0^*) \in ((D^* \times E^*) \cap D_3^*) \times (B_2^* \cap B^*)$ *be such that*

$$\delta J_1^*(\hat{v}_2^*, \hat{v}_1^*, \hat{v}_0^*) = 0$$

and $u_0 \in V_1$, *where*

$$u_0 = \frac{\hat{v}_1^* + \hat{v}_2^* + K_1(-\gamma\nabla^2 + 2v_0^*)f}{K_2 - K - \gamma\nabla^2 + K_1(-\gamma\nabla^2 + 2\hat{v}_0^*)^2}.$$

Under such hypotheses, we have

$$\delta J(u_0) = 0,$$

so that

$$
\begin{aligned}
J(u_0) &= \inf_{u \in V_1}\left\{ J(u) + \frac{K_1}{2}\int_\Omega (-\gamma\nabla^2 u + 2\hat{v}_0^* u - f)^2 \, dx \right\}\\
&= \inf_{v_2^* \in D^*}\left\{ \sup_{(v_1^*, v_0^*) \in E^* \times B^*} J_1^*(v_2^*, v_1^*, v_0^*) \right\}\\
&= J_1^*(\hat{v}_2^*, \hat{v}_1^*, \hat{v}_0^*).
\end{aligned}
\tag{12.26}
$$

Proof 12.3 Observe that $\delta J_1^*(\hat{v}_2^*, \hat{v}_1^*, \hat{v}_0^*) = 0$ so that, since $(\hat{v}_2^*, \hat{v}_1^*) \in D_3^*, \hat{v}_0^* \in B_2^*$ and J_1^* is quadratic in v_2^*, we may infer that

$$
\begin{aligned}
J_1^*(\hat{v}_2^*, \hat{v}_1^*, \hat{v}_0^*) &= \inf_{v_2^* \in Y^*} J_1^*(v_2^*, \hat{v}_1^*, \hat{v}_0^*)\\
&= \sup_{(v_1^*, v_0^*) \in E^* \times B^*} J_1^*(\hat{v}_2^*, v_1^*, v_0^*).
\end{aligned}
\tag{12.27}
$$

Therefore, from a standard saddle point theorem, we have that

$$J_1^*(\hat{v}_2^*, \hat{v}_1^*, \hat{v}_0^*) \inf_{v_2^* \in Y^*}\left\{ \sup_{(v_1^*, v_0^*) \in E^* \times B^*} J_1^*(v_2^*, v_1^*, v_0^*) \right\}.$$

Now we are going to show that

$$\delta J(u_0) = 0.$$

From

$$\frac{\partial J_1^*(\hat{v}_2^*, \hat{v}_1^*, \hat{v}_0^*)}{\partial v_2^*} = 0,$$

we have

$$-u_0 + \frac{\hat{v}_2^*}{K_2} = 0,$$

and thus

$$\hat{v}_2^* = K_2 u_0.$$

From

$$\frac{\partial J_1^*(\hat{v}_2^*, \hat{v}_1^*, \hat{v}_0^*)}{\partial v_1^*} = \mathbf{0},$$

we obtain

$$-u_0 - \frac{\hat{v}_1^* - f}{2\hat{v}_0^* + K} = 0,$$

and thus

$$\hat{v}_1^* = -2\hat{v}_0^* u_0 - K u_0 + f.$$

Finally, denoting

$$D = -\gamma \nabla^2 u_0 + 2\hat{v}_0^* u_0 - f,$$

from

$$\frac{\partial J_1^*(\hat{v}_2^*, \hat{v}_1^*, \hat{v}_0^*)}{\partial v_0^*} = \mathbf{0},$$

we have

$$-2D u_0 + u_0^2 - \frac{\hat{v}_0^*}{\alpha} - \beta = 0,$$

so that

$$\hat{v}_0^* = \alpha(u_0^2 - \beta - 2D u_0). \tag{12.28}$$

Observe now that

$$\hat{v}_1^* + \hat{v}_2^* + K_1(-\gamma \nabla^2 + 2\hat{v}_0^*)f = (K_2 - K - \gamma \nabla^2 + K_1(-\gamma \nabla^2 + 2\hat{v}_0^*)^2)u_0$$

so that

$$\begin{aligned} &K_2 u_0 - 2\hat{v}_0 u_0 - K u_0 + f \\ = {}& K_2 u_0 - K u_0 - \gamma \nabla^2 u_0 + K_1(-\gamma \nabla^2 + 2\hat{v}_0^*)(-\gamma \nabla^2 u_0 + 2\hat{v}_0^* u_0 - f). \end{aligned} \tag{12.29}$$

The solution for this last system of equations (12.28) and (12.29) is obtained through the relations

$$\hat{v}_0^* = \alpha(u_0^2 - \beta)$$

and

$$-\gamma \nabla^2 u_0 + 2\hat{v}_0^* u_0 - f = D = 0,$$

so that

$$\delta J(u_0) = -\gamma \nabla^2 u_0 + 2\alpha(u_0^2 - \beta)u_0 - f = 0$$

and

$$\delta \left\{ J(u_0) + \frac{K_1}{2} \int_\Omega (-\gamma \nabla^2 u_0 + 2\hat{v}_0^* u_0 - f)^2 \, dx \right\} = 0.$$

Moreover, from the Legendre transform properties

$$F_1^*(\hat{v}_2^*, \hat{v}_1^*, \hat{v}_0^*) = \langle u_0, \hat{v}_2^* + \hat{v}_1^* \rangle_{L^2} - F_1(u_0, \hat{v}_0^*),$$

$$F_2^*(\hat{v}_2^*) = \langle u_0, \hat{v}_2^* \rangle_{L^2} - F_2(u_0),$$

$$G^*(\hat{v}_1^*, \hat{v}_0^*) = -\langle u_0, \hat{v}_1^* \rangle_{L^2} - \langle 0, \hat{v}_0^* \rangle_{L^2} - G(u_0, 0),$$

so that

$$
\begin{aligned}
J_1^*(\hat{v}_2^*, \hat{v}_1^*, \hat{v}_0^*) &= -F_1^*(\hat{v}_2^*, \hat{v}_1^*, \hat{v}_0^*) + F_2^*(\hat{v}_2^*) - G^*(\hat{v}_1^*, \hat{v}_0^*) \\
&= F_1(u_0, \hat{v}_0^*) - F_2(u_0) + G(u_0, 0) \\
&= J(u_0).
\end{aligned}
\tag{12.30}
$$

Observe now that

$$
\begin{aligned}
J(u_0) &= J_1^*(\hat{v}_2^*, \hat{v}_1^*, \hat{v}_0^*) \\
&\leq \frac{\gamma}{2} \int_\Omega \nabla u \cdot \nabla u \, dx - \frac{K}{2} \int_\Omega u^2 \, dx \\
&\quad + \frac{K_1}{2} \int_\Omega (-\gamma \nabla^2 u + \hat{v}_0^* u - f)^2 \, dx + \langle u, \hat{v}_1^* \rangle_{L^2} - \langle u, f \rangle_{L^2} \\
&\quad - \frac{1}{2} \int_\Omega \frac{(v_1^*)^2}{2 v_0^* + K} \, dx - \frac{1}{2\alpha} \int_\Omega (v_0^*)^2 \, dx - \beta \int_\Omega v_0^* \, dx \\
&\leq \frac{\gamma}{2} \int_\Omega \nabla u \cdot \nabla u \, dx - \langle u, f \rangle_{L^2} \\
&\quad + \frac{K_1}{2} \int_\Omega (-\gamma \nabla^2 u + \hat{v}_0^* u - f)^2 \, dx \\
&\quad + \sup_{(v_1^*, v_0^*) \in D^* \times B^*} \left\{ + \langle u, \hat{v}_1^* \rangle_{L^2} - \frac{1}{2} \int_\Omega \frac{(v_1^*)^2}{2 v_0^* + K} \, dx \right. \\
&\quad \left. - \frac{1}{2\alpha} \int_\Omega (v_0^*)^2 \, dx - \frac{1}{2\alpha} \int_\Omega (v_0^*)^2 \, dx - \beta \int_\Omega v_0^* \, dx \right\} \\
&= J(u) + \frac{K_1}{2} \int_\Omega (-\gamma \nabla^2 u + 2 \hat{v}_0^* u - f)^2 \, dx,
\end{aligned}
\tag{12.31}
$$

$\forall u \in V_1$.

Hence, we have got

$$J(u_0) = \inf_{u \in V_1} \left\{ J(u) + \frac{K_1}{2} \int_\Omega (-\gamma \nabla^2 u + 2 \hat{v}_0^* u - f)^2 \, dx \right\}.$$

Joining the pieces, we have got

$$
\begin{aligned}
J(u_0) &= \inf_{u \in V} \left\{ J(u) + \frac{K_1}{2} \int_\Omega (-\gamma \nabla^2 u + 2 \hat{v}_0^* u - f)^2 \, dx \right\} \\
&= \inf_{v_2^* \in Y^*} \left\{ \sup_{(v_1^*, v_0^*) \in E^* \times (B^* \cap B_r(\hat{v}_0^*))} J_1^*(\hat{v}_2^*, v_1^*, v_0^*) \right\} \\
&= J_1^*(\hat{v}_2^*, \hat{v}_1^*, \hat{v}_0^*).
\end{aligned}
\tag{12.32}
$$

The proof is complete.

12.6   Another dual variational formulation

In this section, again for $\Omega \subset \mathbb{R}^3$ an open, bounded, connected set with a regular (Lipschitzian) boundary $\partial \Omega$, $\gamma > 0$, $\alpha > 0$, $\beta > 0$ and $f \in L^2(\Omega)$, we denote $F_1 : V \times Y \to \mathbb{R}$, $F_2 : V \to \mathbb{R}$ and $G : Y \to \mathbb{R}$ by

$$
\begin{aligned}
F_1(u, v_0^*) &= \frac{\gamma}{2} \int_\Omega \nabla u \cdot \nabla u \, dx + \langle u^2, v_0^* \rangle_{L^2} \\
&\quad + \frac{K_1}{2} \int_\Omega (-\gamma \nabla^2 u + 2v_0^* u - f)^2 \, dx + \frac{K_2}{2} \int_\Omega u^2 \, dx, \quad (12.33)
\end{aligned}
$$

$$
F_2(u) = \frac{K_2}{2} \int_\Omega u^2 \, dx + \langle u, f \rangle_{L^2},
$$

and

$$
G(u^2) = \frac{\alpha}{2} \int_\Omega (u^2 - \beta)^2 \, dx.
$$

We define also

$$
J_1(u, v_0^*) = F_1(u, v_0^*) - F_2(u) - \langle u^2, v_0^* \rangle_{L^2} + G(u^2),
$$

$$
J(u) = \frac{\gamma}{2} \int_\Omega \nabla u \cdot \nabla u \, dx + \frac{\alpha}{2} \int_\Omega (u^2 - \beta)^2 \, dx - \langle u, f \rangle_{L^2},
$$

$$
A^+ = \{ u \in V : uf > 0, \text{ a.e. in } \Omega \},
$$

$$
V_2 = \{ u \in V : \|u\|_\infty \leq K_3 \},
$$

$$
V_1 = A^+ \cap V_2,
$$

and $F_1^* : [Y^*]^2 \to \mathbb{R}$, $F_2^* : Y^* \to \mathbb{R}$, and $G^* : Y^* \to \mathbb{R}$, by

$$
\begin{aligned}
&F_1^*(v_2^*, v_0^*) \\
&= \sup_{u \in V} \{ \langle u, v_2^* \rangle_{L^2} - F_1(u, v_0^*) \} \\
&= \frac{1}{2} \int_\Omega \frac{(v_2^* + K_1(-\gamma \nabla^2 + 2v_0^*)f)^2}{(-\gamma \nabla^2 + 2v_0^* + K_2 + K_1(-\gamma \nabla^2 + 2v_0^*)^2)} \, dx \\
&\quad - \frac{K_1}{2} \int_\Omega f^2 \, dx, \quad (12.34)
\end{aligned}
$$

$$
\begin{aligned}
F_2^*(v_2^*) &= \sup_{u \in V} \{ \langle u, v_2^* \rangle_{L^2} - F_2(u) \} \\
&= \frac{1}{2K_2} \int_\Omega (v_2^* + f)^2 \, dx, \quad (12.35)
\end{aligned}
$$

and

$$
\begin{aligned}
G^*(v_0^*) &= \sup_{v \in Y} \{ \langle v, v_0^* \rangle_{L^2} - G(v) \} \\
&= \frac{1}{2\alpha} \int_\Omega (v_0^*)^2 \, dx + \beta \int_\Omega v_0^* \, dx \quad (12.36)
\end{aligned}
$$

At this point we define

$$B^* = \{v_0^* \in Y^* \ : \ \|v_0^*\|_\infty \le K/2\},$$

$$B_2^* = \{v_0^* \in Y^* \ : \ -\gamma\nabla^2 + 2v_0^* + K_1(-\gamma\nabla^2 + 2v_0^*)^2 > \mathbf{0}\},$$

$$D_3^* = \{v_2^* \in Y^* \ : \ -1/\alpha + 4K_1[u(v_2^*, v_0^*)^2] + 100/K_2 \le \mathbf{0}, \forall v_0^* \in B^*\},$$

where

$$u(v_2^*, v_0^*) = \frac{\varphi_1}{\varphi},$$

$$\varphi_1 = (v_2^* + K_1(-\gamma\nabla^2 + 2v_0^*)f)$$

and

$$\varphi = (-\gamma\nabla^2 + 2v_0^* + K_1(-\gamma\nabla^2 + 2v_0^*)^2 + K_2),$$

Finally, we also define

$$E_1^* = \{v_2^* \in Y^* \ : \ \|v_2^*\|_\infty \le (5/4)K_2\}.$$

$$E_2^* = \{v_2^* \in Y^* \ : \ fv_2^* > 0, \text{ a.e. in } \Omega\},$$
$$E^* = E_1^* \cap E_2^*,$$

and $J_1^* : E^* \times B^* \to \mathbb{R}$, by

$$J_1^*(v_2^*, v_0^*) = -F_1^*(v_2^*, v_0^*) + F_2^*(v_2^*) - G^*(v_0^*).$$

Moreover, assume

$$K_2 \gg K_1 \gg K \gg K_3 \gg \max\{1, \gamma, \alpha\}.$$

By directly computing $\delta^2 J_1^*(v_2^*, v_0^*)$, recalling that

$$\varphi = (-\gamma\nabla^2 + 2v_0^* + K_1(-\gamma\nabla^2 + 2v_0^*)^2 + K_2),$$
$$\varphi_1 = (v_2^* + K_1(-\gamma\nabla^2 + 2v_0^*)f),$$

and

$$u = \frac{\varphi_1}{\varphi},$$

we obtain

$$\frac{\partial^2 J_1^*(v_2^*, v_0^*)}{\partial(v_2^*)^2} = 1/K_2 - 1/\varphi,$$

and

$$\frac{\partial^2 J_1^*(v_2^*, v_0^*)}{\partial(v_0^*)^2} = 4u^2 K_1 - 1/\alpha + \mathcal{O}(1/K_2) < \mathbf{0},$$

in $E^* \times B^*$.

Considering such statements and definitions, we may prove the following theorem.

Theorem 12.6.1 *Let $(\hat{v}_2^*, \hat{v}_0^*) \in (E^* \cap D_3^*) \times (B^* \cap B_2^*)$ be such that*

$$\delta J_1^*(\hat{v}_2^*, \hat{v}_0^*) = \mathbf{0}$$

and $u_0 \in V_1$ be such that

$$u_0 = \frac{\hat{v}_2^* + K_1(-\gamma \nabla^2 + 2\hat{v}_0^*)f}{K_2 + 2\hat{v}_0^* - \gamma \nabla^2 + K_1(-\gamma \nabla^2 + 2\hat{v}_0^*)^2}.$$

Under such hypotheses, we have

$$\delta J(u_0) = \mathbf{0},$$

so that

$$
\begin{aligned}
J(u_0) &= \inf_{u \in V_1} \left\{ J(u) + \frac{K_1}{2} \int_\Omega (-\gamma \nabla^2 u + 2\hat{v}_0^* u - f)^2 \, dx \right\} \\
&= \inf_{v_2^* \in E^*} \left\{ \sup_{v_0^* \in B^*} J_1^*(v_2^*, v_0^*) \right\} \\
&= J_1^*(\hat{v}_2^*, \hat{v}_0^*).
\end{aligned}
\tag{12.37}
$$

Proof 12.4   Observe that $\delta J_1^*(\hat{v}_2^*, \hat{v}_0^*) = \mathbf{0}$ so that, since $\hat{v}_2^* \in D_3^*$, $\hat{v}_0^* \in B_2^*$ and J_1^* is quadratic in v_2^*, we get

$$\sup_{v_0^* \in B^*} J_1^*(\hat{v}_2^*, v_0^*) = J_1^*(\hat{v}_2^*, \hat{v}_0^*) = \inf_{v_2^* \in E^*} J_1^*(v_2^*, \hat{v}_0^*).$$

Consequently, from this and the Min-Max Theorem, we obtain

$$J_1^*(\hat{v}_2^*, \hat{v}_0^*) = \inf_{v_2^* \in E^*} \left\{ \sup_{v_0^* \in B^*} J_1^*(v_2^*, v_0^*) \right\} = \sup_{v_0^* \in B^*} \left\{ \inf_{v_2^* \in E^*} J_1^*(v_2^*, v_0^*) \right\}.$$

Now we are going to show that

$$\delta J(u_0) = \mathbf{0}.$$

From

$$\frac{\partial J_1^*(\hat{v}_2^*, \hat{v}_0^*)}{\partial v_2^*} = 0,$$

we have

$$-u_0 + \frac{\hat{v}_2^*}{K_2} = 0,$$

and thus

$$\hat{v}_2^* = K_2 u_0.$$

Finally, denoting

$$D = -\gamma \nabla^2 u_0 + 2\hat{v}_0^* u_0 - f,$$

from

$$\frac{\partial J_1^*(\hat{v}_2^*, \hat{v}_0^*)}{\partial v_0^*} = 0,$$

we have

$$-2Du_0 + u_0^2 - \frac{\hat{v}_0^*}{\alpha} - \beta = 0,$$

so that

$$\hat{v}_0^* = \alpha(u_0^2 - \beta - 2Du_0). \tag{12.38}$$

Observe now that

$$\hat{v}_2^* + K_1(-\gamma\nabla^2 + 2\hat{v}_0^*)f = (K_2 - \gamma\nabla^2 + 2\hat{v}_0^* + K_1(-\gamma\nabla^2 + 2\hat{v}_0^*)^2)u_0$$

so that

$$
\begin{aligned}
& K_2 u_0 - 2\hat{v}_0 u_0 - K u_0 + f \\
= \; & K_2 u_0 - K u_0 - \gamma\nabla^2 u_0 + K_1(-\gamma\nabla^2 + 2\hat{v}_0^*)(-\gamma\nabla^2 u_0 + 2\hat{v}_0^* u_0 - f). \quad (12.39)
\end{aligned}
$$

The solution for this last equation is obtained through the relation

$$-\gamma\nabla^2 u_0 + 2\hat{v}_0^* u_0 - f = D = 0,$$

so that from this and (12.38), we get

$$\hat{v}_0^* = \alpha(u_0^2 - \beta).$$

Thus,

$$\delta J(u_0) = -\gamma\nabla^2 u_0 + 2\alpha(u_0^2 - \beta)u_0 - f = 0$$

and

$$\delta\left\{J(u_0) + \frac{K_1}{2}\int_\Omega(-\gamma\nabla^2 u_0 + 2\hat{v}_0^* u_0 - f)^2\,dx\right\} = 0.$$

Moreover, from the Legendre transform properties

$$F_1^*(\hat{v}_2^*, \hat{v}_0^*) = \langle u_0, \hat{v}_2^* \rangle_{L^2} - F_1(u_0, \hat{v}_0^*),$$

$$F_2^*(\hat{v}_2^*) = \langle u_0, \hat{v}_2^* \rangle_{L^2} - F_2(u_0),$$

$$G^*(\hat{v}_0^*) = \langle u_0^2, \hat{v}_0^* \rangle_{L^2} - G(u_0^2),$$

so that

$$
\begin{aligned}
J_1^*(\hat{v}_2^*, \hat{v}_0^*) &= -F_1^*(\hat{v}_2^*, \hat{v}_0^*) + F_2^*(\hat{v}_2^*) - G^*(\hat{v}_0^*) \\
&= F_1(u_0, \hat{v}_0^*) - F_2(u_0) - \langle u_0^2, \hat{v}_0^* \rangle_{L^2} + G(u_0^2) \\
&= J(u_0). \tag{12.40}
\end{aligned}
$$

Observe now that

$$
\begin{aligned}
J(u_0) &= J_1^*(\hat{v}_2^*, \hat{v}_0^*) \\
&\leq \frac{\gamma}{2} \int_\Omega \nabla u \cdot \nabla u \, dx - \langle u^2, \hat{v}_0^* \rangle_{L^2} \\
&\quad + \frac{K_1}{2} \int_\Omega (-\gamma \nabla^2 u + \hat{v}_0^* u - f)^2 \, dx - \langle u, f \rangle_{L^2} \\
&\quad - \frac{1}{2\alpha} \int_\Omega (v_0^*)^2 \, dx - \beta \int_\Omega v_0^* \, dx \\
&\leq \frac{\gamma}{2} \int_\Omega \nabla u \cdot \nabla u \, dx - \langle u, f \rangle_{L^2} \\
&\quad + \frac{K_1}{2} \int_\Omega (-\gamma \nabla^2 u + \hat{v}_0^* u - f)^2 \, dx \\
&\quad + \sup_{v_0^* \in Y^*} \left\{ -\langle u^2, \hat{v}_0^* \rangle_{L^2} - \frac{1}{2\alpha} \int_\Omega (v_0^*)^2 \, dx - \beta \int_\Omega v_0^* \, dx \right\} \\
&= J(u) + \frac{K_1}{2} \int_\Omega (-\gamma \nabla^2 u + 2\hat{v}_0^* u - f)^2 \, dx. \quad (12.41)
\end{aligned}
$$

$\forall u \in V_1$.

Hence, we have got

$$
J(u_0) = \inf_{u \in V_1} \left\{ J(u) + \frac{K_1}{2} \int_\Omega (-\gamma \nabla^2 u + 2\hat{v}_0^* u - f)^2 \, dx \right\}.
$$

Joining the pieces, we have got

$$
\begin{aligned}
J(u_0) &= \inf_{u \in V_1} \left\{ J(u) + \frac{K_1}{2} \int_\Omega (-\gamma \nabla^2 u + 2\hat{v}_0^* u - f)^2 \, dx \right\} \\
&= \inf_{v_2^* \in E^*} \left\{ \sup_{v_0^* \in B^*} J_1^*(v_2^*, v_0^*) \right\} \\
&= J_1^*(\hat{v}_2^*, \hat{v}_0^*). \quad (12.42)
\end{aligned}
$$

The proof is complete.

12.7 A related numerical computation through the generalized method of lines

We start by recalling that the generalized method of lines was originally introduced in the book entitled "Topics on Functional Analysis, Calculus of Variations and Duality" [22], published in 2011.

Indeed, the present results are extensions and applications of previous ones which have been published since 2011, in books and articles such as [22, 17, 12, 13]. About the Sobolev spaces involved, we would mention [1]. Concerning the applications, related models in physics are addressed in [4, 52].

We also emphasize that, in such a method, the domain of the partial differential equation in question is discretized in lines (or more generally, in curves), and the concerning solution is written on these lines as functions of boundary conditions and the domain boundary shape.

In fact, in its previous format, this method consists of an application of a kind of a partial finite differences procedure combined with the Banach fixed point theorem to obtain the relation between two adjacent lines (or curves).

In the present chapter, we propose an improvement concerning the way we truncate the series solution obtained through an application of the Banach fixed point theorem to find the relation between two adjacent lines. The results obtained are very good even as a typical parameter $\varepsilon > 0$ is very small.

In the next lines and sections we develop in details such a numerical procedure.

12.7.1 About a concerning improvement for the generalized method of lines

Let $\Omega \subset \mathbb{R}^2$ where

$$\Omega = \{(r, \theta) \in \mathbb{R}^2 \ : \ 1 \leq r \leq 2, \ 0 \leq \theta \leq 2\pi\}.$$

Consider the problem of solving the partial differential equation

$$\begin{cases} -\varepsilon \left(\frac{\partial^2 u}{\partial r^2} + \frac{1}{r}\frac{\partial u}{\partial r} + \frac{1}{r^2}\frac{\partial^2 u}{\partial \theta^2} \right) + \alpha u^3 - \beta u = f, & \text{in } \Omega, \\ u = u_0(\theta), & \text{on } \partial\Omega_1, \\ u = u_f(\theta), & \text{on } \partial\Omega_2. \end{cases} \tag{12.43}$$

Here

$$\Omega = \{(r, \theta) \in \mathbb{R}^2 \ : \ 1 \leq r \leq 2, \ 0 \leq \theta \leq 2\pi\},$$

$$\partial\Omega_1 = \{(1, \theta) \in \mathbb{R}^2 \ : \ 0 \leq \theta \leq 2\pi\},$$

$$\partial\Omega_2 = \{(2, \theta) \in \mathbb{R}^2 \ : \ 0 \leq \theta \leq 2\pi\},$$

$\varepsilon > 0$, $\alpha > 0, \beta > 0$, and $f \equiv 1$, on Ω.

In a partial finite differences scheme, such a system stands for

$$-\varepsilon \left(\frac{u_{n+1} - 2u_n + u_{n-1}}{d^2} + \frac{1}{t_n}\frac{u_n - u_{n-1}}{d} + \frac{1}{t_n^2}\frac{\partial^2 u_n}{\partial \theta^2} \right) + \alpha u_n^3 - \beta u_n = f_n,$$

$\forall n \in \{1, \cdots, N-1\}$, with the boundary conditions

$$u_0 = 0,$$

and

$$u_N = 0.$$

Here N is the number of lines and $d = 1/N$.

In particular, for $n = 1$ we have

$$-\varepsilon \left(\frac{u_2 - 2u_1 + u_0}{d^2} + \frac{1}{t_1} \frac{(u_1 - u_0)}{d} + \frac{1}{t_1^2} \frac{\partial^2 u_1}{\partial \theta^2} \right) + \alpha u_1^3 - \beta u_1 = f_1,$$

so that

$$u_1 = \left(u_2 + u_1 + u_0 + \frac{1}{t_1}(u_1 - u_0) \, d + \frac{1}{t_1^2} \frac{\partial^2 u_1}{\partial \theta^2} d^2 + (-\alpha u_1^3 + \beta u_1 - f_1) \frac{d^2}{\varepsilon} \right) / 3.0,$$

We solve this last equation through the Banach fixed point theorem, obtaining u_1 as a function of u_2.

Indeed, we may set

$$u_1^0 = u_2$$

and

$$u_1^{k+1} = \left(u_2 + u_1^k + u_0 + \frac{1}{t_1}(u_1^k - u_0) \, d + \frac{1}{t_1^2} \frac{\partial^2 u_1^k}{\partial \theta^2} d^2 \right.$$
$$\left. + (-\alpha (u_1^k)^3 + \beta u_1^k - f_1) \frac{d^2}{\varepsilon} \right) / 3.0, \qquad (12.44)$$

$\forall k \in \mathbb{N}$.

Thus, we may obtain

$$u_1 = \lim_{k \to \infty} u_1^k \equiv H_1(u_2, u_0).$$

Similarly, for $n = 2$, we have

$$u_2 = \left(u_3 + u_2 + H_1(u_2, u_0) + \frac{1}{t_1}(u_2 - H_1(u_2, u_0)) \, d + \frac{1}{t_1^2} \frac{\partial^2 u_2}{\partial \theta^2} d^2 \right.$$
$$\left. + (-\alpha u_2^3 + \beta u_2 - f_2) \frac{d^2}{\varepsilon} \right) / 3.0, \qquad (12.45)$$

We solve this last equation through the Banach fixed point theorem, obtaining u_2 as a function of u_3 and u_0.

Indeed, we may set

$$u_2^0 = u_3$$

and

$$u_2^{k+1} = \left(u_3 + u_2^k + H_1(u_2^k, u_0) + \frac{1}{t_2}(u_2^k - H_1(u_2^k, u_0)) \, d + \frac{1}{t_2^2} \frac{\partial^2 u_2^k}{\partial \theta^2} d^2 \right.$$
$$\left. + (-\alpha (u_2^k)^3 + \beta u_2^k - f_2) \frac{d^2}{\varepsilon} \right) / 3.0, \qquad (12.46)$$

$\forall k \in \mathbb{N}$.

Thus, we may obtain

$$u_2 = \lim_{k \to \infty} u_2^k \equiv H_2(u_3, u_0).$$

Now reasoning inductively, having

$$u_{n-1} = H_{n-1}(u_n, u_0),$$

we may get

$$
\begin{aligned}
u_n &= \left(u_{n+1} + u_n + H_{n-1}(u_n, u_0) + \frac{1}{t_n}(u_n - H_{n-1}(u_n, u_0))\, d + \frac{1}{t_n^2} \frac{\partial^2 u_n}{\partial \theta^2} d^2 \right. \\
&\left. + (-\alpha u_n^3 + \beta u_n - f_n)\frac{d^2}{\varepsilon} \right) /3.0,
\end{aligned}
\tag{12.47}
$$

We solve this last equation through the Banach fixed point theorem, obtaining u_n as a function of u_{n+1} and u_0.

Indeed, we may set

$$u_n^0 = u_{n+1}$$

and

$$
\begin{aligned}
u_n^{k+1} &= \left(u_{n+1} + u_n^k + H_{n-1}(u_n^k, u_0) + \frac{1}{t_n}(u_n^k - H_{n-1}(u_n^k, u_0))\, d + \frac{1}{t_n^2} \frac{\partial^2 u_n^k}{\partial \theta^2} d^2 \right. \\
&\left. + (-\alpha (u_n^k)^3 + \beta u_n^k - f_n)\frac{d^2}{\varepsilon} \right) /3.0,
\end{aligned}
\tag{12.48}
$$

$\forall k \in \mathbb{N}$.

Thus, we may obtain

$$u_n = \lim_{k \to \infty} u_n^k \equiv H_n(u_{n+1}, u_0).$$

We have obtained $u_n = H_n(u_{n+1}, u_0),\ \forall n \in \{1, \cdots, N-1\}$.
In particular, $u_N = u_f(\theta)$, so that we may obtain

$$u_{N-1} = H_{N-1}(u_N, u_0) = H_{N-1}(0) \equiv F_{N-1}(u_N, u_0) = F_{N-1}(u_f(\theta), u_0(\theta)).$$

Similarly,

$$u_{N-2} = H_{N-2}(u_{N-1}, u_0) = H_{N-2}(H_{N-1}(u_N, u_0)) = F_{N-2}(u_N, u_0) = F_{N-1}(u_f(\theta), u_0(\theta)),$$

an so on, up to obtaining

$$u_1 = H_1(u_2) \equiv F_1(u_N, u_0) = F_1(u_f(\theta), u_0(\theta)).$$

The problem is then approximately solved.

12.7.2 Software in Mathematica for solving such an equation

We recall that the equation to be solved is a Ginzburg-Landau type one, where

$$\begin{cases} -\varepsilon\left(\frac{\partial^2 u}{\partial r^2} + \frac{1}{r}\frac{\partial u}{\partial r} + \frac{1}{r^2}\frac{\partial^2 u}{\partial \theta^2}\right) + \alpha u^3 - \beta u = f, & \text{in } \Omega, \\ u = 0, & \text{on } \partial\Omega_1, \\ u = u_f(\theta), & \text{on } \partial\Omega_2. \end{cases} \tag{12.49}$$

Here

$$\Omega = \{(r,\theta) \in \mathbb{R}^2 : 1 \le r \le 2, \, 0 \le \theta \le 2\pi\},$$
$$\partial\Omega_1 = \{(1,\theta) \in \mathbb{R}^2 : 0 \le \theta \le 2\pi\},$$
$$\partial\Omega_2 = \{(2,\theta) \in \mathbb{R}^2 : 0 \le \theta \le 2\pi\},$$

$\varepsilon > 0$, $\alpha > 0, \beta > 0$, and $f \equiv 1$, on Ω. In a partial finite differences scheme, such a system stands for

$$-\varepsilon\left(\frac{u_{n+1} - 2u_n + u_{n-1}}{d^2} + \frac{1}{t_n}\frac{u_n - u_{n-1}}{d} + \frac{1}{t_n^2}\frac{\partial^2 u_n}{\partial \theta^2}\right) + \alpha u_n^3 - \beta u_n = f_n,$$

$\forall n \in \{1, \cdots, N-1\}$, with the boundary conditions

$$u_0 = 0,$$

and

$$u_N = u_f[x].$$

Here N is the number of lines and $d = 1/N$.

At this point, we present the concerning software for an approximate solution. Such a software is for $N = 10$ (10 lines) and $u_0[x] = 0$.

****

1. $m8 = 10$; $(N = 10 \text{ lines})$

2. $d = 1/m8$;

3. $e_1 = 0.1$; $(\varepsilon = 0.1)$

4. $A = 1.0$;

5. $B = 1.0$;

6. $For[i = 1, i < m8, i++, f[i] = 1.0]$; $(f \equiv 1, \text{ on } \Omega)$

7. $a = 0.0$;

8. $For[i = 1, i < m8, i++,$
   $Clear[b, u]$;
   $t[i] = 1 + i * d$;
   $b[x_-] = u[i+1][x]$;

9. $For[k = 1, k < 30, k++,$ (we have fixed the number of iterations)

$$z = \left(u[i+1][x] + b[x] + a + \frac{1}{t[i]}(b[x] - a) * d \right.$$
$$\left. + \frac{1}{t[i]^2} D[b[x], \{x, 2\}] * d^2 + (-A * b[x]^3 + B * u[x] + f[i]) * \frac{d^2}{e_1} \right) / 3.0;$$

$z =$
$Series[z, \{u[i+1][x], 0, 3\}, \{u[i+1]'[x], 0, 1\}, \{u[i+1]''[x], 0, 1\},$
$\{u[i+1]'''[x], 0, 0\}, \{u[i+1]''''[x], 0, 0\}];$

$z = Normal[z],$

$z = Expand[z];$

$b[x\_] = z;$

10. $a_1[i] = z;$

11. $Clear[b];$

12. $u[i+1][x\_] = b[x];$

13. $a = a_1[i]$];

14. $b[x\_] = u_f[x];$

15. $For[i = 1, i < m8, i++,$

$A_1 = a_1[m8 - i];$

$A_1 = Series[A_1, \{u_f[x], 0, 3\}, \{u_f'[x], 0, 1\}, \{u_f''[x], 0, 1\}, \{u_f'''[x], 0, 0\},$
$\{u_f''''[x], 0, 0\}];$

$A_1 = Normal[A_1];$

$A_1 = Expand[A_1];$

$u[m8 - i][x\_] = A_1;$

$b[x\_] = A_1;$

$Print[u[m8/2][x]];$

****

The numerical expressions for the solutions of the concerning $N = 10$ lines are given by

$$u[1][x] = 0.47352 + 0.00691 u_f[x] - 0.00459 u_f[x]^2 + 0.00265 u_f[x]^3 + 0.00039(u_f'')[x]$$
$$-0.00058 u_f[x](u_f'')[x] + 0.00050 u_f[x]^2(u_f'')[x] - 0.000181213 u_f[x]^3(u_f'')[x] \quad (12.50)$$

$$u[2][x] = 0.76763 + 0.01301 u_f[x] - 0.00863 u_f[x]^2 + 0.00497 u_f[x]^3 + 0.00068(u_f'')[x]$$
$$-0.00103 u_f[x](u_f'')[x] + 0.00088 u_f[x]^2(u_f'')[x] - 0.00034 u_f[x]^3(u_f'')[x] \quad (12.51)$$

$$u[3][x] = 0.91329 + 0.02034u_f[x] - 0.01342u_f[x]^2 + 0.00768u_f[x]^3 + 0.00095(u_f'')[x]$$
$$-0.00144u_f[x](u_f'')[x] + 0.00122u_f[x]^2(u_f'')[x] - 0.00051u_f[x]^3(u_f'')[x] \quad (12.52)$$

$$u[4][x] = 0.97125 + 0.03623u_f[x] - 0.02328u_f[x]^2 + 0.01289u_f[x]^3 + 0.00147331(u_f'')[x]$$
$$-0.00223u_f[x](u_f'')[x] + 0.00182uf[x]^2(u_f'')[x] - 0.00074u_f[x]^3(u_f'')[x] \quad (12.53)$$

$$u[5][x] = 1.01736 + 0.09242u_f[x] - 0.05110u_f[x]^2 + 0.02387u_f[x]^3 + 0.00211(u_f'')[x]$$
$$-0.00378u_f[x](u_f'')[x] + 0.00292u_f[x]^2(u_f'')[x] - 0.00132u_f[x]^3(u_f'')[x] \quad (12.54)$$

$$u[6][x] = 1.02549 + 0.21039u_f[x] - 0.09374u_f[x]^2 + 0.03422u_f[x]^3 + 0.00147(u_f'')[x]$$
$$-0.00634u_f[x](u_f'')[x] + 0.00467u_f[x]^2(u_f'')[x] - 0.00200u_f[x]^3(u_f'')[x] \quad (12.55)$$

$$u[7][x] = 0.93854 + 0.36459u_f[x] - 0.14232u_f[x]^2 + 0.04058u_f[x]^3 + 0.00259(u_f'')[x]$$
$$-0.00747373u_f[x](u_f'')[x] + 0.0047969u_f[x]^2(u_f'')[x] - 0.00194u_f[x]^3(u_f'')[x] \quad (12.56)$$

$$u[8][x] = 0.74649 + 0.57201u_f[x] - 0.17293u_f[x]^2 + 0.02791u_f[x]^3 + 0.00353(u_f'')[x]$$
$$-0.00658u_f[x](u_f'')[x] + 0.00407u_f[x]^2(u_f'')[x] - 0.00172u_f[x]^3(u_f'')[x] \quad (12.57)$$

$$u[9][x] = 0.43257 + 0.81004u_f[x] - 0.13080u_f[x]^2 + 0.00042u_f[x]^3 + 0.00294(u_f'')[x]$$
$$-0.00398u_f[x](u_f'')[x] + 0.00222u_f[x]^2(u_f'')[x] - 0.00066u_f[x]^3(u_f'')[x] \quad (12.58)$$

12.7.3 Some plots concerning the numerical results

In this section, we present the lines $2, 4, 6, 8$ related to results obtained in the last section.

Indeed, we present such mentioned lines, in a first step, for the previous results obtained through the generalized method of lines and, in a second step, through a numerical method which is the combination of the Newton's one and the generalized method of lines. In a third step, we also present the graphs by considering the expression of the lines as those also obtained through the generalized method of lines, up to the numerical coefficients for each function term, which are obtained by the numerical optimization of the functional J, below specified. We consider the case in which $u_0(x) = 0$ and $u_f(x) = \sin(x)$.

For the procedure mentioned above as the third step, recalling that $N = 10$ lines, considering that $u_f''(x) = -u_f(x)$, we may approximately assume the following general line expressions:

$$u_n(x) = a(1,n) + a(2,n)u_f(x) + a(3,n)u_f(x)^3 + a(4,n)u_f(x)^3, \ \forall n \in \{1, \cdots N-1\}.$$

Defining

$$W_n = -e_1 \frac{(u_{n+1}(x) - 2u_n(x) + u_{n-1}(x))^2}{d} - \frac{e_1}{t_n}\frac{(u_n(x) - u_{n-1}(x))}{d} - \frac{e_1}{t_n^2}u_n''(x) + u_n(x)^3 - u_n(x) - 1,$$

and

$$J(\{a(j,n)\}) = \sum_{n=1}^{N-1} \int_0^{2\pi} (W_n)^2 \, dx$$

we obtain $\{a(j,n)\}$ by numerically minimizing J.

Hence, we have obtained the following lines for these cases. For such graphs, we have considered 300 nodes in x, with $2\pi/300$ as units in $x \in [0, 2\pi]$.

For the Lines 2, 4, 6, 8, through the generalized method of lines, please see Figures 12.1, 12.4, 12.7, 12.10.

For the Lines 2, 4, 6, 8, through a combination of the Newton's and the generalized method of lines, please see Figures 12.2, 12.5, 12.8, 12.11.

Finally, for the Line 2, 4, 6, 8 obtained through the minimization of the functional J, please see Figures 12.3, 12.6, 12.9, 12.12.

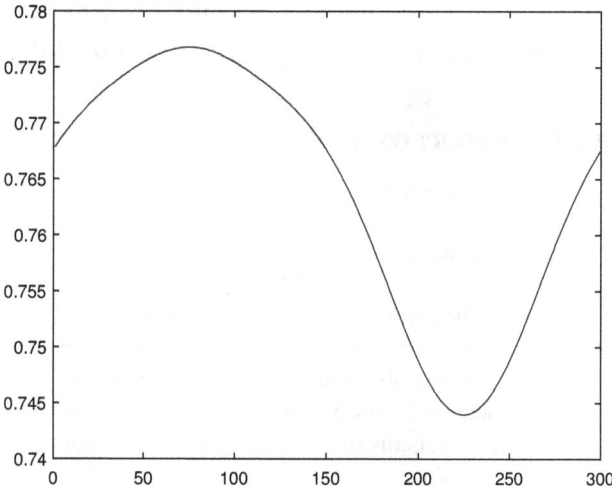

Figure 12.1: Line 2, solution $u_2(x)$ through the general method of lines.

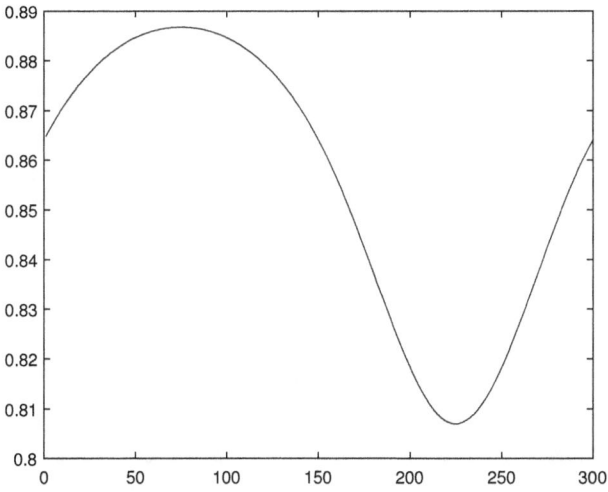

Figure 12.2: Line 2, solution $u_2(x)$ through the Newton's method.

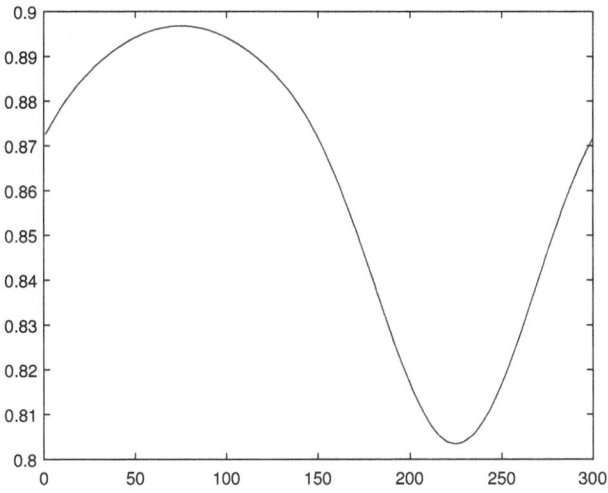

Figure 12.3: Line 2, solution $u_2(x)$ through the minimization of functional J.

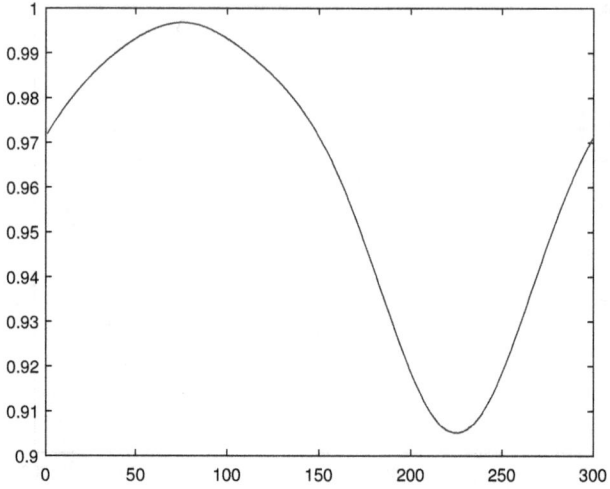

Figure 12.4: Line 4, solution $u_4(x)$ through the general method of lines.

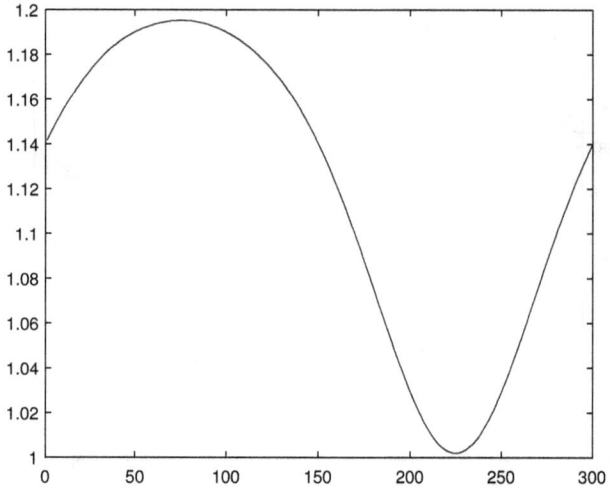

Figure 12.5: Line 4, solution $u_4(x)$ through the Newton's method.

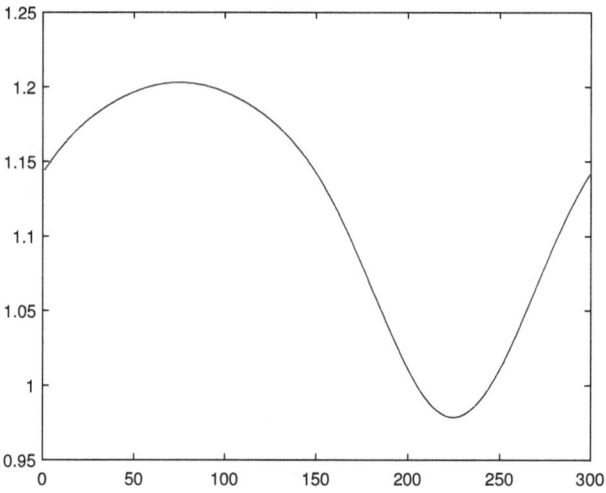

Figure 12.6: Line 4, solution $u_4(x)$ through the minimization of functional J.

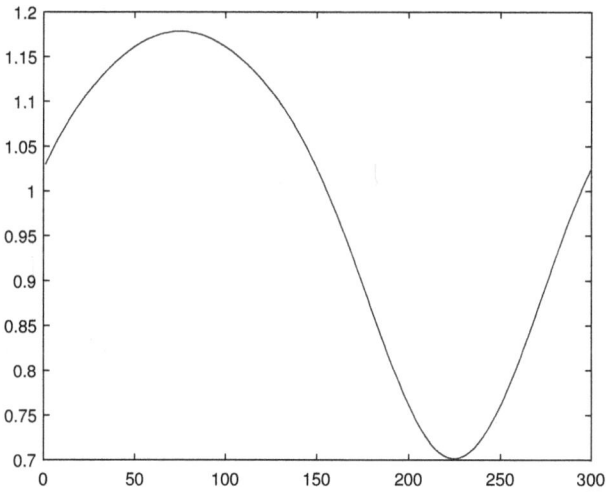

Figure 12.7: Line 6, solution $u_6(x)$ through the general method of lines.

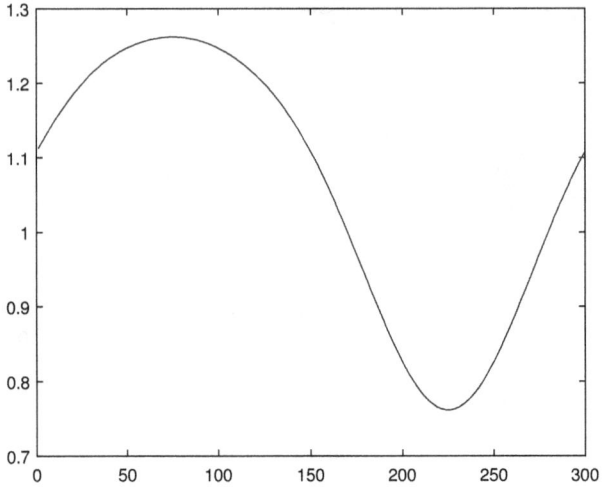

Figure 12.8: Line 6, solution $u_6(x)$ through the Newton's method.

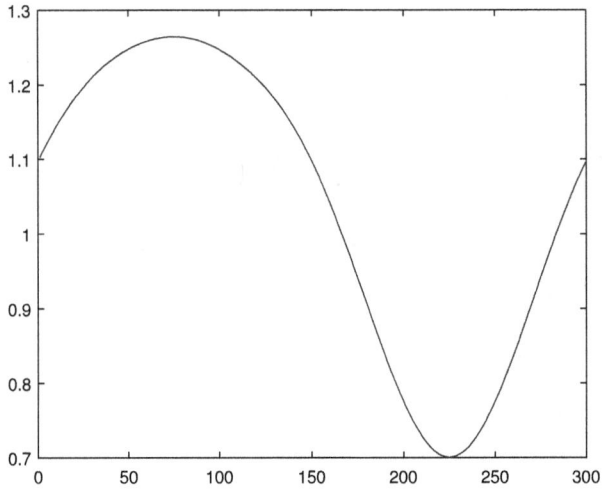

Figure 12.9: Line 6, solution $u_6(x)$ through the minimization of functional J.

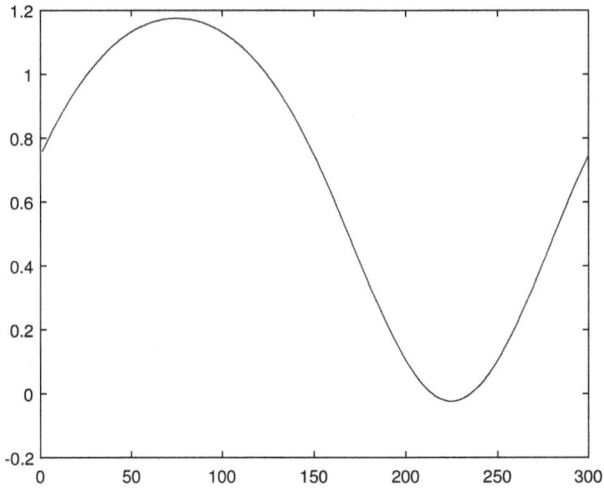

Figure 12.10: Line 8, solution $u_8(x)$ through the general method of lines.

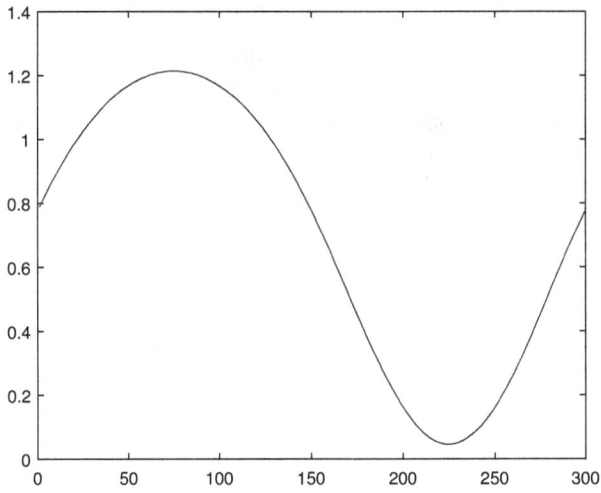

Figure 12.11: Line 8, solution $u_8(x)$ through the Newton's method.

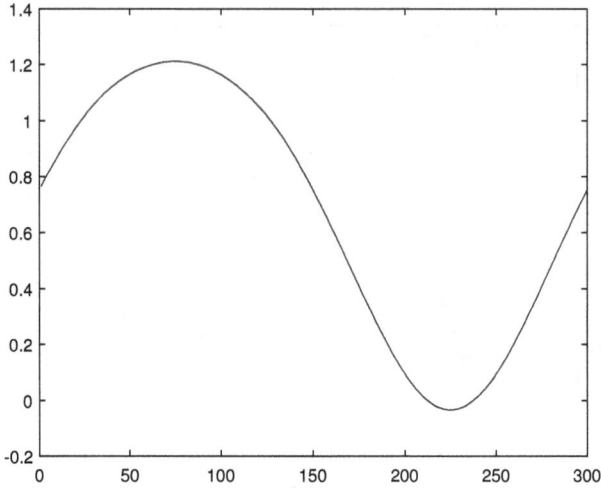

Figure 12.12: Line 8, solution $u_8(x)$ through the minimization of functional J.

12.8 Conclusion

In the first part of this article, we develop duality principles for non-convex variational optimization. In the final concerning sections, we propose dual convex formulations suitable for a large class of models in physics and engineering. In the last article section, we present an advance concerning the computation of a solution for a partial differential equation through the generalized method of lines. In particular, in its previous versions, we used to truncate the series in d^2 however, we have realized the results are much better by taking line solutions in series for $u_f[x]$ and its derivatives, as it is indicated in the present software.

This is a little difference concerning the previous procedure, but with a great result improvement as the parameter $\varepsilon > 0$ is small.

Indeed, with a sufficiently large N (number of lines), we may obtain very good qualitative results even as $\varepsilon > 0$ is very small.

Chapter 13

A Note on the Korn's Inequality in a N-Dimensional Context and a Global Existence Result for a Non-Linear Plate Model

13.1 Introduction

In this article, we present a proof for the Korn inequality in $\mathbb{R}^n$. The results are based on the standard tools of functional analysis and the Sobolev spaces theory.

We emphasize such a proof is relatively simple and easy to follow since it is established in a very transparent and clear fashion.

Remark 13.1.1 *This chapter has been accepted for publication in a similar article format by the MDPI Journal Applied Mathematics, reference [21]:*

F.S. Botelho, A Note on Korn's Inequality in an N-Dimensional Context and a Global Existence Result for a Non-Linear Plate Model. Applied Math 2023, 1: 1–11.

About the references, we highlight, related results in a three dimensional context may be found in [54]. Other important classical results on the Korn's inequality and concerning applications to models in elasticity may be found in [30, 31, 32].

Remark 13.1.2 *Generically throughout the text we denote*

$$\|u\|_{0,2,\Omega} = \left(\int_\Omega |u|^2 \, dx \right)^{1/2}, \ \forall u \in L^2(\Omega),$$

and

$$\|u\|_{0,2,\Omega} = \left(\sum_{j=1}^n \|u_j\|_{0,2,\Omega}^2 \right)^{1/2}, \ \forall u = (u_1, \dots, u_n) \in L^2(\Omega; \mathbb{R}^n).$$

Moreover,

$$\|u\|_{1,2,\Omega} = \left(\|u\|_{0,2,\Omega}^2 + \sum_{j=1}^n \|u_{x_j}\|_{0,2,\Omega}^2 \right)^{1/2}, \ \forall u \in W^{1,2}(\Omega),$$

where we shall also refer throughout the text to the well known corresponding analogous norm for $u \in W^{1,2}(\Omega; \mathbb{R}^n)$.

At this point, we first introduce the following definition.

Definition 13.1.3 *Let $\Omega \subset \mathbb{R}^n$ be an open, bounded set. We say that $\partial\Omega$ is $\hat{C}^1$ if such a manifold is oriented and for each $x_0 \in \partial\Omega$, denoting $\hat{x} = (x_1, \dots, x_{n-1})$ for a local coordinate system compatible with the manifold $\partial\Omega$ orientation, there exist $r > 0$ and a function $f(x_1, \dots, x_{n-1}) = f(\hat{x})$ such that*

$$W = \overline{\Omega} \cap B_r(x_0) = \{ x \in B_r(x_0) \mid x_n \le f(x_1, \dots, x_{n-1}) \}.$$

Moreover $f(\hat{x})$ is a Lipschitz continuous function, so that

$$|f(\hat{x}) - f(\hat{y})| \le C_1 |\hat{x} - \hat{y}|_2, \text{ on its domain,}$$

for some $C_1 > 0$. Finally, we assume

$$\left\{ \frac{\partial f(\hat{x})}{\partial x_k} \right\}_{k=1}^{n-1}$$

is classically defined, almost everywhere also on its concerning domain, so that $f \in W^{1,2}$.

Remark 13.1.4 *This mentioned set Ω is of a Lipschitzian type, so that we may refer to such a kind of sets as domains with a Lipschitzian boundary, or simply as Lipschitzian sets.*

At this point, we recall the following result found in [13], at page 222 in its Chapter 11.

Theorem 13.1.5 *Assume $\Omega \subset \mathbb{R}^n$ is an open bounded set, and that $\partial\Omega$ is $\hat{C}^1$. Let $1 \le p < \infty$, and let V be a bounded open set such that $\Omega \subset\subset V$. Then there exists a bounded linear operator*

$$E : W^{1,p}(\Omega) \to W^{1,p}(\mathbb{R}^n),$$

such that for each $u \in W^{1,p}(\Omega)$ we have:

1. *$Eu = u$, a.e. in Ω,*

2. *Eu has support in V, and*

3. *$\|Eu\|_{1,p,\mathbb{R}^n} \leq C\|u\|_{1,p,\Omega}$, where the constant depend only on p, Ω, and V.*

Remark 13.1.6 *Considering the proof of such a result, the constant $C > 0$ may be also such that*

$$\|e_{ij}(Eu)\|_{0,2,V} \leq C(\|e_{ij}(u)\|_{0,2,\Omega} + \|u\|_{0,2,\Omega}), \ \forall u \in W^{1,2}(\Omega; \mathbb{R}^n), \ \forall i, j \in \{1, \ldots, n\},$$

for the operator $e : W^{1,2}(\Omega; \mathbb{R}^n) \to L^2(\Omega; \mathbb{R}^{n \times n})$ specified in the next theorem.

Finally, as the meaning is clear, we may simply denote $Eu = u$.

13.2   The main results, the Korn inequalities

Our main result is summarized by the following theorem.

Theorem 13.2.1 *Let $\Omega \subset \mathbb{R}^n$ be an open, bounded and connected set with a $\hat{C}^1$ (Lipschitzian) boundary $\partial\Omega$.*
  Define $e : W^{1,2}(\Omega; \mathbb{R}^n) \to L^2(\Omega; \mathbb{R}^{n \times n})$ by

$$e(u) = \{e_{ij}(u)\}$$

where

$$e_{ij}(u) = \frac{1}{2}(u_{i,j} + u_{j,i}), \ \forall i, j \in \{1, \ldots, n\},$$

and where generically, we denote

$$u_{i,j} = \frac{\partial u_i}{\partial x_j}, \ \forall i, j \in \{1, \cdots, n\}.$$

Define also,

$$\|e(u)\|_{0,2,\Omega} = \left(\sum_{i=1}^{n} \sum_{j=1}^{n} \|e_{ij(u)}\|_{0,2,\Omega}^2 \right)^{1/2}.$$

Let $L \in \mathbb{R}^+$ be such $V = [-L, L]^n$ is also such that $\overline{\Omega} \subset V^0$.
Under such hypotheses, there exists $C(\Omega, L) \in \mathbb{R}^+$ such that

$$\|u\|_{1,2,\Omega} \leq C(\Omega, L) \left(\|u\|_{0,2,\Omega} + \|e(u)\|_{0,2,\Omega} \right), \ \forall u \in W^{1,2}(\Omega; \mathbb{R}^n).$$

Proof 13.1   Suppose, to obtain contradiction, the concerning claim does not hold. Hence, for each $k \in \mathbb{N}$ there exists $u_k \in W^{1,2}(\Omega;\mathbb{R}^n)$ such that

$$\|u_k\|_{1,2,\Omega} > k\left(\|u_k\|_{0,2,\Omega} + \|e(u_k)\|_{0,2,\Omega}\right).$$

In particular defining

$$v_k = \frac{u_k}{\|u_k\|_{1,2,\Omega}}$$

we obtain

$$\|v_k\|_{1,2,\Omega} = 1 > k\left(\|v_k\|_{0,2,\Omega} + \|e(v_k)\|_{0,2,\Omega}\right),$$

so that

$$\left(\|v_k\|_{0,2,\Omega} + \|e(v_k)\|_{0,2,\Omega}\right) < \frac{1}{k}, \; \forall k \in \mathbb{N}.$$

From this we have got,

$$\|v_k\|_{0,2,\Omega} < \frac{1}{k},$$

and

$$\|e_{ij}(v_k)\|_{0,2,\Omega} < \frac{1}{k}, \; \forall k \in \mathbb{N},$$

so that

$$\|v_k\|_{0,2,\Omega} \to 0, \text{ as } k \to \infty,$$

and

$$\|e_{ij}(v_k)\|_{0,2,\Omega} \to 0, \text{ as } k \to \infty.$$

In particular

$$\|(v_k)_{j,j}\|_{0,2,\Omega} \to 0, \; \forall j \in \{1,\dots,n\}.$$

At this point we recall the following identity in the distributional sense, found in [31], page 12,

$$\partial_j(\partial_l v_i) = \partial_j e_{il}(v) + \partial_l e_{ij}(v) - \partial_i e_{jl}(v), \; \forall i,j,l \in \{1,\dots,n\}. \tag{13.1}$$

Fix $j \in \{1,\dots,n\}$ and observe that

$$\|(v_k)_j\|_{1,2,v} \le C\|(v_k)_j\|_{1,2,\Omega},$$

so that

$$\frac{C}{\|(v_k)_j\|_{1,2,v}} \ge \frac{1}{\|(v_k)_j\|_{1,2,\Omega}}, \; \forall k \in \mathbb{N}.$$

Hence,

$$\|(v_k)_j\|_{1,2,\Omega}$$

$$= \sup_{\varphi \in C^1(\Omega)} \left\{ \langle \nabla(v_k)_j, \nabla\varphi \rangle_{L^2(\Omega)} + \langle (v_k)_j, \varphi \rangle_{L^2(\Omega)} : \|\varphi\|_{1,2,\Omega} \leq 1 \right\}$$

$$= \left\langle \nabla(v_k)_j, \nabla\left(\frac{(v_k)_j}{\|(v_k)_j\|_{1,2,\Omega}} \right) \right\rangle_{L^2(\Omega)}$$

$$+ \left\langle (v_k)_j, \left(\frac{(v_k)_j}{\|(v_k)_j\|_{1,2,\Omega}} \right) \right\rangle_{L^2(\Omega)}$$

$$\leq C \left(\left\langle \nabla(v_k)_j, \nabla\left(\frac{(v_k)_j}{\|(v_k)_j\|_{1,2,V}} \right) \right\rangle_{L^2(V)} + \left\langle (v_k)_j, \left(\frac{(v_k)_j}{\|(v_k)_j\|_{1,2,V}} \right) \right\rangle_{L^2(V)} \right)$$

$$= C \sup_{\varphi \in C_c^1(V)} \left\{ \langle \nabla(v_k)_j, \nabla\varphi \rangle_{L^2(V)} + \langle (v_k)_j, \varphi \rangle_{L^2(V)} : \|\varphi\|_{1,2,V} \leq 1 \right\}. \qquad (13.2)$$

Here, we recall that $C > 0$ is the constant concerning the extension Theorem 13.1.5. From such results and (13.1), we have that

$$\sup_{\varphi \in C^1(\Omega)} \left\{ \langle \nabla(v_k)_j, \nabla\varphi \rangle_{L^2(\Omega)} + \langle (v_k)_j, \varphi \rangle_{L^2(\Omega)} : \|\varphi\|_{1,2,\Omega} \leq 1 \right\}$$

$$\leq C \sup_{\varphi \in C_c^1(V)} \left\{ \langle \nabla(v_k)_j, \nabla\varphi \rangle_{L^2(V)} + \langle (v_k)_j, \varphi \rangle_{L^2(V)} : \|\varphi\|_{1,2,V} \leq 1 \right\}$$

$$= C \sup_{\varphi \in C_c^1(V)} \left\{ \langle e_{jl}(v_k), \varphi_{,l} \rangle_{L^2(V)} + \langle e_{jl}(v_k), \varphi_{,l} \rangle_{L^2(V)} \right.$$

$$\left. - \langle e_{ll}(v_k), \varphi_{,j} \rangle_{L^2(V)} + \langle (v_k)_j, \varphi \rangle_{L^2(V)}, : \|\varphi\|_{1,2,V} \leq 1 \right\}. \qquad (13.3)$$

Therefore

$$\|(v_k)_j\|_{(W^{1,2}(\Omega))}$$

$$= \sup_{\varphi \in C^1(\Omega)} \left\{ \langle \nabla(v_k)_j, \nabla\varphi \rangle_{L^2(\Omega)} + \langle (v_k)_j, \varphi \rangle_{L^2(\Omega)} : \|\varphi\|_{1,2,\Omega} \leq 1 \right\}$$

$$\leq C \left(\sum_{l=1}^{n} \left\{ \|e_{jl}(v_k)\|_{0,2,V} + \|e_{ll}(v_k)\|_{0,2,V} \right\} + \|(v_k)_j\|_{0,2,V} \right)$$

$$\leq C_1 \left(\sum_{l=1}^{n} \left\{ \|e_{jl}(v_k)\|_{0,2,\Omega} + \|e_{ll}(v_k)\|_{0,2,\Omega} \right\} + \|(v_k)_j\|_{0,2,\Omega} \right)$$

$$< \frac{C_2}{k}, \qquad (13.4)$$

for appropriate $C_1 > 0$ and $C_2 > 0$.

Summarizing,

$$\|(v_k)_j\|_{(W^{1,2}(\Omega))} < \frac{C_2}{k}, \ \forall k \in \mathbb{N}.$$

From this we have got

$$\|v_k\|_{1,2,\Omega} \to 0, \text{ as } k \to \infty,$$

which contradicts

$$\|v_k\|_{1,2,\Omega} = 1, \forall k \in \mathbb{N}.$$

The proof is complete.

Corollary 13.2.2 *Let* $\Omega \subset \mathbb{R}^n$ *be an open, bounded and connected set with a* $\hat{C}^1$ *boundary* $\partial\Omega$. *Define* $e : W^{1,2}(\Omega; \mathbb{R}^n) \to L^2(\Omega; \mathbb{R}^{n \times n})$ *by*

$$e(u) = \{e_{ij}(u)\}$$

where

$$e_{ij}(u) = \frac{1}{2}(u_{i,j} + u_{j,i}), \forall i, j \in \{1, \dots, n\}.$$

Define also,

$$\|e(u)\|_{0,2,\Omega} = \left(\sum_{i=1}^{n} \sum_{j=1}^{n} \|e_{ij}(u)\|_{0,2,\Omega}^2\right)^{1/2}.$$

Let $L \in \mathbb{R}^+$ *be such* $V = [-L, L]^n$ *is also such that* $\overline{\Omega} \subset V^0$. *Moreover, define*

$$\hat{H}_0 = \{u \in W^{1,2}(\Omega; \mathbb{R}^n) : u = \mathbf{0}, \text{ on } \Gamma_0\},$$

where $\Gamma_0 \subset \partial\Omega$ *is a measurable set such that the Lebesgue measure* $m_{\mathbb{R}^{n-1}}(\Gamma_0) > 0$.
Assume also Γ_0 *is such that for each* $j \in \{1, \dots, n\}$ *and each* $x = (x_1, \dots, x_n) \in \Omega$ *there exists* $x_0 = ((x_0)_1, \dots, (x_0)_n) \in \Gamma_0$ *such that*

$$(x_0)_l = x_l, \forall l \neq j, l \in \{1, \dots, n\},$$

and the line

$$A_{x_0, x} \subset \overline{\Omega}$$

where

$$A_{x_0, x} = \{(x_1, \dots, (1-t)(x_0)_j + tx_j, \dots, x_n) : t \in [0, 1]\}.$$

Under such hypotheses, there exists $C(\Omega, L) \in \mathbb{R}^+$ *such that*

$$\|u\|_{1,2,\Omega} \leq C(\Omega, L) \|e(u)\|_{0,2,\Omega}, \forall u \in \hat{H}_0.$$

Proof 13.2 Suppose, to obtain contradiction, the concerning claim does not hold. Hence, for each $k \in \mathbb{N}$ there exists $u_k \in \hat{H}_0$ such that

$$\|u_k\|_{1,2,\Omega} > k \|e(u_k)\|_{0,2,\Omega}.$$

In particular defining

$$v_k = \frac{u_k}{\|u_k\|_{1,2,\Omega}}$$

similarly to the proof of the last theorem, we may obtain

$$\|(v_k)_{j,j}\|_{0,2,\Omega} \to 0, \text{ as } k \to \infty, \forall j \in \{1,\ldots,n\}.$$

From this, the hypotheses on Γ_0 and from the standard Poincaré inequality proof we obtain

$$\|(v_k)_j\|_{0,2,\Omega} \to 0, \text{ as } k \to \infty, \forall j \in \{1,\ldots,n\}.$$

Thus, also similarly as in the proof of the last theorem, we may infer that

$$\|v_k\|_{1,2,\Omega} \to 0, \text{ as } k \to \infty,$$

which contradicts

$$\|v_k\|_{1,2,\Omega} = 1, \forall k \in \mathbb{N}.$$

The proof is complete.

13.3   An existence result for a non-linear model of plates

In the present section, as an application of the results on the Korn's inequalities presented in the previous sections, we develop a new global existence proof for a Kirchhoff-Love thin plate model. Previous results on existence in mathematical elasticity and related models may be found in [30, 31, 32].

At this point we start to describe the primal formulation.

Let $\Omega \subset \mathbb{R}^2$ be an open, bounded, connected set which represents the middle surface of a plate of thickness h. The boundary of Ω, which is assumed to be regular (Lipschitzian), is denoted by $\partial\Omega$. The vectorial basis related to the cartesian system $\{x_1,x_2,x_3\}$ is denoted by $(\mathbf{a}_\alpha,\mathbf{a}_3)$, where $\alpha = 1,2$ (in general Greek indices stand for 1 or 2), and where $\mathbf{a}_3$ is the vector normal to Ω, whereas $\mathbf{a}_1$ and $\mathbf{a}_2$ are orthogonal vectors parallel to Ω. Also, $\mathbf{n}$ is the outward normal to the plate surface.

The displacements will be denoted by

$$\hat{\mathbf{u}} = \{\hat{u}_\alpha, \hat{u}_3\} = \hat{u}_\alpha\mathbf{a}_\alpha + \hat{u}_3\mathbf{a}_3.$$

The Kirchhoff-Love relations are

$$\hat{u}_\alpha(x_1,x_2,x_3) = u_\alpha(x_1,x_2) - x_3 w(x_1,x_2)_{,\alpha}$$
$$\text{and } \hat{u}_3(x_1,x_2,x_3) = w(x_1,x_2). \tag{13.5}$$

Here $-h/2 \leq x_3 \leq h/2$ so that we have $u = (u_\alpha, w) \in U$ where

$$U = \{(u_\alpha, w) \in W^{1,2}(\Omega;\mathbb{R}^2) \times W^{2,2}(\Omega),$$
$$u_\alpha = w = \frac{\partial w}{\partial \mathbf{n}} = 0 \text{ on } \partial\Omega\}$$
$$= W_0^{1,2}(\Omega;\mathbb{R}^2) \times W_0^{2,2}(\Omega).$$

It is worth emphasizing that the boundary conditions here specified refer to a clamped plate.

We define the operator $\Lambda : U \to Y \times Y$, where $Y = Y^* = L^2(\Omega; \mathbb{R}^{2 \times 2})$, by

$$\Lambda(u) = \{\gamma(u), \kappa(u)\},$$

$$\gamma_{\alpha\beta}(u) = \frac{u_{\alpha,\beta} + u_{\beta,\alpha}}{2} + \frac{w_{,\alpha} w_{,\beta}}{2},$$

$$\kappa_{\alpha\beta}(u) = -w_{,\alpha\beta}.$$

The constitutive relations are given by

$$N_{\alpha\beta}(u) = H_{\alpha\beta\lambda\mu} \gamma_{\lambda\mu}(u), \tag{13.6}$$

$$M_{\alpha\beta}(u) = h_{\alpha\beta\lambda\mu} \kappa_{\lambda\mu}(u), \tag{13.7}$$

where: $\{H_{\alpha\beta\lambda\mu}\}$ and $\{h_{\alpha\beta\lambda\mu} = \frac{h^2}{12} H_{\alpha\beta\lambda\mu}\}$, are symmetric positive definite fourth order tensors. From now on, we denote $\{\overline{H}_{\alpha\beta\lambda\mu}\} = \{H_{\alpha\beta\lambda\mu}\}^{-1}$ and $\{\overline{h}_{\alpha\beta\lambda\mu}\} = \{h_{\alpha\beta\lambda\mu}\}^{-1}$.

Furthermore, $\{N_{\alpha\beta}\}$ denote the membrane force tensor and $\{M_{\alpha\beta}\}$ the moment one. The plate stored energy, represented by $(G \circ \Lambda) : U \to \mathbb{R}$, is expressed by

$$(G \circ \Lambda)(u) = \frac{1}{2} \int_\Omega N_{\alpha\beta}(u) \gamma_{\alpha\beta}(u)\, dx + \frac{1}{2} \int_\Omega M_{\alpha\beta}(u) \kappa_{\alpha\beta}(u)\, dx \tag{13.8}$$

and the external work, represented by $F : U \to \mathbb{R}$, is given by

$$F(u) = \langle w, P \rangle_{L^2(\Omega)} + \langle u_\alpha, P_\alpha \rangle_{L^2(\Omega)}, \tag{13.9}$$

where $P, P_1, P_2 \in L^2(\Omega)$ are external loads in the directions $\mathbf{a}_3$, $\mathbf{a}_1$ and $\mathbf{a}_2$ respectively. The potential energy, denoted by $J : U \to \mathbb{R}$, is expressed by:

$$J(u) = (G \circ \Lambda)(u) - F(u)$$

Finally, we also emphasize that from now on, as their meaning are clear, we may denote $L^2(\Omega)$ and $L^2(\Omega; \mathbb{R}^{2 \times 2})$ simply by L^2, and the respective norms by $\| \cdot \|_2$. Moreover derivatives are always understood in the distributional sense, $\mathbf{0}$ may denote the zero vector in appropriate Banach spaces and, the following and relating notations are used:

$$w_{,\alpha} = \frac{\partial w}{\partial x_\alpha},$$

$$w_{,\alpha\beta} = \frac{\partial^2 w}{\partial x_\alpha \partial x_\beta},$$

$$u_{\alpha,\beta} = \frac{\partial u_\alpha}{\partial x_\beta},$$

$$N_{\alpha\beta,1} = \frac{\partial N_{\alpha\beta}}{\partial x_1},$$

and

$$N_{\alpha\beta,2} = \frac{\partial N_{\alpha\beta}}{\partial x_2}.$$

13.4 On the existence of a global minimizer

At this point we present an existence result concerning the Kirchhoff-Love plate model.

We start with the following two remarks.

Remark 13.1 Let $\{P_\alpha\} \in L^\infty(\Omega; \mathbb{R}^2)$. We may easily obtain by appropriate Lebesgue integration $\{\tilde{T}_{\alpha\beta}\}$ symmetric and such that

$$\tilde{T}_{\alpha\beta,\beta} = -P_\alpha, \text{ in } \Omega.$$

Indeed, extending $\{P_\alpha\}$ to zero outside Ω if necessary, we may set

$$\tilde{T}_{11}(x,y) = -\int_0^x P_1(\xi,y) \, d\xi,$$

$$\tilde{T}_{22}(x,y) = -\int_0^y P_2(x,\xi) \, d\xi,$$

and

$$\tilde{T}_{12}(x,y) = \tilde{T}_{21}(x,y) = 0, \text{ in } \Omega.$$

Thus, we may choose a $C > 0$ sufficiently big, such that

$$\{T_{\alpha\beta}\} = \{\tilde{T}_{\alpha\beta} + C\delta_{\alpha\beta}\}$$

is positive definite in Ω, so that

$$T_{\alpha\beta,\beta} = \tilde{T}_{\alpha\beta,\beta} = -P_\alpha,$$

where

$$\{\delta_{\alpha\beta}\}$$

is the Kronecker delta.

So, for the kind of boundary conditions of the next theorem, we do NOT have any restriction for the $\{P_\alpha\}$ norm.

Summarizing, the next result is new and it is really a step forward concerning the previous one in Ciarlet [31]. We emphasize, this result and its proof through such a tensor $\{T_{\alpha\beta}\}$ are new, even though the final part of the proof is established through a standard procedure in the calculus of variations.

Finally, more details on the Sobolev spaces involved may be found in [1, 12, 13, 35]. Related duality principles are addressed in [34, 12, 13].   ■

At this point we present the main theorem in this section.

Theorem 13.4.1 *Let $\Omega \subset \mathbb{R}^2$ be an open, bounded, connected set with a Lipschitzian boundary denoted by $\partial\Omega = \Gamma$. Suppose $(G \circ \Lambda) : U \to \mathbb{R}$ is defined by*

$$G(\Lambda u) = G_1(\gamma(u)) + G_2(\kappa(u)), \ \forall u \in U,$$

where

$$G_1(\gamma u) = \frac{1}{2} \int_\Omega H_{\alpha\beta\lambda\mu} \gamma_{\alpha\beta}(u) \gamma_{\lambda\mu}(u) \, dx,$$

and

$$G_2(\kappa u) = \frac{1}{2} \int_\Omega h_{\alpha\beta\lambda\mu} \kappa_{\alpha\beta}(u) \kappa_{\lambda\mu}(u) \, dx,$$

where

$$\Lambda(u) = (\gamma(u), \kappa(u)) = (\{\gamma_{\alpha\beta}(u)\}, \{\kappa_{\alpha\beta}(u)\}),$$

$$\gamma_{\alpha\beta}(u) = \frac{u_{\alpha,\beta} + u_{\beta,\alpha}}{2} + \frac{w_{,\alpha} w_{,\beta}}{2},$$

$$\kappa_{\alpha\beta}(u) = -w_{,\alpha\beta},$$

and where

$$\begin{aligned} J(u) &= W(\gamma(u), \kappa(u)) - \langle P_\alpha, u_\alpha \rangle_{L^2(\Omega)} \\ &\quad - \langle w, P \rangle_{L^2(\Omega)} - \langle P_\alpha^t, u_\alpha \rangle_{L^2(\Gamma_t)} \\ &\quad - \langle P^t, w \rangle_{L^2(\Gamma_t)}, \end{aligned} \tag{13.10}$$

where,

$$\begin{aligned} U &= \{u = (u_\alpha, w) = (u_1, u_2, w) \in W^{1,2}(\Omega; \mathbb{R}^2) \times W^{2,2}(\Omega) : \\ & u_\alpha = w = \frac{\partial w}{\partial \mathbf{n}} = 0, \text{ on } \Gamma_0\}, \end{aligned} \tag{13.11}$$

where $\partial\Omega = \Gamma_0 \cup \Gamma_t$ *and the Lebesgue measures*

$$m_\Gamma(\Gamma_0 \cap \Gamma_t) = 0,$$

and

$$m_\Gamma(\Gamma_0) > 0.$$

We also define,

$$\begin{aligned} F_1(u) &= -\langle w, P \rangle_{L^2(\Omega)} - \langle u_\alpha, P_\alpha \rangle_{L^2(\Omega)} - \langle P_\alpha^t, u_\alpha \rangle_{L^2(\Gamma_t)} \\ &\quad - \langle P^t, w \rangle_{L^2(\Gamma_t)} + \langle \varepsilon_\alpha, u_\alpha^2 \rangle_{L^2(\Gamma_t)} \\ &\equiv -\langle u, \mathbf{f} \rangle_{L^2} + \langle \varepsilon_\alpha, u_\alpha^2 \rangle_{L^2(\Gamma_t)} \\ &\equiv -\langle u, \mathbf{f}_1 \rangle_{L^2} - \langle u_\alpha, P_\alpha \rangle_{L^2(\Omega)} + \langle \varepsilon_\alpha, u_\alpha^2 \rangle_{L^2(\Omega)}, \end{aligned} \tag{13.12}$$

where

$$\langle u, \mathbf{f}_1 \rangle_{L^2} = \langle u, \mathbf{f} \rangle_{L^2} - \langle u_\alpha, P_\alpha \rangle_{L^2(\Omega)},$$

$\varepsilon_\alpha > 0$, $\forall \alpha \in \{1, 2\}$ *and*

$$\mathbf{f} = (P_\alpha, P) \in L^\infty(\Omega; \mathbb{R}^3).$$

Let $J : U \to \mathbb{R}$ be defined by

$$J(u) = G(\Lambda u) + F_1(u), \ \forall u \in U.$$

Assume there exists $\{c_{\alpha\beta}\} \in \mathbb{R}^{2 \times 2}$ such that $c_{\alpha\beta} > 0, \ \forall \alpha, \beta \in \{1,2\}$ and

$$G_2(\kappa(u)) \geq c_{\alpha\beta} \|w_{,\alpha\beta}\|_2^2, \ \forall u \in U.$$

Under such hypotheses, there exists $u_0 \in U$ such that

$$J(u_0) = \min_{u \in U} J(u).$$

Proof 13.3   Observe that we may find $\mathbf{T}_\alpha = \{(T_\alpha)_\beta\}$ such that

$$div\mathbf{T}_\alpha = T_{\alpha\beta,\beta} = -P_\alpha,$$

and also such that $\{T_{\alpha\beta}\}$ is positive definite and symmetric (please, see Remark 13.1).
    Thus defining

$$v_{\alpha\beta}(u) = \frac{u_{\alpha,\beta} + u_{\beta,\alpha}}{2} + \frac{1}{2}w_{,\alpha}w_{,\beta}, \tag{13.13}$$

we obtain

$$
\begin{aligned}
J(u) &= G_1(\{v_{\alpha\beta}(u)\}) + G_2(\kappa(u)) - \langle u, \mathbf{f} \rangle_{L^2} + \langle \varepsilon_\alpha, u_\alpha^2 \rangle_{L^2(\Gamma_t)} \\
&= G_1(\{v_{\alpha\beta}(u)\}) + G_2(\kappa(u)) + \langle T_{\alpha\beta,\beta}, u_\alpha \rangle_{L^2(\Omega)} - \langle u, \mathbf{f}_1 \rangle_{L^2} + \langle \varepsilon_\alpha, u_\alpha^2 \rangle_{L^2(\Gamma_t)} \\
&= G_1(\{v_{\alpha\beta}(u)\}) + G_2(\kappa(u)) - \left\langle T_{\alpha\beta}, \frac{u_{\alpha,\beta} + u_{\beta,\alpha}}{2} \right\rangle_{L^2(\Omega)} \\
&\quad + \langle T_{\alpha\beta}n_\beta, u_\alpha \rangle_{L^2(\Gamma_t)} - \langle u, \mathbf{f}_1 \rangle_{L^2} + \langle \varepsilon_\alpha, u_\alpha^2 \rangle_{L^2(\Gamma_t)} \\
&= G_1(\{v_{\alpha\beta}(u)\}) + G_2(\kappa(u)) - \left\langle T_{\alpha\beta}, v_{\alpha\beta}(u) - \frac{1}{2}w_{,\alpha}w_{,\beta} \right\rangle_{L^2(\Omega)} - \langle u, \mathbf{f}_1 \rangle_{L^2} + \langle \varepsilon_\alpha, u_\alpha^2 \rangle_{L^2(\Gamma_t)} \\
&\quad + \langle T_{\alpha\beta}n_\beta, u_\alpha \rangle_{L^2(\Gamma_t)} \\
&\geq c_{\alpha\beta}\|w_{,\alpha\beta}\|_2^2 + \frac{1}{2}\langle T_{\alpha\beta}, w_{,\alpha}w_{,\beta} \rangle_{L^2(\Omega)} - \langle u, \mathbf{f}_1 \rangle_{L^2} + \langle \varepsilon_\alpha, u_\alpha^2 \rangle_{L^2(\Gamma_t)} + G_1(\{v_{\alpha\beta}(u)\}) \\
&\quad - \langle T_{\alpha\beta}, v_{\alpha\beta}(u) \rangle_{L^2(\Omega)} + \langle T_{\alpha\beta}n_\beta, u_\alpha \rangle_{L^2(\Gamma_t)}. \tag{13.14}
\end{aligned}
$$

From this, since $\{T_{\alpha\beta}\}$ is positive definite, clearly J is bounded below.
Let $\{u_n\} \in U$ be a minimizing sequence for J. Thus there exists $\alpha_1 \in \mathbb{R}$ such that

$$\lim_{n \to \infty} J(u_n) = \inf_{u \in U} J(u) = \alpha_1.$$

From (13.14), there exists $K_1 > 0$ such that

$$\|(w_n)_{,\alpha\beta}\|_2 < K_1, \forall \alpha, \beta \in \{1,2\}, \ n \in \mathbb{N}.$$

Therefore, there exists $w_0 \in W^{2,2}(\Omega)$ such that, up to a subsequence not relabeled,

$$(w_n)_{,\alpha\beta} \rightharpoonup (w_0)_{,\alpha\beta}, \text{ weakly in } L^2,$$

$\forall \alpha, \beta \in \{1,2\}$, as $n \to \infty$.

Moreover, also up to a subsequence not relabeled,

$$(w_n)_{,\alpha} \to (w_0)_{,\alpha}, \text{ strongly in } L^2 \text{ and } L^4, \tag{13.15}$$

$\forall \alpha, \in \{1,2\}$, as $n \to \infty$.

Also from (13.14), there exists $K_2 > 0$ such that,

$$\|(v_n)_{\alpha\beta}(u)\|_2 < K_2, \forall \alpha, \beta \in \{1,2\}, n \in \mathbb{N},$$

and thus, from this, (13.13) and (13.15), we may infer that there exists $K_3 > 0$ such that

$$\|(u_n)_{\alpha,\beta} + (u_n)_{\beta,\alpha}\|_2 < K_3, \forall \alpha, \beta \in \{1,2\}, n \in \mathbb{N}.$$

From this and Korn's inequality, there exists $K_4 > 0$ such that

$$\|u_n\|_{W^{1,2}(\Omega;\mathbb{R}^2)} \leq K_4, \ \forall n \in \mathbb{N}.$$

So, up to a subsequence not relabeled, there exists $\{(u_0)_\alpha\} \in W^{1,2}(\Omega, \mathbb{R}^2)$, such that

$$(u_n)_{\alpha,\beta} + (u_n)_{\beta,\alpha} \rightharpoonup (u_0)_{\alpha,\beta} + (u_0)_{\beta,\alpha}, \text{ weakly in } L^2,$$

$\forall \alpha, \beta \in \{1,2\}$, as $n \to \infty$, and,

$$(u_n)_\alpha \to (u_0)_\alpha, \text{ strongly in } L^2,$$

$\forall \alpha \in \{1,2\}$, as $n \to \infty$.

Moreover, the boundary conditions satisfied by the subsequences are also satisfied for w_0 and u_0 in a trace sense, so that

$$u_0 = ((u_0)_\alpha, w_0) \in U.$$

From this, up to a subsequence not relabeled, we get

$$\gamma_{\alpha\beta}(u_n) \rightharpoonup \gamma_{\alpha\beta}(u_0), \text{ weakly in } L^2,$$

$\forall \alpha, \beta \in \{1,2\}$, and

$$\kappa_{\alpha\beta}(u_n) \rightharpoonup \kappa_{\alpha\beta}(u_0), \text{ weakly in } L^2,$$

$\forall \alpha, \beta \in \{1,2\}$.

Therefore, from the convexity of G_1 in γ and G_2 in κ we obtain

$$
\begin{aligned}
\inf_{u \in U} J(u) &= \alpha_1 \\
&= \liminf_{n \to \infty} J(u_n) \\
&\geq J(u_0). \tag{13.16}
\end{aligned}
$$

Thus,

$$J(u_0) = \min_{u \in U} J(u).$$

The proof is complete.

13.5 Conclusion

In this chapter, we have developed a new proof for the Korn inequality in a specific n-dimensional context. In the second text part, we present a global existence result for a non-linear model of plates. Both results represent some new advances concerning the present literature. In particular, the results for the Korn's inequality so far known are for a three dimensional context such as in [54], for example, whereas we have here addressed a more general n-dimensional case.

In a future research, we intend to address more general models, including the corresponding results for manifolds in $\mathbb{R}^n$.

References

[1] Adams, R.A. 1975. *Sobolev Spaces*. Academic Press, New York.

[2] Adams, R.A. and Fournier, J.F. 2003. *Sobolev Spaces, Second Edition*. Elsevier.

[3] Allaire, G. 2002. *Shape Optimization by the Homogenization Method*. Springer-Verlag, New York.

[4] Annet, J.F. 2010. *Superconductivity, Superfluids and Condensates*. Oxford Master Series in Condensed Matter Physics, Oxford University Press, Reprint.

[5] Attouch, H., Buttazzo, G. and Michaille, G. 2006. *Variational Analysis in Sobolev and BV Spaces*. MPS-SIAM Series in Optimization, Philadelphia.

[6] Bachman, G. and Narici, L. 2000. *Functional Analysis*. Dover Publications, Reprint.

[7] Ball, J.M. and James, R.D. 1987. Fine mixtures as minimizers of energy. Archive for Rational Mechanics and Analysis, 100: 15–52.

[8] Bendsoe, M.P. and Sigmund, O. 2003. *Topology Optimization, Theory Methods and Applications*. Springer, Berlin.

[9] Bethuel, F., Brezis, H. and Helein, F. 1994. *Ginzburg-Landau Vortices*. Birkhäuser, Basel.

[10] Bielski, W.R., Galka, A. and Telega, J.J. 1988. The complementary energy principle and duality for geometrically nonlinear elastic shells. I. Simple case of moderate rotations around a tangent to the middle surface. Bulletin of the Polish Academy of Sciences, Technical Sciences, 38(7–9).

[11] Bielski, W.R. and Telega, J.J. 1985. A contribution to contact problems for a class of solids and structures. Arch. Mech., 37: (4-5): 303–320, Warszawa.

[12] Botelho, F. 2014. *Functional Analysis and Applied Optimization in Banach Spaces*. Springer Switzerland.

[13] Botelho, F.S. 2020. *Functional Analysis, Calculus of Variations and Numerical Methods in Physics and Engineering*. CRC Taylor and Francis, Florida.

[14] Botelho, F. 2009. *Variational Convex Analysis*. Ph.D. Thesis, Virginia Tech, Blacksburg, VA-USA.

[15] Botelho, F. 2010. *Variational Convex Analysis, Applications to non-Convex Models*. Lambert Academic Publishing, Berlin.

[16] Botelho, F. 2010. Dual variational formulations for a non-linear model of plates. Journal of Convex Analysis, 17(1): 131–158.

[17] Botelho, F. 2012. Existence of solution for the Ginzburg-Landau system, a related optimal control problem and its computation by the generalized method of lines. Applied Mathematics and Computation, 218: 11976–11989.

[18] Botelho, F.S. 2022. An approximate proximal numerical procedure concerning the generalized method of lines mathematics, 10(16): 2950. https://doi.org/10.3390/math10162950.

[19] Botelho, F.S. 2023. On Lagrange Multiplier Theorems for non-smooth optimization for a large class of variational models in Banach spaces. Contemporary Mathematics, to appear. DOI: https://doi.org/10.37256/cm.3420221711.

[20] Botelho, F.S. 2023. Dual variational formulations for a large class of non-convex models in the calculus of variations mathematics, 11(1): 63. https://doi.org/10.3390/math11010063 - 24 Dec 2022.

[21] Botelho, F.S. 2023. A note on Korn's inequality in an N-dimensional context and a global existence result for a non-linear plate model. Applied Math, 1: 1–11.

[22] Botelho, F. 2011. *Topics on Functional Analysis, Calculus of Variations and Duality*. Academic Publications, Sofia.

[23] Botelho, F. 2012. On duality principles for scalar and vectorial multi-well variational problems. Nonlinear Analysi,s 75: 1904–1918.

[24] Botelho, F. 2013. On the Lagrange multiplier theorem in Banach spaces. Computational and Applied Mathematics, 32: 135–144.

[25] Botelho, F. 2017. *A Classical Description of Variational Quantum Mechanics and Related Models*. Nova Science Publishers, New York.

[26] Brezis, H. 1987. *Analyse Fonctionnelle*. Masson.

[27] Chenchiah, I.V. and Bhattacharya, K. 2008. The relaxation of two-well energies with possibly unequal moduli. Arch. Rational Mech. Anal., 187: 409–479.

[28] Chipot, M. 1993. Approximation and oscillations. Microstructure and Phase Transition, the IMA Volumes in Mathematics and Applications, 54: 27–38.

[29] Choksi, R., Peletier, M.A. and Williams, J.F. 2009. On the phase diagram for microphase separation of Diblock copolymers: An approach via a nonlocal Cahn-Hilliard functional. to appear in SIAM J. Appl. Math.

[30] Ciarlet, P. 1988. *Mathematical Elasticity*, Vol. I—Three Dimensional Elasticity. North Holland Elsevier.

[31] Ciarlet, P. 1997. *Mathematical Elasticity*, Vol. II—Theory of Plates. North Holland Elsevier.

[32] Ciarlet, P. 2000. *Mathematical Elasticity*, Vol. III—Theory of Shells. North Holland Elsevier.

[33] Dacorogna, B. 1989. *Direct methods in the Calculus of Variations*. Springer-Verlag.

[34] Ekeland, I. and Temam, R. 1976. *Convex Analysis and Variational Problems*. North Holland.

[35] Evans, L.C. 1998. *Partial Differential Equations*. Graduate Studies in Mathematics, 19, AMS.

[36] Fidalgo, U. and Pedregal, P. 2007. A general lower bound for the relaxation of an optimal design problem in conductivity with a quadratic cost functional. Pre-print.

[37] Firoozye, N.B. and Khon, R.V. 1993. Geometric parameters and the relaxation for multiwell energies. Microstructure and Phase Transition, the IMA Volumes in Mathematics and Applications, 54: 85–110.

[38] Fonseca, I. and Leoni, G. 2007. *Modern Methods in the Calculus of Variations, L^p Spaces*. Springer, New York.

[39] Constantin, P. and Foias, C. 1989. *Navier-Stokes Equation*. University of Chicago Press, Chicago.

[40] Hamouda, M., Han, D., Jung, C.Y. and Temam, R. 2018. Boundary layers for the 3D primitive equations in a cube: The zero-mode. Journal of Applied Analysis and Computation, 8(3): 873–889, DOI:10.11948/2018.873.

[41] Giorgini, A., Miranville, A. and Temam, R. 2019. Uniqueness and regularity for the Navier-Stokes-Cahn-Hilliard system. SIAM J. of Mathematical Analysis (SIMA), 51(3): 2535–2574. https://doi.org/10.1137/18M1223459.

[42] Foias, C., Rosa, R.M.S. and Temam, R.M. 2019. Properties of stationary statistical solutions of the three-dimensional Navier-Stokes equations. J. of Dynamics and Differential Equations, Special Issue in Memory of George Sell, 31(3): 1689–1741. https://doi.org/10.1007/s10884-018-9719-2.

[43] Giaquinta, M. and Hildebrandt, S. 1996. *Calculus of Variations I*. A Series of Comprehensive Studies in Mathematics, Vol. 310, Springer.

[44] Giaquinta, M. and Hildebrandt, S. 1996. *Calculus of Variations I*. A Series of Comprehensive Studies in Mathematics, Vol. 311, Springer.

[45] Giorgi, T. and Smits, R.T. 2003. Remarks on the existence of global minimizers for the Ginzburg-Landau energy functional. Nonlinear Analysis, Theory Methods and Applications, 53(147): 155.

[46] Giusti, E. 2005. *Direct Methods in the Calculus of Variations*. World Scientific, Singapore, Reprint.

[47] Isakov, V. 2006. *Inverse Problems for Partial Differential Equations*. Series Applied Mathematical Sciences 127, Second Edition, Springer, New York.

[48] Izmailov, A. and Solodov, M. 2009. *Otimização* Volume 1, Second Edition. IMPA, Rio de Janeiro.

[49] Izmailov, A. and Solodov, M. 2007. *Otimização* Volume 2. IMPA, Rio de Janeiro.

[50] James, R.D. and Kinderlehrer, D. 1990. Frustration in ferromagnetic materials. Continuum Mech. Thermodyn., 2: 215–239.

[51] Ito, K. and Kunisch, K. 2008. *Lagrange Multiplier Approach to Variational Problems and Applications*. Advances in Design and Control, SIAM, Philadelphia.

[52] Landau, L.D. and Lifschits, E.M. 2008. *Course of Theoretical Physics, Vol. 5—Statistical Physics, Part 1*. Butterworth-Heinemann, Elsevier, Reprint.

[53] Lifschits, E.M. and Pitaevskii, L.P. 2002. *Course of Theoretical Physics, Vol. 9—Statistical Physics, Part 2*. Butterworth-Heinemann, Elsevier, Reprint.

[54] Lebedev, L.P. and Cloud, M.J. 2020. Korn's Inequality. *In*: Altenbach, H. and Öchsner, A. (eds.). Encyclopedia of Continuum Mechanics. Springer, Berlin, Heidelberg. https://doi.org/10.1007/978-3-662-55771-6 217.

[55] Luenberger, D.G. 1969. *Optimization by Vector Space Methods*. John Wiley and Sons, Inc.

[56] Milton, G.W. 2002. *Theory of Composites*. Cambridge Monographs on Applied and Computational Mathematics. Cambridge University Press, Cambridge.

[57] Molter, A., Silveira, O.A.A., Fonseca, J. and Bottega, V. 2010. Simultaneous piezoelectric actuator and sensor placement optimization and control design of manipulators with flexible links using SDRE method. Mathematical Problems in Engineering, 2010: 1–23.

[58] Pedregal, P. 1997. *Parametrized Measures and Variational Principles*. Progress in Nonlinear Differential Equations and their Applications, 30, Birkhauser.

[59] Pedregal, P. and Yan, B. 2007. *On Two Dimensional Ferromagnetism*. Pre-print.

[60] Reed, M. and Simon, B. 2003. *Methods of Modern Mathematical Physics, Volume I, Functional Analysis*. Reprint Elsevier, Singapore.

[61] Robinson, S.M. 1980. Strongly regular generalized equations. Math. of Oper. Res., 5(1980): 43–62.

[62] Rockafellar, R.T. 1970. *Convex Analysis*. Princeton Univ. Press.

[63] Rogers, R.C. 1991. A nonlocal model of the exchange energy in ferromagnet materials. Journal of Integral Equations and Applications, 3(1) (Winter).

[64] Rogers, R.C. 1988. Nonlocal variational problems in nonlinear electromagneto-elastostatics. SIAM J. Math. Anal., 19(6) (November).

[65] Rudin, W. 1991. *Functional Analysis*. Second Edition, McGraw-Hill.

[66] Rudin, W. 1987. *Real and Complex Analysis*. Third Edition, McGraw-Hill.

[67] Royden, H. 2006. *Real Analysis*. Third Edition, Prentice Hall, India.

[68] Sigmund, O. 2001. A 99 line topology optimization code written in MATLAB®. Struc. Muldisc. Optim., 21: 120–127, Springer-Verlag.

[69] Silva, J.J.A., Alvin, A., Vilhena, M., Petersen, C.Z. and Bodmann, B. 2011. On a closed form solution of the point cinetics equations with reactivity feedback of temperature. International Nuclear Atlantic Conference-INAC-ABEN, Belo Horizonte, MG, Brazil, October 24–28, ISBN: 978-85-99141-04-05.

[70] Stein, E.M. and Shakarchi, R. 2005. *Real Analysis*. Princeton Lectures in Analysis III, Princeton University Press.

[71] Strikwerda, J.C. 2004. *Finite Difference Schemes and Partial Differential Equations*. SIAM, Second Edition.

[72] Talbot, D.R.S. and Willis, J.R. 2004. Bounds for the effective contitutive relation of a nonlinear composite. Proc. R. Soc. Lond., 460: 2705–2723.

[73] Temam, R. 2001. *Navier-Stokes Equations*. AMS Chelsea, Reprint.

[74] Telega, J.J. 1989. *On the complementary energy principle in non-linear elasticity. Part I: Von Karman plates and three dimensional solids*, C.R. Acad. Sci. Paris, Serie II, 308, 1193–1198; Part II: Linear elastic solid and non-convex boundary condition. Minimax approach, ibid, pp. 1313–1317.

[75] Galka, A. and Telega, J.J. 1995. Duality and the complementary energy principle for a class of geometrically non-linear structures. Part I. Five parameter
shell model; Part II. Anomalous dual variational principles for compressed elastic beams. Arch. Mech., 47(677–698): 699–724.

[76] Börgens, E., Kanzow, C. and Steck, D. 2019. Local and global analysis of multiplier methods in constrained optimization in Banach spaces. SIAM Journal on
Control and Optimization, 57(6).

[77] Aubin, J.P. and Ekeland, I. 1984. *Applied Non-linear Analysis.* Wiley Interscience, New York.

[78] Clarke, F.H. 1983. *Optimization and Non-Smooth Analysis.* Wiley Interscience,
New York.

[79] Clarke, F. 2013. *Functional Analysis, Calculus of Variations and Optimal Control.* Springer New York.

[80] Kanzow, C., Steck, D. and Wachsmuth, D. 2018. An augmented Lagrangian
method for optimization problems in Banach spaces. SIAM Journal on Control
and Optimization, 56(1).

[81] Toland, J.F. 1979. A duality principle for non-convex optimisation and the calculus of variations. Arch. Rath. Mech. Anal., 71(1): 41–61.

[82] Troutman, J.L. 1996. *Variational Calculus and Optimal Control.* Second Edition, Springer.

[83] Vinh, N.X. 1993. *Flight Mechanics of High Performance Aircraft.* Cambridge
University Press, New York.

[84] Weinberg, S. 1972. *Gravitation and Cosmology.* Principles and Applications of
the General Theory of Relativity. Wiley and Sons (Cambridge, Massachusetts).

Index

For Product Safety Concerns and Information please contact our EU
representative GPSR@taylorandfrancis.com
Taylor & Francis Verlag GmbH, Kaufingerstraße 24, 80331 München, Germany

www.ingramcontent.com/pod-product-compliance
Lightning Source LLC
Chambersburg PA
CBHW060328220326
41598CB00023B/2636

9 781032 192109